ANATOMY and PHYSIOLOGY

Volume 1: Cells; Tissues; Integument; Skeletal, Muscular, and Digestive Systems; Blood, Lymph, Circulatory System

About the Authors

Edwin B. Steen is Professor Emeritus of Biology at Western Michigan University where he taught anatomy, physiology, histology, embryology, and related subjects for over thirty years. He received his A.B. degree from Wabash College, his A.M. degree from Columbia University, and his Ph.D. degree from Purdue University. He has served on the faculties of Wabash College, College of the City of New York, New York University, University of Cincinnati, and Purdue University. He was the principal contributing editor for the revision of *Taber's Cyclopedic Medical Dictionary* and a contributor to the eighth edition of *Acronyms, Initialisms, & Abbreviations Dictionary.* He is the author of a *Laboratory Manual and Study Guide for Anatomy and Physiology,* a *Dictionary of Biology,* and, with J. H. Price, *Human Sex and Sexuality with a Dictionary of Sexual Terms.* For several years he was a member of the Michigan Board of Examiners in the Basic Sciences and he has served as an editorial consultant for a number of publishers in the fields of anatomy and physiology.

Ashley Montagu studied at the universities of London and Florence and also at Columbia University where he received his Ph.D. degree. He has taught anatomy and physical anthropology for many years to undergraduate and graduate students at medical and dental schools, including New York University, the Hahnemann Medical College and Hospital, Philadelphia, Rutgers University, University of California, and Princeton University. Professor Ashley Montagu is the author of many notable works on anatomy, physical anthropology, and other biological sciences.

ANATOMY and PHYSIOLOGY

Volume 1: Cells; Tissues; Integument; Skeletal, Muscular, and Digestive Systems; Blood, Lymph, Circulatory System

Second Edition
Revised and Enlarged

Edwin B. Steen
Ashley Montagu

BARNES & NOBLE BOOKS
A DIVISION OF HARPER & ROW, PUBLISHERS
New York, Cambridge, Philadelphia
San Francisco, London, Mexico City
São Paulo, Sydney

 For information address Harper & Row, Publishers, Inc., 10 East 53rd Street, New York, N.Y. 10022. Published simultaneously in Canada by Fitzhenry & Whiteside Limited, Toronto.

Library of Congress Cataloging in Publication Data

Steen, Edwin Benzel, 1901–
Anatomy and physiology.

(College outline series ; COS/190)
Includes index.
Contents: v. 1. Cells, tissues, integument—skeletal, muscular, digestive, and circulatory systems.
1. Human physiology—Outlines, syllabi, etc. 2. Anatomy, Human—Outlines, syllabi, etc. I. Montagu, Ashley, 1905– II. Title. III. Series. [DNLM: 1. Anatomy—Outlines. 2. Physiology—Outlines. QS 18 S814a]
QP41.S75 1984 612 83–47599
ISBN 0–06–460190–0 (pbk. : v. 1)

---

84 85 86 87 10 9 8 7 6 5 4 3 2 1

CONTENTS

PREFACE

This new revised edition of *Anatomy and Physiology* is designed to meet the needs of students and others who wish to learn or review the essentials of human anatomy and physiology. It provides a comprehensive summary for students of the biological sciences, nursing, occupational therapy, and physical education. It should be most helpful to medical and dental students at all stages in their education—undergraduate, graduate, and postgraduate—and to students preparing to be physician's assistants, medical technologists, dental hygienists, physical therapists, and others whose activities require a knowledge of the human body and how it works. It should also serve as a useful reference book for laymen who desire information about the human body.

Since the authors have for many years taught students in all the groups mentioned above, they are acutely aware of the fact that in this field of study there is no satisfactory substitute for work in the laboratory. No one can acquire a thorough, functional knowledge of anatomy and physiology by merely reading a book. Consequently, this volume is not intended as a substitute for laboratory research but rather as a supplement to it, as a remembrancer following laboratory work, and, finally, as a refresher of the memory long after introductory studies of the subject have been concluded. For laboratory work, the following manual is available: Edwin B. Steen. *Laboratory Manual and Study Guide for Anatomy and Physiology*. 3rd ed. Dubuque, Iowa: William C. Brown Co., 1976.

The Outline constitutes a completely integrated textbook of anatomy and physiology, covering the fundamentals of these inextricably intertwined subjects. The reader is asked to bear in mind always that anatomic structure and physiological function are but different aspects of the same thing, just as physics and chemistry are but two different ways of looking at matter. Anatomy and physiology are the names we give to two ways in which we look at organic matter, structure being a function of function, and function being a function of structure. There is rarely a structure without function and no function without structure, the two being inseparably associated.

In this new, revised edition, the authors have attempted to bring the subject matter on all topics up to date and to incorporate the latest findings and developments throughout. Many new illustrations have been added in order to make the text more understandable. The sections at the end of each chapter that list pathological conditions involving the various organ systems have been enlarged and amplified.

The authors wish to express their special thanks and appreciation to Jeanne Flagg, Editor, Barnes & Noble Books, for her work in the preparation of this new revised edition, to Carol Miller for her editorial review of the manuscript, and to Bruce Emmer for his careful work in copy editing. They also wish to thank the publishers who kindly granted them permission to reproduce a number of the figures. Acknowledgment of sources of these figures is made in the legends. To the staff of Harper & Row the authors express their thanks for their work and cooperation in the preparation and production of this revision.

The authors also thank Louise Yorke, librarian at the Princeton Medical Center, and Helen Zimmerberg and Louise Schaeffer of the Biology Library, Princeton University, for their always gracious assistance.

While the authors have made every effort to insure accuracy in their writing, errors and mistakes sometimes occur. They will welcome any information calling attention to possible errors or inaccuracies.

EDWIN B. STEEN
ASHLEY MONTAGU

1: INTRODUCTION TO ANATOMY AND PHYSIOLOGY

Anatomy is the science of the structure of organisms. It is concerned with the discovery and the systematic statement of facts about the structure of the organs and organ systems which make up the machinery of the complete living organism. (A fact is a "verifiable datum of experience.") Originally restricted to the gathering and classification of facts by means of dissection, anatomy (from the Greek, *ana,* apart, and *temnein,* to cut) now comprises many branches.

BRANCHES OF ANATOMY

The branches of anatomy are known by terms that indicate either the nature of their procedures, their objectives, or the parts or regions of the body under investigation.

Gross anatomy is the macroscopic study of structure through dissection, in which inspection is achieved with the naked eye. In it the regions and parts of the body are studied with regard to their general form, external features, and main divisions. The study of form is known as *morphology.* The study of the regions of the body by dissection and by means of sections taken through the body at different planes, and resulting in a sort of descriptive geography of the body, is called *regional* or *topographic anatomy.*

Microscopic anatomy, also termed *histology,* is the study of the minute structure of organs and tissues, accomplished by means of the microscope. In addition to the conventional light microscope (LM), the utilization of dark-field, fluorescence, phase-contrast, interference, polarizing, and ultraviolet microscopes, and finally the electron microscope (EM) has greatly increased our knowledge of tissues. *Electron microscopy,* with magnifications of several hundred thousand times, reveals extremely minute details of structure comprising *ultrastructure. Scanning electron microscopy* (SEM) provides detailed, three-dimensional surface views of structures.

Cytology is the study of the structure, function, and development of cells, especially with reference to their relationship to the body as a whole.

Developmental anatomy is the study of growth and development during the entire life of the organism. Treating a more restricted area, *embryology* is concerned with the origin and development of the organism from the fertilized egg to birth.

Comparative anatomy is the study of the organ systems of various classes of organisms with emphasis on their structural relationships to each other and to the body.

Genetics is the study of heredity. It deals with the transmission of traits from one generation to the next and with the nature and causes of variations.

Systematic anatomy is the study of the various organ systems as structural and functional units. The study of the skeletal system is called *osteology;* of joints, *arthrology;* of the muscular system, *myology;* the integumentary system, *dermatology;* the nervous system, *neurology;* the circulatory system, *angiology;* the endocrine system, *endocrinology;* the internal organs collectively comprising the digestive, respiratory, urinary, and reproductive systems, *splanchnology.*

Some special fields are *applied anatomy, pathologic anatomy, radiographic anatomy,* and *surgical anatomy.*

ANATOMIC TERMINOLOGY

The need for accurate reference and economy of expression demands a special and uniform terminology for communication of anatomic data. Such a terminology serves at least two specific purposes: (1) to give names to structures so that they may be readily identified and distinguished from all others, and (2) through terms that refer to location, position, direction, and plane, to describe unambiguously the relationships between structures.

Terms Referring to Location or Position. The following terms assume the body to be in the "anatomic position," in which the individual is standing erect, with face forward, arms at the sides, and palms forward (Fig. 1-1).

Anterior toward the front of the body
Posterior toward the back of the body

(In quadrupeds, *anterior* means "toward the head end"; *posterior,* "toward the caudal or tail end"; *ventral,* toward the underneath or the belly side; *dorsal,* toward the back or the uppermost side.)

Ventral toward the anterior side
Dorsal toward the posterior side

Superior above, upper
Inferior below, lower

Superficial on or near the surface
Deep remote from the surface

Internal within, inside
External without, outside

(*Inside* or *interior* and *outside* or *exterior* are reserved for reference to body cavities and hollow organs.)

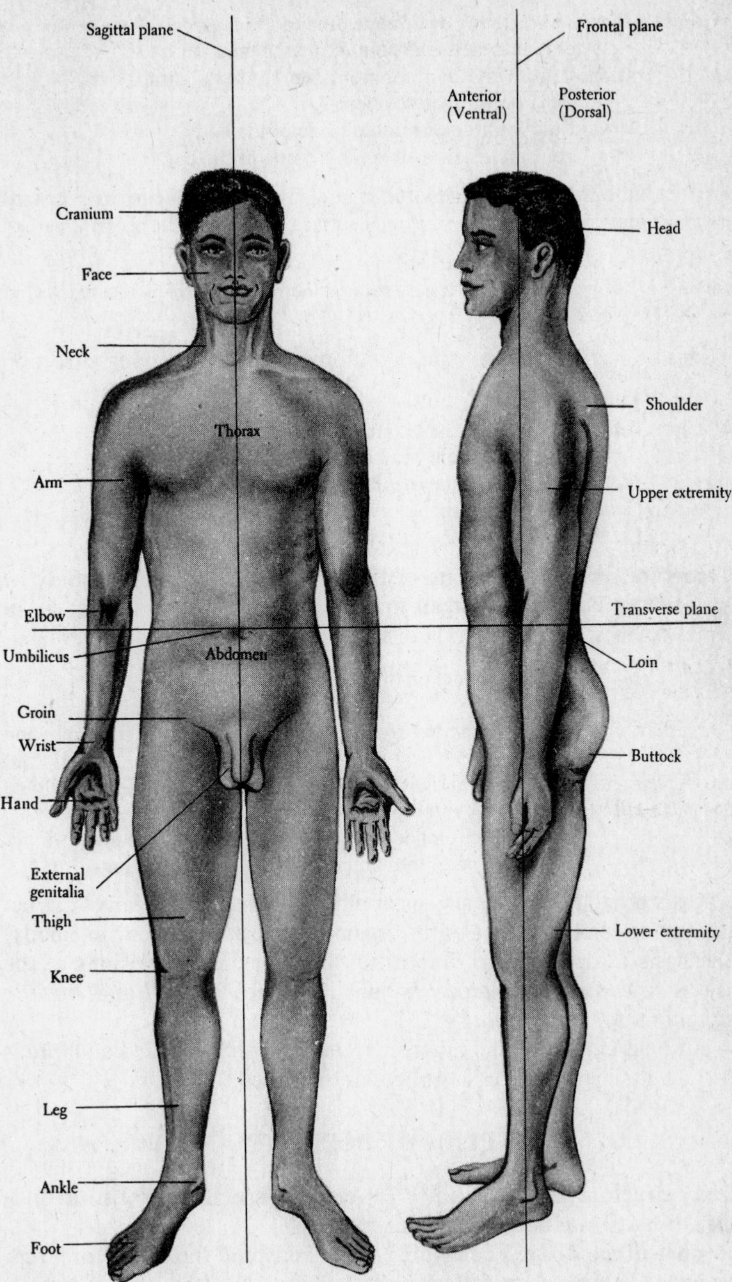

Fig. 1-1. The body in anatomic position.

Proximal	nearest to the body or to some other point regarded as the center of a system; nearest to the point of attachment
Distal	farthest from the body or from some other central point; farthest from the point of attachment
Medial	toward the midline or middle plane of the body
Lateral	away from the midline or middle plane of the body
Central	toward the center of or mid-region; toward the principal axis; inward
Peripheral	toward the periphery or outer surface; away from the center; outward
Parietal	of, pertaining to, arising from, or located on a wall
Visceral	of or pertaining to the viscera or the internal organs located within one of the body cavities

Terms Referring to Direction. Anatomic terms specifying direction are:

Craniad	toward the cranium
Cephalad	toward the head end
Mesiad	toward the median plane
Caudad	toward the tail end; away from the head
Laterad	toward the side; away from the median plane

Terms Referring to Sections of the Body. For purposes of study or description, a body or an organ may be cut into "sections" that lie in various "planes" (Fig. 1-2). These sections are:

Sagittal	a vertical section that divides the body into right and left portions
Midsagittal	a medial, sagittal section that divides the body into right and left halves
Transverse, cross, or horizontal	a cut made at right angles to the long axis of the body, dividing the body into upper and lower portions
Frontal or coronal	a vertical cut made at right angles to the sagittal plane, dividing the body into anterior and posterior portions

The foregoing terms can also be applied to individual organs or structures. In this sense, the axis of the organ (not the axis of the whole body) is the basis of description. Frequently, therefore, in discussions of the body as a whole, *longitudinal* is substituted for *sagittal* and *median longitudinal* for *midsagittal.*

A cut made in any plane other than the foregoing is called an *oblique section,* but such a section is not commonly used.

GENERAL PLAN OF BODY STRUCTURE

Body structure may be studied in terms of specific regions, or divisions, with definite boundaries.

Regions of the Body. The regions of the body and the organs or structures they contain are as follows:

Region	*Organs or Structures*
Head	Brain and sense organs and some of the food-getting and respiratory structures
Neck	Larynx, trachea, esophagus, thyroid and parathyroid glands
Thorax*	Heart and large blood vessels, lungs, trachea, bronchi, esophagus, thymus, thoracic duct
Abdomen*	Digestive, urinary, and reproductive organs; spleen and adrenal glands
Extremities	Arms, forearms, hands, thighs, legs, feet

*The thorax and the abdomen together constitute the *trunk.*

A general idea of the structure of the body can be obtained by a study of midsagittal, transverse, and frontal sections or views.

Midsagittal Section. This is a section made through the midsagittal plane (Fig. 1-2A). It reveals that the body contains two cavities, a *dorsal cavity* and a *ventral cavity.* These cavities with their divisions and the principal organs contained within them are listed in the following table.

Cavity	Division	Subdivision
Dorsal cavity	Cranial cavity (brain)	
	Vertebral canal (spinal cord)	
Ventral cavity	Thoracic cavity	Pleural cavities (lungs)
		Pericardial cavity (heart) (This cavity is located within the *mediastinum,* the region between the pleural cavities, which also contains the thymus gland.)
	Abdominopelvic cavity	Abdominal cavity (small intestine and most of large intestine, liver, gallbladder, pancreas, spleen, adrenals, and kidneys)
		Pelvic cavity (bladder, sigmoid colon, rectum, male organs—seminal vesicles, prostate, and a portion of the vas deferens—and female organs—ovaries, uterus, uterine tubes, vagina)

Transverse Section. This cross section (Fig. 1-3) shows that the body is essentially a "tube within a tube." The body wall is the *outer tube;* the digestive tract, the *inner tube.* The two are separated by a potential space, the *ventral cavity.*

In the region of the abdomen, the body wall consists of the following layers: (1) the outer covering, the *skin* or *integument;* (2) the underlying *subcutaneous tissue* or *fascia;* (3) a *muscular layer* of striated muscles; and (4) the innermost layer or *peritoneum,* a serous membrane.

Observe how the peritoneum of one side meets that of the other side in the dorsal midline to form a double-layered membrane, the *mesen-*

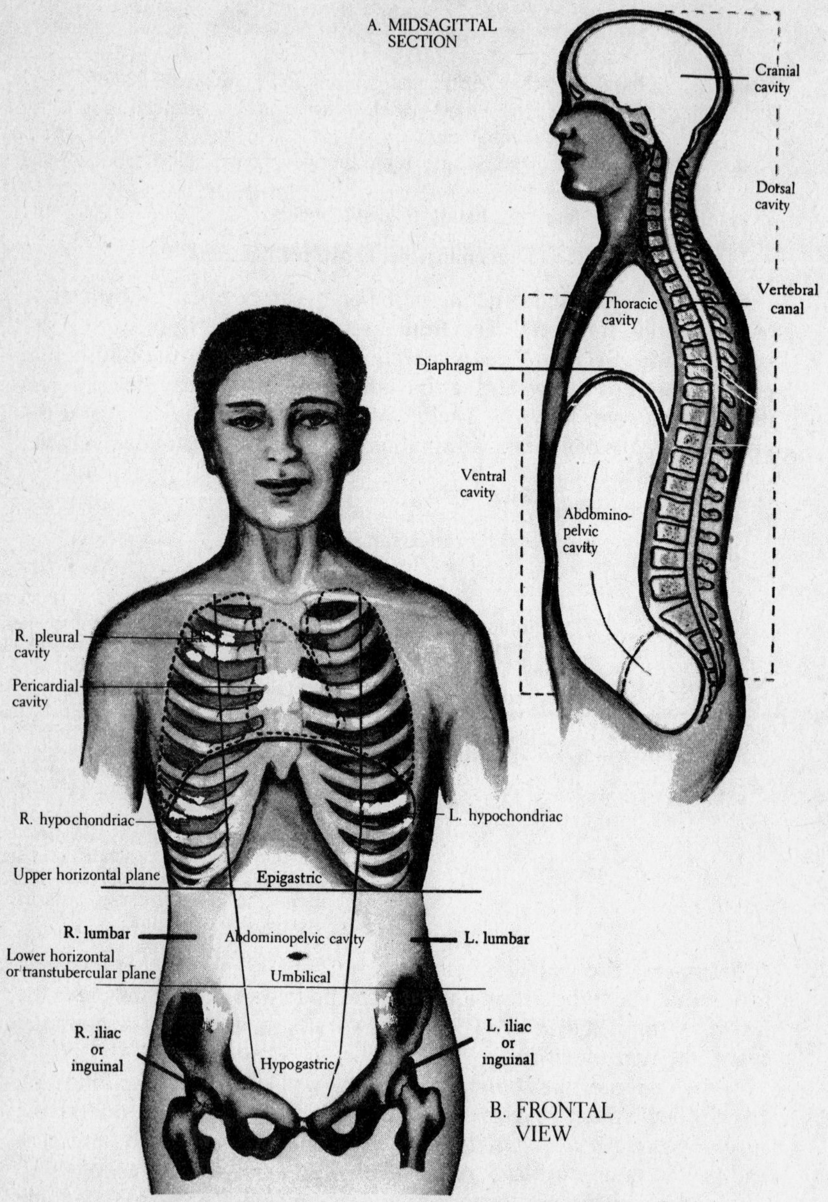

Fig. 1-2. Diagram of the body to show planes, sections, regions, and cavities.

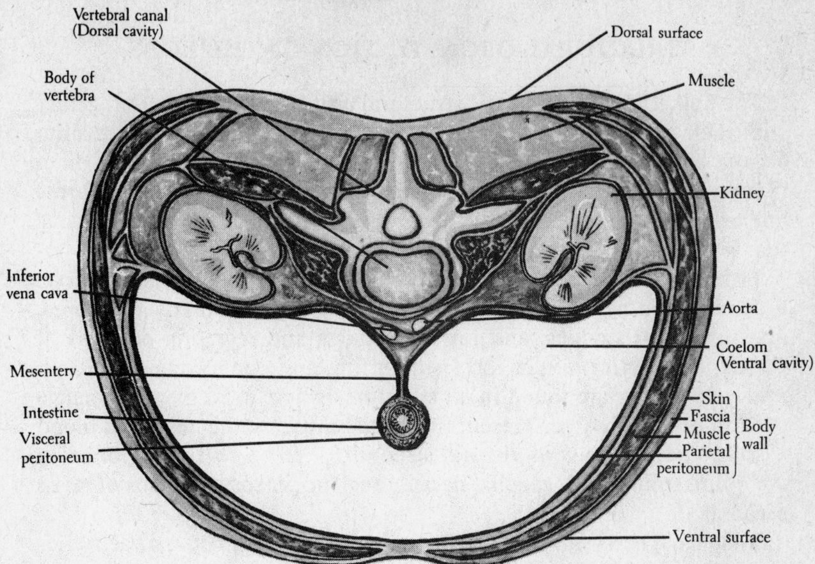

Fig. 1-3. Transverse section through the abdomen showing the plan of the body.

tery, which supports the intestine and encloses the blood vessels, the lymphatic vessels, and the nerves supplying it. The peritoneum lining the body wall is the *parietal peritoneum;* that covering the internal organs is the *visceral peritoneum.*

The dorsal and ventral cavities are seen in this section. Note that the kidneys are not actually within the body cavity but lie dorsal to the peritoneum. In this position they are said to be *retroperitoneal.*

The vertebral column is also shown as the main supporting axis of the body. Within it is the dorsal cavity (*vertebral canal*), which contains the spinal cord.

In the region of the thorax, the body wall contains some skeletal structures (the *ribs*) and encloses the thoracic cavity, which consists of three cavities, the two *pleural cavities,* lined with *pleura,* and, between them, the *pericardial cavity,* lined with *pericardium.*

Note that the body cavities are *potential cavities* rather than large empty spaces. The body organs almost completely fill the spaces of the various cavities, the slight space remaining between the organs being occupied by a serous fluid.

Frontal Section. This view of the trunk (Fig. 1-3) shows the subdivisions of the ventral cavity and some of the internal organs contained within them. The thoracic and abdominal cavities are separated by a muscular partition, the *diaphragm.*

STRUCTURAL ORGANIZATION OF THE BODY

The Cell. The cell is the basic structural unit of the body. Cells, together with their intercellular material, are organized into *tissues,* tissues into *organs,* and organs into *systems.* An *organ* is a structure of two or more tissues that has a more or less definite form and structure and performs one or more specific functions. A *system* is a group of associated organs that work together in performing a series of related functions.

Tissues. The primary tissues of the body are epithelial, connective, muscular, and nervous. *Epithelial tissues* are those that cover surfaces, line tubes and cavities, and form the ducts and secreting portions of glands. *Connective tissues* form supporting and connecting structures. *Muscular tissues* are found in all structures where movement or change of form occur. They are present in skeletal muscles, the heart and blood vessels, and the walls of visceral structures. *Nervous tissue* is found in the brain, spinal cord, ganglia, nerves, and the sensory portions of sense organs.

Systems. The systems of the body, in which all functional activities occur, are the following:

INTEGUMENTARY SYSTEM. This includes the skin and its derivatives (hair, nails, and glands). Its principal functions are protection, prevention of dehydration, and regulation of body temperature. Serving as a base for sensory receptors, it provides information about our environment.

SKELETAL SYSTEM. This includes the bones and cartilage, which form supporting structures, and the ligaments, which bind the bones together at joints. The skeleton functions in support, protects vital organs, and serves for the attachment of muscles by which movement is accomplished. The soft tissue (bone marrow) in the hollow spaces of bones is the seat of the manufacture of blood cells.

MUSCULAR SYSTEM. This includes the contractile tissues of the body, namely skeletal, cardiac, and smooth muscle. These muscles are responsible for the maintenance of posture and all active movements of the body, including locomotion, change in position of body parts, and the movement of blood, food, and other substances through tubes.

DIGESTIVE SYSTEM. This includes the alimentary canal from mouth to anus and its associated glands. The *alimentary canal* functions in the ingestion, digestion, and absorption of food and water and in the elimination of undigested food and some metabolic wastes. The glands (salivary, gastric, intestinal) are the source of enzymes essential for digestion. Other glands, the pancreas and liver, are essential in the digestion and metabolism of carbohydrates and fats. The pancreas is the source of the hormones insulin and glucagon; the liver is the source of bile.

CIRCULATORY SYSTEM. This includes the heart and all the blood vessels (arteries, arterioles, veins, venules, and capillaries) that are essential for circulation of the blood to and from the tissues. *Blood* is the medium through which water, nutrients, oxygen, minerals, and other essential substances, such as hormones and vitamins, are carried to all cells and through which the products of cellular activity, such as waste and metabolic products and internal secretions, are carried away.

Also included in the circulatory system is the *lymphatic system,* which includes the lymph vessels (thoracic ducts, lymphatics, capillaries, lacteals) that convey lymph, and the lymphatic organs (lymph nodules, nodes, tonsils, thymus gland, and spleen), which serve various functions. This system functions in the return of tissue fluid to the blood and in the protection of the body from disease-causing agents. The latter is accomplished principally through the action of phagocytes and the production of antibodies.

RESPIRATORY SYSTEM. This system includes the respiratory passageways (nasal cavities, pharynx, larynx, trachea, bronchi, bronchioles, and alveolar sacs) and the lungs. It functions in the exchange of gases (oxygen and carbon dioxide) between the air and the blood and between the blood and the tissues. It also functions in vocalization, in the elimination of excess heat and water, and in the regulation of acid-base balance of the body fluids.

URINARY SYSTEM. This includes the kidneys, ureters, urinary bladder, and urethra. In the production of urine, the kidneys remove from the blood toxic waste products of protein metabolism, such as urea and uric acid, and also nontoxic materials, such as water and inorganic salts. The kidneys thus eliminate waste products and, in the process, function in the maintenance of water, acid-base, and electrolyte balance. In the male, the urethra functions in the transport of the seminal fluid, which contains spermatozoa.

NERVOUS SYSTEM. This includes the brain, spinal cord, ganglia, nerves, and sensory receptors. Impulses originating in sense organs or receptors convey information concerning the nature of the environment, both external and internal, to the spinal cord, the brain, or both, where, either consciously or by reflex action, appropriate action is taken to initiate responses (movement or secretion) by which proper adjustment to the environment is made. Through nervous impulses, the various activities of the body are integrated and coordinated. In addition, the brain is the seat of such higher mental functions as perception, thinking, reasoning, understanding, judgment, and memory; it is also the seat of the emotions and the source of certain hormones.

ENDOCRINE SYSTEM. This includes the glands of internal secretion (hypophysis or pituitary gland, thyroid gland, parathyroid glands, adrenal glands, pancreatic islets of Langerhans, pineal gland, thymus gland, ovaries, testes, and other secreting tissue). These organs secrete *hormones,* chemical agents that are transported by the bloodstream to other organs or tissues (target organs) on which they exert their effects, either stimulating or inhibiting activity. Through hormones, various activities of the body, especially those involved in growth, development, and reproduction, are regulated and coordinated. Hormones are also involved in the regulation of such basic metabolic processes as the utilization of oxygen, sugar, and various minerals.

REPRODUCTIVE SYSTEM. This system is primarily concerned with the production of offspring. The *male reproductive organs* include the testes, which produce spermatozoa; the ducts (ductus deferens, ejaculatory ducts, and urethra), which convey the sperm outside the body; the accessory glands (seminal vesicles, prostate gland, and bulbourethral glands), which contribute to the seminal fluid; and the penis, the copulatory organ. The *female reproductive organs* include the ovaries, which produce ova; the uterine tubes, which serve as a site for fertilization and for conduction of the fertilized egg, or zygote, to the uterus; the uterus, in which the embryo or fetus develops; the vagina, which serves as the female copulatory organ, as a birth canal, and as a passageway for menstrual fluid; and the external genitalia or vulva (the clitoris, major and minor labia, and mons pubis). The mammary glands are sometimes regarded as reproductive organs.

The testes and ovaries also function as endocrine organs producing hormones that regulate the development of secondary sex characteristics and various physiological processes. The penis is also an excretory organ.

SUMMARY. The integrated functioning of all these organ systems is essential for the maintenance of life, for physical and mental health, for adjustment to the environment, and for the continuation of the species. The detailed study of the anatomy and physiology of each of these systems constitutes the subject matter of this text.

METHODS OF STUDYING LIFE ACTIVITIES

Physiology, the *study of function,* encompasses all the activities of an organism—not only the activities of the organism as a whole but also those of individual organs, tissues, and cells. It is concerned with all the physical and chemical changes that take place within the organism. Physiologists study man as a physiochemical machine and seek to explain life processes in the body on the basis of physical and chemical changes that occur within it and among its parts. It should be borne in mind, however, that *function is an aspect of structure* and that *all functions imply structure.*

Anatomy is primarily an observational science, and its origins go back into the early history of humankind. Modern anatomy dates from the work of Andreas Vesalius (1514–1564). *Physiology* is primarily an experimental science, and its origins as such date from the work of William Harvey (1578–1657) on circulation. With the advent of the electronic and nuclear age, new and sophisticated instruments and techniques have enabled physiologists to study processes that were considered impossible a few years ago. The following are some of the methods and procedures used in the study of life activities.

Direct Observation. This is examination of bodily structures and functions under both normal and abnormal conditions, noting changes that occur.

Microscopic Examination. In addition to the light microscope, a number of specialized microscopes (see page 1) have been developed. The use of these instruments, especially the electron microscope, has revealed details of cellular fine structure (ultrastructure) previously unknown.

Chemical Analysis. This is examination of the chemical makeup of the substances used and products produced by the body in its metabolic activities in health and disease. Special types of apparatus and techniques utilized include the use of the spectrophotometer, flame photometry, laser beam, ultrasonic waves, and various separation techniques such as use of molecular sieves, paper and thin-layer chromatography, electrophoresis, ion exchange, fraction collection, density gradient determination, and ultracentrifugation. By the use of modern staining procedures and specialized techniques, such as radioimmunoassay (RIA) and enzymeimmunoassay (EIA), for the identification of enzymes and other substances, many of the metabolic processes that take place within the cells can be determined with great accuracy and the localization of function in specific cells can be established.

Tissue Culture. Tissue culture is the growth of body tissues or cells in a culture medium outside the body.

Grafting and Transplantation. This is the transference of tissues from one part of the body to another region or from one organism to another.

Perfusion Methods. These involve the removal of organs from the body and their observation and study while they are supplied with blood or nutrient fluids.

Injection Techniques. Here, specific substances such as air, dyes, radioactive isotopes, and hormones are introduced into the body so that their course can be traced and their effects noted. This technique is employed in obtaining arteriograms and venograms of various organs, such as the brain, heart, and kidney. The cavities of various organs, such as the gallbladder, gastrointestinal tract, lungs, and brain, can be revealed by employing such procedures.

Computerized Axial Tomography (CAT). This procedure combines the techniques of radiology, scintillation counting, computer processing, and cathode ray tube display to produce a view of a transverse section of the body or a part of it that is remarkably similar to an anatomic cross section. It is of special value in depicting intracranial and intra-abdominal abnormalities. CAT provides clear outlines of structures not visible by other techniques.

Nuclear Magnetic Resonance (NMR). A great advance on and development of CAT, the images produced by this method do not depend upon ionizing radiation, but on magnetic fields and radio frequency pulses, making it possible in the brain to distinguish between white and gray matter, and thus the diagnosis of demyelinating disorders in the living, such as multiple sclerosis, etc. Tumors are easily visualized as well as other lesions, thus avoiding invasive techniques such as angiography and pneumoencephalography. It is possible to produce longitudinal and sagittal scans of the brain and spinal cord. The technique can show the myocardium without the use of contrast medium, and in the kidney can differentiate between cortex and medulla. In the liver parenchymal disorders of various kinds can be distinguished, and primary biliary cirrhosis from ordinary cirrhosis.

Positron Emission Tomography (PET). The PET reveals the biochemical changes occurring as the living brain interacts with the world around it. Deoxyglucose is labeled with a positron-emitting isotope (18-fluorine) and the uptake of the radioactive compound can be measured with special detectors in the awake subject's brain. By breathing air containing a trace of carbon monoxide, after a few minutes PET detectors can determine the amount of isotope present in different areas of the brain, and also measure the metabolism of the brain. Basic information on how the normal brain functions in healthy people, such, for example, as the changes occurring in the brain while listening to a story, or in *thinking* about writing as also in writing, or in tasks requiring verbal as compared with spatial analysis. By this method the binding of a neurotransmitter to its receptor may also be seen—and much else.

Electronic Instrumentation. This is the application of electronics to the study of life activities. It has led to the development of *bioinstrumentation systems* that are capable of detecting extremely minute physiological changes, converting them into electrical impulses, increasing their amplitude, and presenting them to the observer in an understandable form, either visual or auditory. Basically such a system involves an *input transducer,* a sensing device that responds to a biological activity, such as a nerve impulse, muscle contraction, heartbeat, or sound wave; an *amplifier* that increases the signal; and an *output transducer* that converts the electric signal into a form that can be detected and interpreted. This may be in the form of a graphic recorder in which a writing device records on paper the changes that occur; a cathode ray oscilloscope, in which changes can be noted visually; or a loudspeaker in which audible changes are noted. Other output transducers, such as panel meters, counting devices, and combined audiovisual monitors, are employed.

The preceding principles are employed in the study of the heart by an *electrocardiograph* (ECG), the brain by an *electroencephalograph* (EEG), eye muscles by an *electrooculograph* (EOG), retina of the eye by an *electroretinograph* (ERG), and body muscles by an *electromyograph* (EMG).

Special Physiological Devices. Special types of apparatus, such as respirometers, calorimeters, kymographs, and others, make possible an understanding of various physiological activities. The development of microelectrodes and microdissection devices, which have made microsurgery possible, have also provided much information about what goes on within the individual cells of tissues.

Summary. Through the use of these methods and procedures and the utilization of special techniques such as the use of ultrasound, artery catheterization, and others, a vast amount of information has been accumulated about the body. This has resulted in a better understanding of the factors responsible for the normal or abnormal functioning of organs. Progress in physiology has enabled physicians to develop more reliable procedures for diagnosis of disease and to develop more effective therapeutic measures, which have done much to alleviate pain and suffering and to prolong life.

BASIC LIFE PROCESSES

The basic life processes can be better understood if they are first studied in a simple organism, such as an ameba. The *ameba* is a one-celled animal consisting of protoplasm, a jellylike substance of which every plant and animal cell is composed. The form of an ameba, though roughly the same for each species, is variable among individuals. As a living entity, however, the ameba exhibits characteristics that are generally attributable to all animal life: irritability, conductivity, contractility, metabolic activities, reproduction, and adaptation.

Irritability and Conductivity. *Irritability* is the ability of an organism to respond to a stimulus (i.e., an internal or external environmental change). If an ameba is disturbed, as by a weak electric shock or by a touch, it will respond with movement accomplished by a flow of its protoplasm and the formation of blunt, rootlike processes called *pseudopodia.* The ability to respond at a point remote from the stimulus depends upon *conductivity,* in which a state of excitation is transmitted, a property possessed by nerve cells to a high degree.

Contractility. *Contractility* is the ability of an organism to change its form by shortening. Such a reaction may be due either to an external stimulus or to factors within the animal. In the human body certain cells (the white blood cells, or leucocytes) have this ability, a reaction termed *ameboid movement.* All muscle activity depends on the ability of protoplasm to shorten or change shape.

Metabolic Activities. The foregoing processes require the expenditure of energy and for the supply of energy the ameba, like all other living organisms, depends on food. Although the term *metabolism* in its broadest sense includes all physical and chemical processes involved in life activities, it is commonly applied only to the processes by which food substances are converted into energy or body substance. Metabolism has two aspects, *anabolism*, the building-up processes involving synthesis of complex molecules from simple ones, and *catabolism*, the breaking-down processes, in particular the breakdown of food, and the resultant liberation of energy.

The detailed processes of *food getting and utilization* are (1) ingestion; (2) digestion; (3) utilization; (4) respiration; (5) excretion; and (6) egestion.

INGESTION. The ameba extends its pseudopodia and engulfs a food particle, which consists of organic material. The particle becomes enclosed in a cavity, the *food vacuole.*

DIGESTION. Within the vacuole, the food, which is composed of complex chemical compounds, is broken down, by the action of catalytic agents (enzymes), into simple, diffusible substances. The enzymes, produced by the ameba's protoplasm through *secretion*, pass from the protoplasm into the vacuole by means of *diffusion.* The products of digestion diffuse from the vacuole into the surrounding protoplasm, a process called *absorption.*

UTILIZATION. Within the protoplasm the food is utilized in the following ways: (1) to *increase the amount* of protoplasm (growth by assimilation); (2) for *repair* (rebuilding of protoplasm broken down through cellular activities or injury); (3) for *regulation* of cellular activities; and (4) as a *source of energy.*

RESPIRATION. The energy in a food molecule is released as a result of *oxidation.* Oxygen is taken in from the surrounding water (the ameba's life milieu) by diffusion through the cell membrane. It reacts with the food molecules, which are broken down principally into carbon dioxide (CO_2) and water (H_2O), and energy is released. This exchange of gases between the organism and its environment is called *respiration*, or, more specifically, *external respiration*; the oxidative processes within the cell constitute *internal respiration.*

EXCRETION. As a result of metabolic activities, waste products such as water, urea, uric acid, and carbon dioxide are produced. Excretion is the process of eliminating these substances through the surface of the cell.

EGESTION. Ingested food usually contains some indigestible or unusable material which has taken no part in metabolism. In the ameba the food vacuole eventually ruptures, and this material is discharged from the cell.

Reproduction. While metabolic processes continue, the ameba is a living and growing cell. Eventually, however, growth would cease and degenerative changes occur, and the ameba would die, but for reproduction. The process of reproduction is accomplished by *cell division,* or *mitosis,* in which the nucleus and cytoplasm are divided equally between each of the two daughter amebas. This process, in which only one parent is involved, is *asexual reproduction.* The two new cells, young and rejuvenated, begin life anew.

Adaptation. The ameba lives in an environment, fresh or salt water, which contains many types of chemicals, some favorable to life processes and others unfavorable. Furthermore, the environment is constantly changing; to survive, the ameba must adapt itself to these changes. Life processes, then, are those activities resulting from interaction of the life substance (protoplasm) with the environment. Indeed, life has been defined *as the continuous adjustment of an organism to its environment.*

UNICELLULAR AND MULTICELLULAR ORGANISMS

The continuity between unicellular and multicellular organisms stretches in a long, unbroken chain from the beginnings of life on earth. The differences that have come into being between such organisms are the results of the process of *evolution.* In evolution, progressively adaptive variations have occurred, leading to the development of the highly complex specialized forms that inhabit the earth today.

Life Processes in a Multicellular Organism. A complex organism, such as a human being, is composed of trillions of cells, each consisting of protoplasm, the same type of substance as that found in the ameba. The same fundamental processes described for the ameba take place in each cell of the organism: Food and oxygen are taken in, energy transformations occur, protoplasm is built up and broken down, waste products are formed and eliminated, and cell division occurs, or cells may die. As compared with unicellular organisms, however, the more complex living forms have certain *additional qualities:* specialization of activities, interdependence of cells, and sexual reproduction.

SPECIALIZATION. In a multicellular organism, division of labor prevails. Cells are structurally differentiated into diverse types, each adapted for the performance of specific functions. A fundamental specialization and division of labor is that between cells reserved and later specialized for reproduction and the remaining cells of the body. The former constitute the *reproductive* or *germ cells* and the latter, the *soma.* The soma includes collectively all the cells that make up the structure of the body and share in the maintenance of its activities, such as muscle, gland, nerve, blood, and connective tissue cells; the reproductive cells include only the eggs and the sperm. Their sole function is the production of a succeeding generation.

INTERDEPENDENCE. Whereas the ameba is an independent cell carrying on all the necessary life activities, in a complex organism each cell is dependent on other cells. The ameba (one cell) is capable of moving about more or less freely, but the cells of a complex organism are generally *fixed in position.* For this reason, special mechanisms are required for the circulation of substances to and from cells that are remote from the surface of the organism. In addition, highly developed *correlating* and *integrating mechanisms* are required to coordinate activities in various parts of the body.

SEXUAL REPRODUCTION. The simplest organisms reproduce by asexual means. Most of the single-celled organisms, such as the ameba, and single cells of the body reproduce by means of *binary fission* (asexual cell division): The parent cell divides and becomes a pair of daughter cells. In a majority of multicellular animals, including man, reproduction is sexual: Two parent organisms produce sex cells, which unite to form a third individual. From a single cell (the zygote) resulting from the union of two sex cells (the egg and the sperm), all the succeeding cells of the new organism develop by repeated cell division. The two original sex cells contain the hereditary "packets" (chromosomes containing the genes) that influence the expression of the responses of the tissues to the environment and transmit these potentialities from one generation to the next.

The Body as a Machine. A machine is an apparatus or device capable of releasing energy and performing work. By this definition, all organisms may be regarded as "living machines." If an organism is compared with a nonliving machine, certain similarities and differences are noted.

SIMILARITIES. Nonliving machines and organisms have the following similarities: (1) Each requires fuel as a source of energy; (2) the nature of the fuel is the same for each—i.e., it consists of combustible, organic substances; (3) oxidation or combustion occurs within each, resulting in the release of energy; (4) the released energy is used for the performance of mechanical work or in the production of heat; (5) waste products formed must be eliminated; and (6) various parts have specialized functions, and the failure of one part may cause stoppage of the entire machine (death of organism).

DIFFERENCES. Organisms differ from most nonliving machines in the following respects: (1) They are automatic, or self-operating; (2) they are self-repairing; (3) they have the capacity for growth; (4) they are capable of reproducing their kind; (5) they are adaptable and, within limits, have the ability to alter their structure; (6) they exhibit an awareness of their environment; (7) they exhibit purpose and foresight; and (8) they cannot be set in motion again after prolonged stoppage (death) has occurred.

THEORIES OF THE NATURE OF LIFE

In striving to understand the nature of life, the physiologist makes use of the *mechanistic theory.* This theory holds that all life processes are the result of the chemical and physical activities within an organism and that death is the cessation of these processes. Notwithstanding the foregoing, another theory, the *vitalistic theory*, holds that life is something more than merely an organization of chemical compounds and the physical and chemical reactions taking place in protoplasm. It maintains that life is a "vital something," the nature of which is not yet known, and that this life does not stop at death. The vitalistic theory embodies essentially the spiritual viewpoint on the nature of life.

The cornerstone of our modern knowledge of life processes, and of the mechanistic theory, is the work of Lavoisier (1743–1794), who discovered the nature of burning (oxidation) and applied this to the study of living things. When it was learned that the oxidative processes within the cells of living organisms were basically the same as those occurring outside such cells, the foundation was laid for interpreting living phenomena in physicochemical terms.

Although a vast body of facts has been amassed about fundamental processes involved in bodily functions, there is still much to be learned. Physiology is an experimental science; with each new generation of students, teachers, and researchers, more contributions are made and a better understanding of life is obtained.

2: THE CELL AND PROTOPLASM

The cell is the basic unit of the human body. It is a *unit of structure,* since all parts of the body are composed of cells or the products of cells. It is a *unit of function,* since all activities of the body depend on the activities of cells. It is a *unit of heredity,* since it is through the reproductive cells, the egg and the sperm, that the units of genetic materials (genes) are passed from one generation to the next. And, finally, it is a *unit of development,* since it is through processes of cell multiplication and differentiation that new tissues and organs are formed and a new individual develops to maturity.

A *cell* may be defined as a mass of protoplasm consisting of a *nucleus,* or nuclear material, and *cytoplasm,* the outer border of which constitutes the *cell membrane.* A cell is the smallest unit of living substance that is capable of carrying on life functions. In some instances cell membranes are absent, and the result is a multinucleated cell called a *syncytium* or *coenocyte,* such as a striated muscle cell. Some cells may lack a nucleus, such as red blood cells in humans, which lose their nucleus during development.

THE PROTOPLASMIC DOCTRINE

The cell was first seen in plants and so named by Robert Hooke (1635–1703) and described in his *Micrographia* in 1665. Observing a thin slice of cork under a microscope, he saw what reminded him of the "hexangular cells" of honeycomb; hence, the name *cells.* With great prescience Hooke wrote of the possible "passages," or interactions, between the cells: ". . . Me thinks, it seems very probable, that Nature has in these passages, as well as in those of Animal bodies, very many appropriated Instruments and contrivances, whereby to bring her designs and end to pass, which 'tis not improbable, but that some diligent Observer, if help'd with better *Microscopes,* may in time detect." This appears to have been the first surmise that the materials, the "Instruments and contrivances," of development are contained in the cell.

It was not until much additional work had been done by "diligent Observers" that the cell theory was finally formulated on firmer foundations than were available to Hooke. In 1839 Theodor Schwann (1810–1882) stated the view that "all organisms are composed of essentially like parts, namely, of cells," and in 1859 Rudolf Virchow (1821–1902), with the statement that "Every cell arises from a preexisting cell," clinched the argument for the cell theory. The theory stimulated investigation into the nature of cells and led to Max Schultze's formulation of the *protoplasmic doctrine* in 1861. This doctrine regarded the jellylike substance (protoplasm) within the cell membrane as the living substance of all animals and plants.

SIZE AND SHAPES OF CELLS

Animal cells range in *size* from 2 μm* (protozoan blood parasites) to 5 cm or more (ostrich egg yolks) in diameter. Red blood cells of the body have a diameter of 7.7 μm and egg cells in the ovary 120 μm. Some multinucleated cells, such as those of striated muscle, may attain a length of 6 cm. When free and suspended in a fluid, cells tend to assume a spherical shape. In the body, however, they may have almost any shape, depending on their function and their relationship to other cells.

METHODS OF STUDY

Living cells may be studied in the body by the use of a transparent chamber of metal and glass or by observing cells in a natural transparent chamber such as the eye. They may also be studied immediately upon their removal from the body, as in the case of blood cells or living tissue taken in biopsies for the purpose of diagnosis. Living cells are also sometimes grown outside the body in tissue cultures.

In order to obtain detailed information concerning the structure of cells, it is necessary to examine fixed, stained preparations. Briefly, the steps in preparing a tissue for examination are as follows. The tissue is first immersed in a *fixing solution.* This kills the cells and fixes or coagulates the protein, thus preserving the general structure of the nucleus and the cytoplasm. Then the tissue is cut by use of a *microtome* into extremely thin sections in order to permit light rays or electrons to pass through it; this is essential for microscopic examination. In order that the tissue may be cut clearly, it must be infiltrated by an *embedding agent,* usually celloidin or paraffin for light microscopy or an acrylic or epoxy resin for electron microscopy. Preparatory to embedding, *dehydration* is accomplished by passing the tissues through a graded series of alcohols. After sectioning, the tissue is *mounted* on a glass slide and *stained.* By use of selective stains, parts can be readily distinguished and information about their chemical constituents obtained.

Other methods of tissue preparation are sometimes employed. When speed is essential, as in a biopsy during surgery, the tissue may be *frozen* and then sectioned, stained, and examined in a matter of minutes.

*A *micrometer* (formerly, *micron*) = 0.001 millimeter, or approximately 1/25,000 of an inch.

CELL STRUCTURE

A typical cell (Fig. 2-1) may be represented by a *nucleus* surrounded by a mass of *cytoplasm,* the whole enclosed within a *plasma membrane.* The structure of each is discussed in the following paragraphs.

Nucleus. The nucleus of a nondividing cell is usually a single spherical or ovoid body bounded by a *nuclear membrane.* It usually contains a clear, semifluid material called *nuclear sap* or *karyolymph,* in which are found *chromatin granules* and one or more *nucleoli.*

NUCLEAR MEMBRANE. The *nuclear membrane* is a *unit membrane* (two layers separated by a space, the *perinuclear space*). The outer layer of the membrane is continuous with the endoplasmic reticulum of the

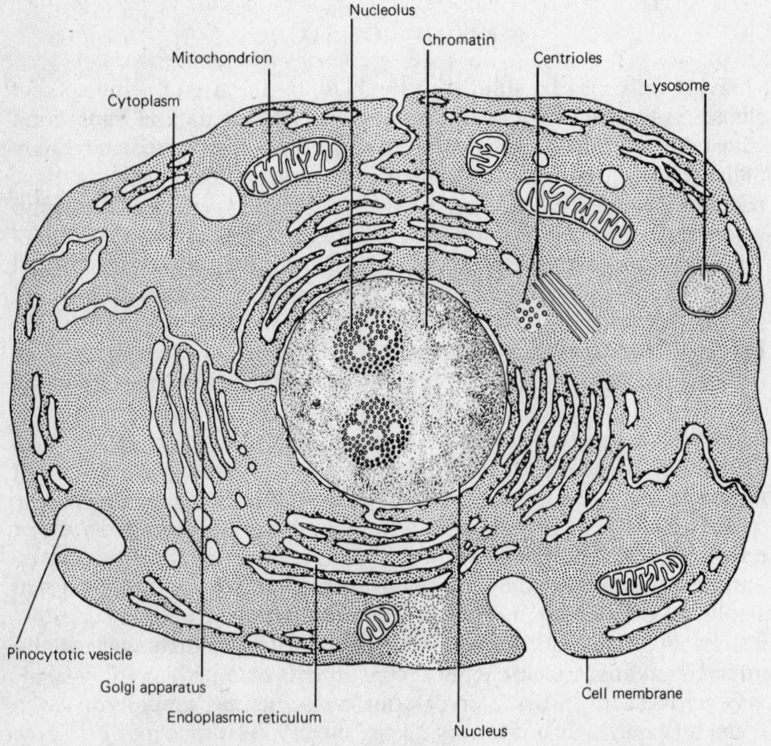

Fig. 2-1. Diagram of a typical cell. (Reprinted with permission of J. B. Lippincott Co. from E. M. Greisheimer and M. P. Wiedeman, *Physiology and Anatomy,* 9th ed., 1972.)

cytoplasm and may bear small projections, the *ribosomes;* the inner layer bears minute *pores.*

CHROMATIN. *Chromatin* is a darkly staining material appearing in the form of granules scattered throughout the nucleus. It is concentrated on the inner surface of the nuclear membrane and also forms a shell about the nucleolus. The basophilic staining property of the chromatin is due to the presence of *deoxyribonucleic acid* (DNA).

NUCLEOLUS. The *nucleolus* is a basophilic body composed principally of *ribonucleic acid* (RNA). Cells may possess one or more nucleoli. Nucleoli disappear during cell division.

NUCLEAR SAP. *Nuclear sap,* or *karyolymph,* is a clear, semifluid matrix that fills the spaces about the chromatin granules and the nucleoli.

During cell division, the chromatin granules become concentrated into densely staining bodies called *chromosomes.* The number of chromosomes is constant in the cells of a given species (46 in the somatic cells of humans). The chromosomes carry the *genes* or hereditary determinants.

FUNCTIONS OF THE NUCLEUS. The nucleus is essential for the life of a cell. Through the nucleic acids, DNA and RNA, it controls all the metabolic activities of the cell; it is a directing and organizing unit of the cell and a repository of the genetic materials, the genes. It is essential for cellular reproduction.

Cytoplasm. The cytoplasm is the portion of a cell that lies outside the nucleus. It is primarily involved in the functional activities of a cell, such as secretion, absorption, conduction, and contraction.

The cytoplasm consists of a semifluid matrix in which are suspended a number of organelles and inclusions. *Membranous organelles* play an active role in the metabolic activities of a cell; *nonmembranous organelles* are not involved in metabolic activities but have special functions. *Inclusions* are substances such as stored foods, pigments, and secretion granules.

MEMBRANOUS ORGANELLES. These include the cell membrane, endoplasmic reticulum, Golgi apparatus, lysosomes, and mitochondria.

Cell Membrane or Plasmalemma. This plasma membrane is invisible under an ordinary light microscope, but the electron microscope shows it to be about 75Å* in thickness and consisting of two dense layers enclosing a third, clear, central layer. It is a lipoprotein complex, the outer and inner layers or lamella being composed of proteins (about 60 percent) and the middle layer of lipids (about 40 percent). It is a selectively permeable membrane and thus controls the passage of fluids and various substances into and out of cells. Modifications of the cell membrane are *microvilli, desmosomes, terminal bars,* and *intercalated discs,* which serve to increase the absorbing surface or to hold the cells togeth-

* 1 Å = 1 ångstrom = 10^{-8} cm (1/100,000,000 of a centimeter)

er. Invaginations or pockets capable of being pinched off and passing into the interior of the cell may form. These allow *pinocytosis,* the taking in of fluid droplets, or *phagocytosis,* the taking in of particulate matter. A reverse process called *budding,* in which cytoplasmic protrusions of the cell are pinched off, allows cells to discharge fluid or solid material. In addition to the cell membrane, some cells have a *cell coat,* a very thin layer, consisting principally of sialic acid–containing glycoproteins, that assist in holding cells together and help separated cells to reorganize and associate with others of their own kind.

Endoplasmic Reticulum (ER). Also called the *ergastoplasm,* this is a lacelike network of hollow, membranous *tubules* or bladderlike *vesicles.* Large, flattened vesicles are sometimes called *cisternae.* In a part of the endoplasmic reticulum, the walls of the tubules or vesicles bear minute bodies called *ribosomes* on their outer surface. This is called *rough* or *granular reticulum;* that lacking ribosomes is called *smooth* or *agranular reticulum.* The tubular endoplasmic reticulum, together with the other membranous structures (cell membrane, nuclear membrane, Golgi apparatus) forms a continuous, interconnecting network within the cell.

The *ribosomes,* which are attached to the rough endoplasmic reticulum, synthesize proteins, such as digestive enzymes of the digestive glands or antibodies of plasma cells, which are destined for transport. The smooth endoplasmic reticulum is abundant in cells involved in lipid metabolism, especially those that secrete steroid hormones. It is also abundant in liver and muscle cells, which are involved in glycogen metabolism, and in cells of the stomach in which hydrochloric acid (HCl) is produced.

The Golgi Apparatus. This netlike structure is usually located near or surrounding the cell center, or *centrosome.* It is prominent in secretory cells, where it acts on proteins secreted in the rough endoplasmic reticulum, concentrating them by withdrawing salt and water and refining them by adding carbohydrates. It then encloses these substances in small, membranous containers and delivers them to the cytoplasm, where they constitute *secretory granules,* ready to be utilized by the cell or discharged as a secretory product. The Golgi apparatus is variable in size, becoming larger during periods of secretory activity. The Golgi apparatus is also involved in the formation of lysosomes.

Lysosomes. These are small vesicles, each containing a number of hydrolyzing enzymes, especially acid phosphatase. The membrane enclosing the vesicle acts as a barrier that prevents the enzymes from acting on the contents of the cytoplasm of the cell. Normally enzymes act on substances engulfed by the cell in the processes of pinocytosis and phagocytosis. The rapid destruction of tissues after death (postmortem degeneration) is largely the result of the breakdown of lysosomes and the liberation of their enzymes, which cause the digestion or *autolysis* of cell constituents. Lysosomes are present in nearly all cells and are especially abundant in neutrophils and other phagocytes. They originate in the Golgi apparatus and, when first formed, comprise *dense bodies,* which are inactive.

Mitochondria. These membranous organelles are present in all cells. They are especially numerous in cells with high energy requirements, such as liver cells, which may contain as many as 1000. Each mitochondrion is a hollow cylindrical or ovoid body possessing two membranes, an outer one that is smooth and an inner one that is folded, with projections called *cristae.* The space between the cristae comprises the *matrix.* Within the mitochondria are enzymes of the Krebs cycle and of the cytochrome electron-transfer system by which energy obtained from the oxidation of foods is incorporated by oxidative phosphorylation into adenosine triphosphate (ATP) and made available to cells. Mitochrondria are consequently regarded as the powerhouses of the cell.

NONMEMBRANOUS ORGANELLES. These include a number of structures that do not engage actively in metabolic processes but play a role in a number of specialized activities. Among them are free ribosomes, microtubules, filaments and fibrils.

Free Ribosomes. Free ribosomes are those not attached to the endoplasmic reticulum. They synthesize proteins of the cytoplasmic matrix—for example, hemoglobin in red blood cells.

Microtubules. Microtubules are extremely minute tubules present in all cells. They may be dispersed in the cytoplasm, where they serve a supportive function and facilitate the transport of various substances. They aid in the movement of secretion granules in gland cells, contractile proteins in muscle cells, and proteins and neurotransmitters in nerve cells. They are concentrated in parallel fashion in *cilia* and *flagella* and are responsible for their movement. In dividing cells, microtubules form spindle fibers and facilitate the movement of chromosomes. *Centrioles* are two short, cylindrical bodies located in a *central body* (centrosome) near the nucleus. During mitosis they form the poles of the mitotic spindle. They are also involved in the formation of cilia and flagella.

Filaments. Filaments are present in nearly all cells, especially muscle and nerve cells. They serve for support and are involved in contraction processes. A bundle of filaments comprises a *fibril.*

INCLUSIONS. These include certain structures and materials that are not considered to be an integral part of the protoplasm of a cell and are not involved in metabolic functions. These include stored foods, pigments, secretions, and various crystalline substances.

Stored Foods. These include *glycogen* stored in muscle and liver cells and *fat* stored in cells of adipose tissue. Proteins are not stored in animal cells. The protein present in a cell is that which forms an integral part of its structure, and it is not available for use elsewhere in the body except under such extreme conditions as starvation.

Pigments. Coloring matter or pigment occurring in cells may be *exogenous* (arising outside the body) or *endogenous* (arising within the body). Exogenous pigments include *carotene,* the yellow pigment of carrots, egg yolk, and butter; *dust,* such as coal dust, which is inspired; and *mineral matter,* such as lead or silver. Endogenous pigments include *hemoglobin* and *hemosiderin, bilirubin, biliverdin, melanin,* and *lipochromes.*

Secretions. The secretory products of glands are usually stored in the form of secretory granules. Enzymes are produced as *zymogen granules* and stored in cells before their discharge.

Crystalline Substances. These are found in various cells, such as the interstitial cells of the testes, luteal cells of the ovary, and liver cells. Their chemical nature and function are uncertain.

CELL DIVISION

Cells originate by means of cell division, a process in which a parent cell divides and gives rise to two daughter cells. The methods by which cell division is accomplished are amitosis, mitosis, and meiosis.

Amitosis. Amitosis is a simple, direct method of division in which the nucleus and the cytoplasm both constrict, the two resulting masses forming two new cells. Amitosis rarely occurs in the human body.

Mitosis. (Fig. 2-2). Mitosis is a complex, indirect method of cell division. In this process, the chromatin filaments of the nucleus are rearranged into chromosomes, each of which divides into duplicate parts; these identical parts (*chromatids*) then line up opposite each other. Thus the chromosomes, which contain the genes, are divided equally, one-half of the split parts being passed on to each daughter cell. Since all cells in the human body are derived through successive cell divisions from the original fertilized egg cell, it follows that, by mitosis, the nuclear material possessed by all cells of the body is identical.

Mitosis is a continuous process, but for convenience it is usually described as occurring in four stages, or *phases.* The stage between two successive divisions is called the *interphase.* During this phase, the chromosomes lose their condensed state and generally become invisible. Shortly before mitosis begins, the chromosomes duplicate themselves; the two replicas are known as *chromatids.* The four phases are prophase, metaphase, anaphase, and telophase.

PROPHASE. The *centrioles* separate and, with the developing *astral rays* (microtubules), move to opposite poles of the cell. A *spindle* of microtubules develops between the centrioles. Chromatin granules condense and form linear units, the *chromosomes,* which divide longitudinally into chromatids. The chromatid pairs become arranged on the *equatorial plane* of the spindle. The nuclear membrane and the nucleolus disappear.

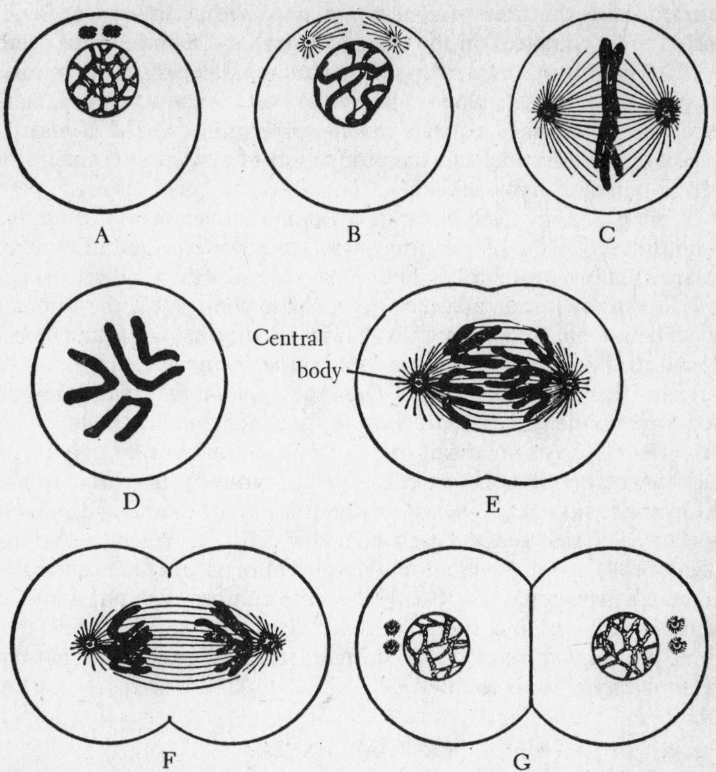

Fig. 2-2. Mitosis in animal cells (diagrammatic). (*A*) Interphase (resting cell). (*B*) Prophase. (*C*) Beginning of metaphase—chromosomes at equator of spindle. (*D*) Polar view of chromosomes at equator of spindle. (*E*) Anaphase. (*F*) Early telophase. (*G*) Daughter cells in interphase. (From G. Alexander, *Biology,* Barnes & Noble, Inc., 1954.)

METAPHASE. The chromatid pairs lie suspended on the spindle midway between the centrioles.

ANAPHASE. The chromatid pairs separate and move, or are drawn, along the spindle fibers toward the two centrioles.

TELOPHASE. Each set of daughter chromosomes reconstructs itself into a nucleus. The chromosomes begin to break up into irregular chromatin granules. Nucleoli reappear and the nuclear membrane re-forms. The two centrioles may divide in preparation for the next division. The cytoplasm constricts and the two daughter cells become distinct from each other and enter interphase.

The division of nuclear material, or nuclear division, is called *karyokinesis;* the division of the cytoplasm, *cytokinesis.*

New cells are produced in enormous numbers throughout a person's lifetime. From the time of conception until the period when growth appears to be complete (in the early twenties) and then on until death, the development of any new tissue or structure depends upon mitosis. The production of new blood cells by the bone marrow and lymphatic tissue, the replacement of cells in the epidermis of the skin and the mucosa of the intestine, and the production of new cells to replace injured or damaged tissue all depend upon the process of mitosis.

Mitosis is also involved in the development of *neoplasms,* or *tumors,* uncontrolled growths of abnormal tissue that serve no useful function and are usually harmful to the host. When the growths are localized and the cells do not invade adjacent tissue, as in most cysts, the tumor is said to be *benign.* Neoplastic cells that invade normal tissue and spread through the body (*metastasize*) are said to be *malignant.* Malignant tumors are *cancers.* Cancers are of two types: *carcinomas* if they are derived from epithelial cells; *sarcomas,* if from nonepithelial cells.

Meiosis. This is a modified form of mitosis that occurs only in the development of reproductive cells. In this process the chromosomes line up as pairs (*synaptic mates*). In the division of the cell, one chromosome of each pair goes to each daughter cell. The result is that the daughter cells possess one-half the normal number of chromosomes.

Through meiosis (see Fig. 8-1, Vol. 2), the number of chromosomes in the functional egg and sperm is reduced to one-half, which is then termed the *haploid number* (23 in man). At fertilization, the normal, *diploid number* (46) is restored.

PROTOPLASM

Protoplasm is the *living substance,* constituting, as T. H. Huxley put it, the *physical basis of life.* All living things are composed of protoplasm or its products. Nonliving things do not contain protoplasm.

Physical Nature of Protoplasm. *Consistency:* viscid, jellylike; a colloidal mixture that may have at one time a fluid consistency (*sol* state) and at another a semisolid gluelike consistency (*gel* state). *Color:* usually semiopaque, grayish in appearance. *Structure:* As seen under the light microscope, the protoplasmic substance may be (1) *granular,* containing granules of various sizes; (2) *alveolar,* having the appearance of soapsuds or bubbles; (3) *fibrillar,* consisting of fine fiberlike structures; or (4) *reticular,* consisting of an interlacing net of threadlike structures.

Chemical Nature of Protoplasm. Protoplasm, as such, cannot be analyzed chemically; when an analysis is attempted, the organization of the cell or of a tissue is disrupted and the life processes stop. The substance is thereupon nonliving, and whether it is of the same nature as it was in the living cell cannot be readily determined. This suggests that "life" depends on a specific chemical and physical organization of protoplasm.

It is possible, however, to determine which elements and compounds are present in a cell or an organism at the time of its death or shortly afterward. The table that follows shows the results of such an analysis of man, but the percentile distribution applies generally to most living organisms.

ELEMENTS COMPRISING THE HUMAN BODY

Element	*Percent by Weight*	*Element*	*Percent by Weight*
Oxygen (O)	65	Potassium (K)	0.35
Carbon (C)	18	Sulfur (S)	0.25
Hydrogen (H)	10	Chlorine (Cl)	0.15
Nitrogen(N)	3	Sodium (Na)	0.15
Calcium (Ca)	2	Magnesium (Mg)	0.05
Phosphorus (P)	1	Iron (Fe)	0.004
		Other elements*	0.046
	99		1.000

*Iodine (I), Copper (Cu), Manganese (Mn), Rubidium (Rb), Cobalt (Co), Nickel (Ni), Zinc (Zn), and Cadmium (Cd), among others.

Compounds in Protoplasm. Only the gases (oxygen and nitrogen) are found in the body in their elemental state. All the other elements are present in the form of compounds, which are either *inorganic* (lacking carbon) or *organic* (containing carbon):

Inorganic Compounds	*Organic Compounds*
Water	Proteins
Inorganic salts	Carbohydrates
	Fats (lipids)

The foregoing compounds are common to all protoplasm. Other compounds, both inorganic and organic, may occur in the protoplasm of specific types of cells.

The percentile distribution of the common compounds in different kinds of living tissue is variable, as shown in the following table:

DISTRIBUTION OF COMPOUNDS IN THE HUMAN BODY

Compound	*Striated Muscle*	*White Matter of Brain*	*Entire Body*
Water	75.0	70.7	66.0
Proteins	20.0	10.0	16.0
Fats	2.0	18.5	13.0
Carbohydrates and extractives	2.0	.....	0.6
Inorganic salts	1.0	0.8	5.0

WATER. Water is the principal chemical constituent of protoplasm. It is indispensable in metabolic processes, because enzymatic activity can take place only in the presence of water. In the body, water exists as *free water,* in which form it acts as the chief solvent and a medium for metabolic processes, and *bound water,* in which form it is bonded to proteins by hydrogen bonds.

Water plays the following roles in protoplasm:

1. It serves as a *solvent.* More substances are soluble in water than in any other solvent. Substances in solution are free to move about within the cell or from one cell to another. Most chemical reactions take place only in solutions, and enzymes act only in a fluid medium.

2. In water, substances such as acids, bases, and salts readily *ionize;* that is, their molecules separate into ions bearing electrical charges. For example, sodium chloride ($NaCl$) separates into a positive sodium ion and a negative chlorine ion ($NaCl \rightarrow Na^{+} + Cl^{-}$). Such a substance is called an *electrolyte,* and its solution conducts an electric current.

3. The properties of water make it important in the *regulation of body temperature.* These properties are its *high specific heat* (i.e., its capacity to take up and give off large amounts of heat with little change in its own temperature) and its *high latent heat of vaporization.*

4. Water has a *high chemical stability.* This makes it possible for chemical reactions to occur in water without involving the water itself.

5. Water has a *high surface tension,* which is of significance in the production of protoplasmic membranes.

6. Water serves as a reacting agent; it is essential in all hydrolytic reactions.

INORGANIC SALTS (MINERAL SUBSTANCES). These are substances that, upon burning, leave an ash. The common salts are compounds of the following elements.

Cations (+)		*Anions (−)*		
Sodium Potassium Calcium Magnesium Iron	which combine with	Chlorine Carbon dioxide Sulfur Phosphorus	to form	Chlorides Carbonates Sulfates Phosphates

Some of the important activities or conditions in the body in which salts play an important role are:

1. Proper osmotic conditions, to maintain proper concentration of water in cells.
2. Proper acid-base balance (buffer salts in the blood and body fluids).
3. Development of bone and teeth (calcium and phosphorus).
4. Normal coagulation of the blood (calcium).
5. Formation of red blood cells (iron for hemoglobin).
6. Proper state of irritability (sodium, potassium, and calcium ions).
7. Formation of certain endocrine secretions (iodine for thyroxine).

8. Function of certain enzymes or enzyme systems (copper, zinc, cobalt, iron, and magnesium in coenzymes).

PROTEINS. Proteins are compounds of carbon, hydrogen, oxygen, and nitrogen (and sometimes sulfur and phosphorus). They are built up from *amino acids,* compounds consisting of a carboxyl (—COOH) group and an amino group (—NH_2) group. Examples are albumen and hemoglobin. In the body, proteins serve the following roles.

1. They form the framework of the protoplasm of cells and hence are essential for growth and repair of tissues.
2. They are sources of energy.
3. They are constituents of all enzymes and some hormones.
4. Through their colloidal properties they are of importance in maintaining the fluid balance in the protoplasm of cells and various body fluids.
5. They are the transporting agents for certain minerals and hormones.
6. They participate in immune reactions.
7. They comprise the contractile elements in muscles.

LIPIDS OR FATS. Lipids are compounds of carbon, hydrogen, and oxygen. Examples are neutral fats, phospholipids, and cholesterol. In the body, lipids serve the following roles.

1. They are the body's most concentrated source of energy.
2. They provide a reserve supply of energy.
3. They support and protect various body organs.
4. They provide an insulating layer in subcutaneous tissues.
5. They are an integral part of cell membranes and thus influence cell permeability.
6. They contain certain essential fatty acids and serve as transport agents for certain vitamins.

CARBOHYDRATES. Carbohydrates are compounds of carbon, hydrogen, and oxygen, with the latter two nearly always in the ratio of 2 to 1. An example is glucose ($C_6H_{12}O_6$). Carbohydrates in the body serve the following functions.

1. They are the most readily available source of energy.
2. They are a storage form of food (e.g., glycogen).
3. They have a regulating effect on protein and lipid metabolism.
4. They are essential for the normal functioning of the central nervous system.

CELL PERMEABILITY

For the maintenance of life processes, which are going on unceasingly in cells, substances are moving more or less continuously from the outside into the cell and from the cell into the surrounding environment.

Food, oxygen, salts, hormones, and other substances must enter the cell, and carbon dioxide, water, products of secretion, and waste materials must be discharged from the cell. These substances pass through the cell membrane, some readily, others with difficulty.

Physiochemical Processes of Cell Permeability. The processes important to bodily activities involved in the passage of substances through cell membranes are *diffusion, filtration,* and *osmosis.*

DIFFUSION. Any kind of chemical activity involves the movement of atoms, ions, molecules, or groups of molecules. Where particles are free to move, there is a general tendency for them to move from a region of high concentration to one of lower concentration. This process is called *diffusion* (Fig. 2-3).

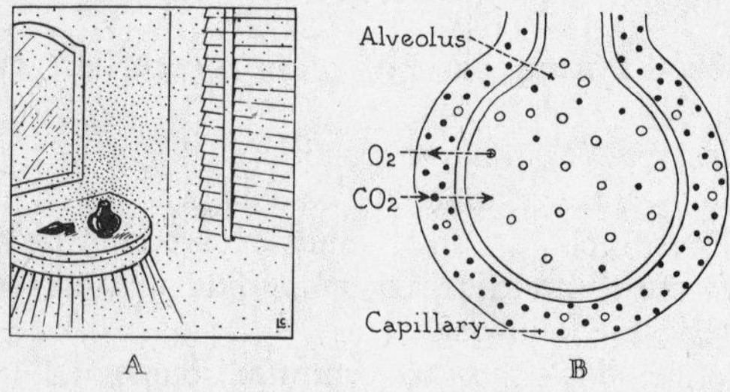

Fig. 2-3. Diffusion. (*A*) Molecules of perfume spread throughout enclosure. (*B*) Diffusion in air sac of lung. CO_2, represented as black dots, passes from capillary into alveolus. O_2 molecules, shown as circles, pass from air sac into blood capillary. (Reprinted with permission of W. B. Saunders Co., Philadelphia, from B. G. King and M. J. Showers, *Human Anatomy and Physiology,* 6th ed., 1969.)

The consequence of diffusion is a uniform distribution of the particles of the diffused substance. Thus, if two gases are admitted to a container, the molecules of each gas will diffuse among those of the other until they are uniformly distributed. Two liquids will behave in the same way, but their diffusion takes place at a slower rate. Solids diffuse extremely slowly, if at all.

The Membrane as a Factor. Diffusion is modified by the presence of a membrane. If the membrane is *permeable* to the substance that is diffusing, the particles will pass through the membrane as though it were not there. If the particles are larger than the pores of the membrane, they will not pass through; that is, diffusion will not take place. Semipermeable membranes allow selective diffusion.

Selective Diffusion (Dialysis). The passage of a substance *in solution* through a membrane is referred to as *dialysis.* Colloids do not dialyze; they do not pass through animal membranes or parchment paper. Crystalloid substances will pass through such membranes. Indeed, this property makes it possible to differentiate colloids from crystalloids and to separate them one from the other.

Examples of Diffusion in the Body. Examples of diffusion are the passage of oxygen from the alveoli of the lungs into the blood and from the blood to the tissue cells; the passage of carbon dioxide from tissue cells into the blood and from the blood into the lungs; and the passage of molecules of digested food from the digestive tract into the blood and from the blood into the tissue cells. In addition, within each cell, diffusion is constantly occurring as particles move about within the protoplasm of the cell.

FILTRATION. This process (Fig. 2-4) involves the passage of a substance through a barrier or filter as a result of the difference in mechanical pressure on the two sides of the filter. The principle can be illustrated by filtering a mixture of salt, sand, and water through filter paper. The water and the salt will pass through; the sand will not.

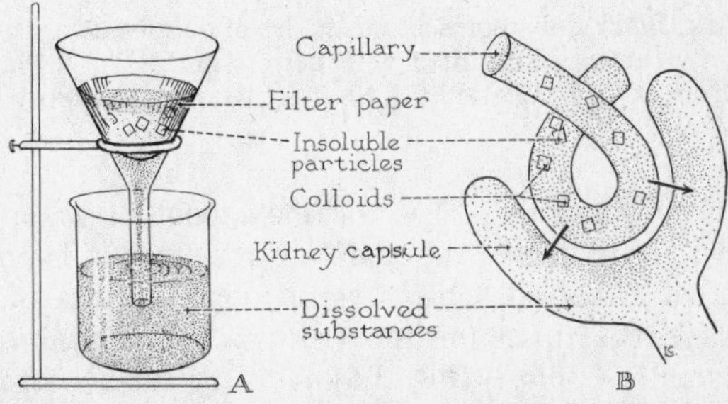

Fig. 2-4. Filtration. (*A*) Water and dissolved substances pass through the filter paper into the beaker. (*B*) Water and dissolved substances in the blood pass from the capillaries into the kidney tubule. Colloids of the blood that do not pass through the capillary wall correspond to insoluble substances in the funnel. (Reprinted with permission of W. B. Saunders Co., Philadelphia, from B. G. King and M. J. Showers, *Human Anatomy and Physiology,* 6th ed., 1969.)

Examples of Filtration in the Body. The action in the glomerulus of the kidney and that in the blood capillaries provide excellent examples of filtration in the body.

In the *glomerulus of the kidney* the mechanical force is supplied by the blood pressure; the wall of the glomerulus (a capillary net) and Bowman's capsule constitute the filter. Water and crystalloid solutes (sugar, salts, urea) pass through. Blood proteins do not pass through, because they are colloids and do not dialyze.

In the *blood capillaries* the mechanical force, again, is the blood pressure. Blood plasma is forced through the capillary wall into the tissue spaces. Blood proteins may get through by passing between the cells rather than through the cell membranes. Red blood cells do not pass through the capillary wall.

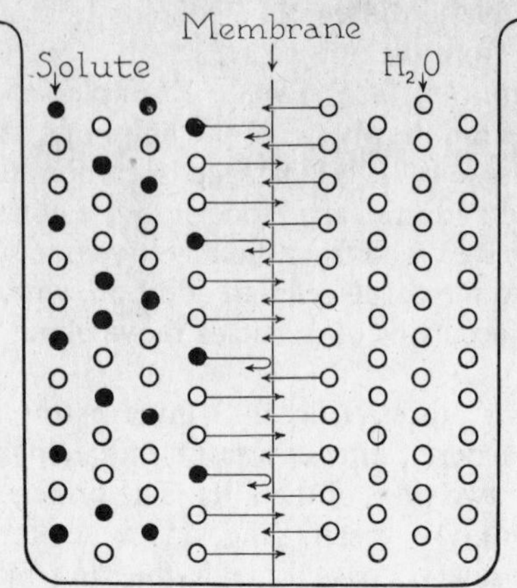

Fig. 2-5. Osmosis. Black dots represent molecules of dissolved substances; circles indicate water molecules. (Reprinted with permission of W. B. Saunders Co., Philadelphia, from B. G. King and M. J. Showers, *Human Anatomy and Physiology*, 6th ed., 1969.)

Osmosis. Osmosis (Fig. 2-5) is the movement of water molecules through a semipermeable membrane as a result of differences in the concentration of dissolved substances on the two sides of the membrane. The force that causes this movement is called *osmotic pressure.*

Demonstration of Osmosis (Fig. 2-6). A thistle tube is set up with its large end covered by a semipermeable membrane, such as an animal bladder. If the tube is filled with water and the large end is placed in a beaker of water so that the levels in the tube and the beaker are the same, it will be noted after a period of time that the level in the beaker does not change but remains the same as that in the beaker. But if the thistle tube is filled with a sugar solution and then placed in a beaker of water with the levels the same, after a short period of time the fluid in the tube will begin to rise and may go upward for a considerable distance, sometimes as much as several feet. From this action it becomes obvious that water molecules are moving through the membrane from the beaker into the tube. By the force of osmotic pressure the movement of the water molecules is tending to equalize the concentration of the two solutions.

In the first of the foregoing demonstrations, the levels remain constant because as many water molecules are passing through the membrane out of the tube as into it. In the second instance, however, both water molecules and sugar molecules are striking against the upper surface of the membrane, but the membrane is permeable only to the water molecules, so that these pass out into the beaker while the sugar molecules are retained in the tube. At the same time that the water molecules

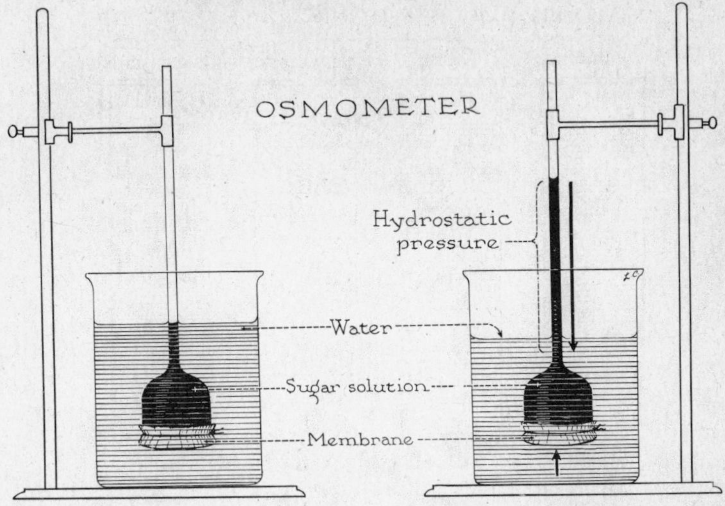

Fig. 2-6. Demonstration of osmosis. More water molecules strike and pass through the semipermeable membrane from the side in contact with pure water than from the side in contact with the sugar solution. Since more water is entering than leaving the thistle tube through the membrane, the height of the column of the sugar solution increases. (Reprinted with permission of W. B. Saunders Co., Philadelphia, from B. G. King and M. J. Showers, *Human Anatomy and Physiology,* 6th ed., 1969.)

in the sugar solution are striking against the membrane, the molecules of the dissolved substance (sugar) are also striking the surface area of the membrane. It is apparent that in a given period of time, fewer water molecules will strike a given area of this surface on the inside (that is, on the side of the tube) than will strike on the outside (the side of the beaker). Consequently, more water molecules enter the tube than leave, and the level rises.

Application of the Principle of Osmosis to Living Tissues. In essence, all cells in the body are small "sacs" of water containing dissolved substances, such as sugars and salts. The surface layer of protoplasm is a semipermeable membrane. Water tends to move through this membrane as the concentration of the fluids on either side of the membrane changes. This process can be illustrated with body cells in the following manner (see Fig. 2-7):

1. If red blood cells are placed in a salt solution with a concentration of 0.85 percent, no change in size is observed. The salt solution exerts an osmotic pressure exactly equal to that exerted within the cell. Such a solution is referred to as *isotonic.*

2. If red blood cells are placed in a very dilute salt solution or in distilled water (water containing no salt at all), water molecules will

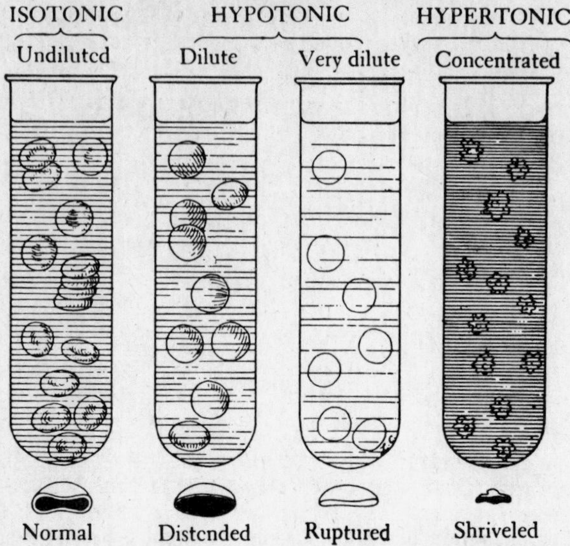

Fig. 2-7. Illustration of the effect of solutions on red blood cells. (Courtesy of I. Gersh. Reprinted with permission of W. B. Saunders Co., Philadelphia, from B. G. King and M. J. Showers, *Human Anatomy and Physiology,* 6th ed., 1969.)

move into the cell, which will increase in size. This condition is called *turgor.* The membrane of the blood cell may become distended, even to the point of bursting *(cytolysis).* Such a solution is said to be *hypotonic.*

3. If red blood cells are placed in a highly concentrated salt solution, say 2 percent, water will leave the cells and the cells will shrink. This phenomenon is known as *plasmolysis.* Ultimately the cells present a shriveled appearance *(crenated cells).* Such a solution is said to be *hypertonic.*

From paragraphs 2 and 3 it can be seen that it is important, in the transfusion of a fluid such as blood plasma or a saline solution, that the fluids transfused exert the same osmotic pressure as that of the fluids the cells live in (in this case, the blood plasma).

PHYSIOLOGIC SALINE SOLUTION. When living cells are studied or examined outside the body, they will retain their normal form and carry on their usual activities only if they are kept in an isotonic solution. For humans, such a solution is a 0.85 percent saline solution. In this solution there are 8.5 grams of salt (sodium chloride) to 1 liter of distilled water. This is a *physiologic saline solution.* Another physiologic saline solution is known as *Ringer-Locke solution.* For the study of human cells, each liter of such a solution would contain 8.5 g sodium chloride, 0.42 g potassium chloride, 0.24 g calcium chloride, 0.2 g sodium bicarbonate, and 2.0 g glucose. The salts provide sodium, potassium, and calcium ions in the same proportion as they exist in the blood plasma. Glucose is added to provide energy, and sodium bicarbonate is added to maintain constant hydrogen ion concentration (a buffering action).

The Passage of Substance Across Cell Membranes. The ways substances enter or leave cells may involve passive transport, facilitated diffusion, active transport, pinocytosis, and phagocytosis. The ability of a cell to accept or reject materials is called *selective permeability*, which is dependent upon various transport systems.

PASSIVE TRANSPORT. In passive transport, substances move through openings or pores in the cell membrane as a result of the random movements of molecules, atoms, or ions by means of diffusion and osmosis. No cellular energy is expended in the process. The size of the molecules, the charge on the ion or atom, and the concentration of substances on the two sides of the membrane are factors that determine the rate and direction of movement. Water and salts are substances that enter and leave cells in this manner.

FACILITATED DIFFUSION. Molecules of sugars and amino acids above 8 Å in size would diffuse through cell membranes much too slowly to meet the needs of a cell; hence, a special mechanism is required to move these substances into cells. Such a mechanism involves a *carrier system*. A substance entering a cell acts on a binding site in the membrane. At this site, a specific molecule called a *carrier* combines with the substance and moves it across the membrane. Then it detaches from the substance, which enters the cytoplasm of the cell. Characteristics of this system involve specificity, saturation, and competition.

Specificity means that a carrier for one type of substance (e.g., sugars) will not act as a carrier for another type (e.g., amino acids), and vice versa. A carrier will combine with only a specific group of chemical substances. *Saturation* means that there are only a limited number of binding sites on the membrane or that there are a limited number of carriers. Hence, increasing the concentration of the substance outside the cell will not necessarily increase the intake of that substance into a cell. *Competition* occurs when two similar substances, such as two amino acids, attempt to enter at the same transport site. As transport sites are limited, the presence of two or more types of competing molecules limits the number of any one type that can enter a cell. In this method of transport, no energy is required.

ACTIVE TRANSPORT. In active transport, molecules move across membranes much more rapidly than by other methods, and they also may be moved against a concentration gradient, that is, from a region of low concentration to one of high concentration. This involves the utilization of energy, which is provided by chemical reactions occurring within the cell. The energy is usually provided by ATP, which is activated by an enzyme, ATPase. In active transport, a carrier is utilized, but it is postulated that instead of acting symmetrically on both sides of the membrane, its actions are asymmetric, the solute or substance molecules combining more readily with it on one surface than on the other. This would necessitate a change in the carrier molecule in its passage from one surface to the other.

Probably the most important example of active transport is the movement of sodium and potassium ions across the cell membrane, as occurs in the functioning of muscle and nerve cells. In resting cells there is a high concentration of sodium ions (Na^+) outside the cell and a high concentration of potassium ions (K^+) within the cell. These concentrations are maintained by active transport of potassium ions into the cell and sodium ions out of the cell (the *sodium pump*). Sometimes the two systems of transport that utilize a carrier are referred to as *mediated transport*.

PINOCYTOSIS. This is a process in which small, fluid-filled vesicles form on the surface of a cell and then move into the cell interior. The cell membrane invaginates, forming a small pocket that pinches off from the surface as a small, isolated, fluid-containing vesicle called a *pinosome.* Within the cell, the contents of the pinosome leak out into the intracellular fluid, or the small vesicle, with many others, is incorporated within a lysosome to form a *multivesicular body,* the contents of which are digested as in phagocytosis.

PHAGOCYTOSIS. This process is similar to pinocytosis except that the contents of the vesicle are solid, such as large protein molecules, microorganisms, or tissue debris resulting from the breakdown of cells. Within the cell, this phagocytic vesicle is called a *phagosome.* It unites with a dense body (an inactive lysosome), which now becomes active, its enzymes digesting the ingested material. Any undigested material becomes a *residual body,* which is extruded from the cell by *budding* or *reverse phagocytosis.* Phagocytosis is especially pronounced in white blood cells and macrophages and constitutes an important defense mechanism against the spread of pathogenic microorganisms.

THE FLUIDS OF THE BODY

Water, the most abundant constituent of animal tissues, constitutes about 60 percent of the total body weight in humans. Body fluids, of which water is the principal constituent, include the following:

Intracellular Fluid. This fluid is found within the cells of the body; it is the fluid within which all metabolic activities take place. It comprises about 40 percent of body weight.

Extracellular Fluid. This includes the interstitial fluid, blood plasma, and lymph. It comprises about 20 percent of body weight.

INTERSTITIAL FLUID. This fluid, commonly called *tissue fluid,* fills the spaces between cells and provides the medium in which cells live. All substances entering and leaving cells pass through this fluid. The interstitial fluid constitutes the *internal environment* of the body.

BLOOD PLASMA AND LYMPH. These fluids are confined to vessels, the blood plasma with its corpuscles, filling the heart and blood vessels, and the lymph filling the lymphatic vessels. Both circulate throughout the body. Blood picks up oxygen from the lungs, nutrients from the digestive tract, and hormones from endocrine glands and transports these substances to cells where they are utilized. From the cells, waste products and products of some glands are picked up and transported to disposal areas or organs where they are utilized. Lymph returns to circulation from the tissue spaces by way of lymphatic vessels.

HOMEOSTASIS

The ability of cells to carry on life processes depends upon the maintenance of a fairly constant internal environment. The function of every organ system is basically to carry on those activities that will enable the body to maintain a more or less stable internal environment.

The body is constantly taking in gases, fluids, and solids, which, when added to the body fluids, tend to alter their composition. At the same time, the metabolic activities of the cells result in the production and discharge into the body fluids of substances that also act to alter their composition. In spite of these processes, which are continuous, the various body fluids remain remarkably constant in their composition. This state of constancy of body fluids, especially the interstitial fluid bathing the cells, is referred to as *homeostasis*. It is the result of a number of compensating mechanisms, particularly the functioning of the kidneys and the lungs, which keep changes of a physical and chemical nature at a minimum.

Three important conditions must be maintained within the body with respect to body fluids: water balance, acid-base balance (pH), and electrolyte balance. In addition the body temperature must be maintained within narrow limits.

Water Balance. The body acquires water through fluid intake and from the metabolism of foods. Water is lost through urination, perspiration, respiration, and defecation. Through the work of the kidneys, skin, lungs, and digestive tract, the water content of the body is maintained at a constant level. An excess of water results in *overhydration*; a deficiency, in *dehydration*. An excess of interstitial fluid is called *edema*.

Acid-Base Balance. This depends upon the concentrations of hydrogen ions (H^+) and hydroxyl ions (OH^-). These ions are constantly entering the body in foods, drinks, and often, self-administered medicines, or they are produced in various metabolic activities. Through respiration, with the excretion of CO_2 and through the production of an acid urine, the H^+ concentration of arterial blood is kept fairly constant at a pH of 7.3 to 7.45 (slightly alkaline). An increase in pH of the blood (alkalosis) or a decrease (acidosis) can lead to serious malfunctioning of the vital organs of the body and possibly death.

Electrolyte Balance. This involves a rather delicate balance between sodium, potassium, and calcium ions, these being the principal ions in the body fluids. An imbalance in these affects osmotic relations between cells and body fluids and also affects the irritability of muscle and nerve cells. Intricate neurohormonal mechanisms are involved in maintaining proper electrolyte balance.

Body Temperature. The temperature of the body is maintained by a balance between heat production and heat loss. Heat is constantly being lost through radiation, conduction, evaporation, and expired air. Heat is produced principally from oxidative processes within cells. Mechanisms for maintaining a constant temperature are regulated principally by the hypothalamus in the brain.

Stability of the internal environment within narrow limits is essential to good health and to life itself. Many diseases and pathologic disorders are the result of factors or conditions that alter this constancy. In general, all functions of the various organ systems are directed toward maintaining homeostasis.

3: TISSUES

A *tissue* is an aggregate of cells, together with their intercellular substance, which forms the structural materials of the body.

The four primary types of tissues are (1) *epithelial tissue,* which covers surfaces, lines cavities, and forms tubes; (2) *connective tissue,* which forms protective, supporting, and binding structures; (3) *muscle,* which comprises the contractile elements; and (4) *nervous tissue,* which includes the sensory, coordinating, and conducting structures. Tissues are differentiated from each other on the basis of the types of cells of which they are composed, the nature and amount of intercellular material, and the functions they perform.

EPITHELIAL TISSUE

Epithelial tissue, or *epithelium,* forms the covering of the outer surfaces of the body and of most internal organs; it lines the digestive and respiratory tracts, the serous cavities, blood vessels, excretory ducts, and reproductive ducts and organs. It comprises the secreting portions and ducts of glands and the sensory portions of sense organs.

Structural Characteristics of Epithelial Tissue. Epithelial cells are arranged in a continuous sheet, usually of one layer. They are packed together closely, with little intercellular substance. Blood vessels are absent, but nerve endings are usually abundant. Epithelial cells lie on a *basement membrane* (*basal lamina*), a thin layer of amorphous substance resting on a layer of reticular fibers.

Types of Epithelium (Fig. 3-1). The types of epithelium are differentiated on the basis of the shape of the cells (squamous, cuboidal, columnar) and their arrangement in the epithelial sheet (simple or stratified).

SQUAMOUS EPITHELIUM. This consists of thin, flat cells with regular or irregular outlines. *Simple squamous epithelium* (cells in a single layer) is found lining the body cavities and in Bowman's capsule of the kidney. That lining the serous cavities (pleural, peritoneal, pericardial) is called *mesothelium;* that lining blood and lymph vessels, *endothelium;* that lining cavities in connective tissue (such as the subarachnoid cavity), *mesenchymal epithelium. Stratified squamous epithelium* (cells in several layers) is found in the epidermis, cornea, esophagus, and vagina. On dry surfaces such as the skin, it is usually *keratinized;* that is, the surface cells metamorphose into a nonliving layer of *keratin,* a tough, proteinaceous, waterproofing material.

CUBOIDAL EPITHELIUM. This type consists of cube-shaped cells or

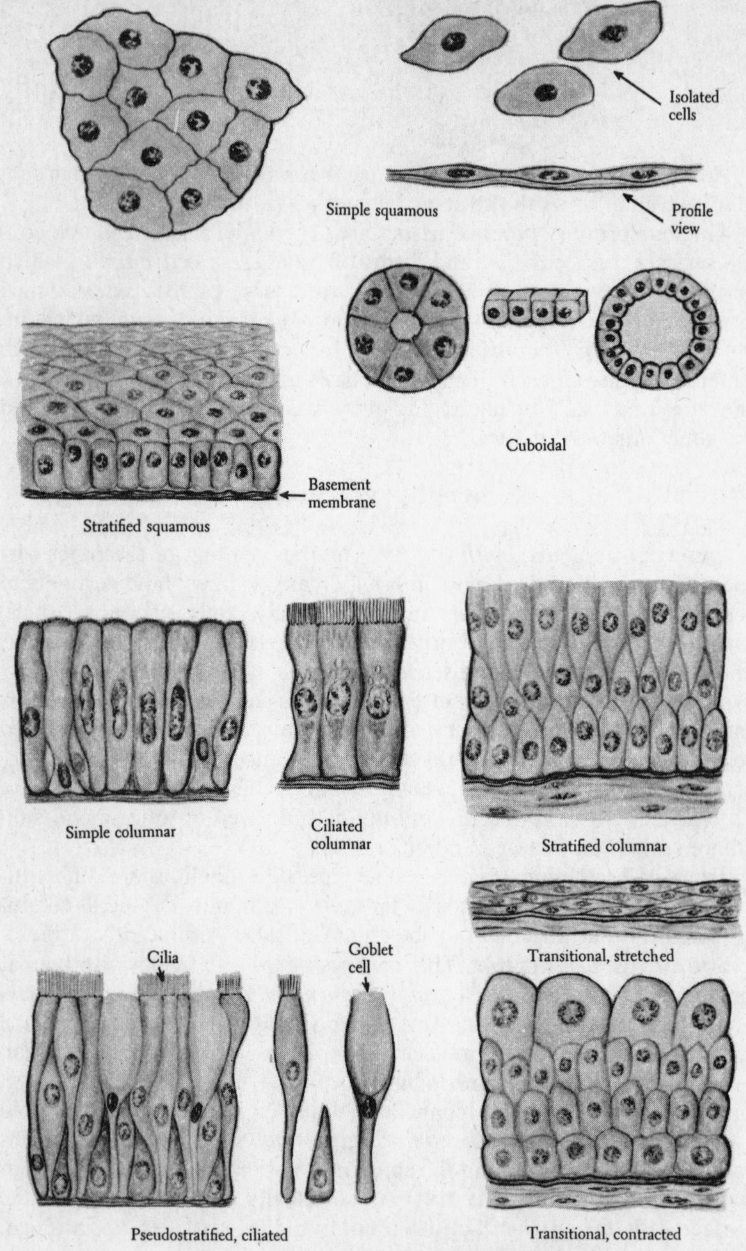

Fig. 3-1. Types of epithelial tissue.

CLASSIFICATION OF EPITHELIAL TISSUE

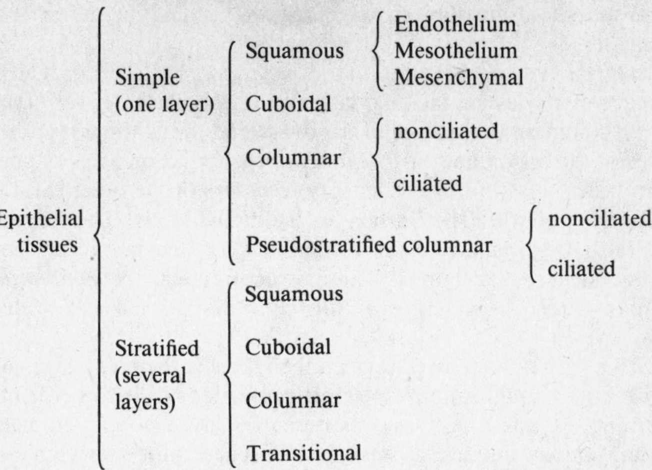

cells in the form of truncated pyramids. It is found in the liver, thyroid gland, kidney ducts and tubules, the ducts of glands, and on the free surface of the ovary. *Stratified cuboidal epithelium* is found in the conjunctiva and in the ovarian follicle.

COLUMNAR EPITHELIUM. Columnar epithelium consists of cells that are long and cylindrical, with nuclei located in their basal ends. It is found lining the stomach, intestine, and cervical canal of the uterus. In the intestine, the cells have absorptive and secretory functions. The secretory cells are specialized, mucus-secreting cells called *goblet cells. Ciliated columnar epithelium* is found in the oviduct, uterus, and nasal sinuses. *Stratified columnar epithelium* is found infrequently in regions where stratified squamous epithelium changes abruptly into *pseudostratified columnar epithelium,* as in the larynx, nasolarynx, and orolarynx. It is also found in the cavernous urethra. Sometimes the surface cells are ciliated.

PSEUDOSTRATIFIED COLUMNAR EPITHELIUM. This epithelium consists of short basal cells interspersed with taller columnar cells. Because the nuclei lie at different levels, the tissue appears to be stratified; however, all the cells rest on a basement membrane and actually constitute a single layer. Pseudostratified columnar epithelium lines the larger excretory ducts of many glands and is present in the male urethra. *Ciliated pseudostratified columnar epithelium* containing many goblet cells is found in the trachea, larger bronchi, and parts of the pharynx.

TRANSITIONAL EPITHELIUM. The transitional type, also called *uroepithelium,* resembles stratified epithelium in that it consists of several layers of cells, but its superficial surface cells are large and rounded instead of squamous in shape. Thus the membrane can stretch without breaking apart the surface cells. This type is well adapted for lining organs and tubes that are subject to expansion, such as the ureter and urinary bladder.

Specialized Types of Epithelium. *Syncytial epithelium,* in which intercellular boundaries are lacking, is found in the placenta; *germinal epithelium* is found on the surface of the ovary and lining the seminiferous tubules of the testes; and *neuroepithelium* is found on sensory surfaces, as in taste buds, and in the sensory receptors of the inner ear.

Specializations of the Free Surface of Epithelial Cells. The free or luminal surface of epithelial cells in certain locations bears minute specialized processes or projections. These structures may be *nonmotile* (e.g., microvilli and stereocilia) or *motile* (e.g., cilia and flagella). A description of these structures follows.

Nonmotile Processes. In certain epithelial cells, there is a layer of extremely minute, nonmotile cytoplasmic projections called *microvilli.* On cells lining the intestine, this constitutes a *striated border;* on cells lining certain kidney tubules, a *brush border.* These processes increase the absorbing surface of the cells. The cells lining certain male reproductive ducts, especially the epididymis, bear long processes called *stereocilia,* which differ from microvilli in being longer, flexible, sinuous, and sometimes branching.

Motile Processes. These include cilia and flagella, both of which are fundamentally alike in structure. Distinctions between cilia and flagella are of a superficial nature. *Cilia* are usually numerous (270 per cell in the trachea), and their motion is wavelike; *flagella* are few in number (one, two, or a few per cell), are usually much longer than cilia, and their motion is whiplike.

Cilia. Cilia are minute, hairlike projections, each of which arises from a *basal body* or *corpuscle,* which is a modified centriole. All cilia have a characteristic internal structure. When viewed in cross section (Fig. 3-2), each shows two *central fibrils* (also called *filaments* or *microtubules*)

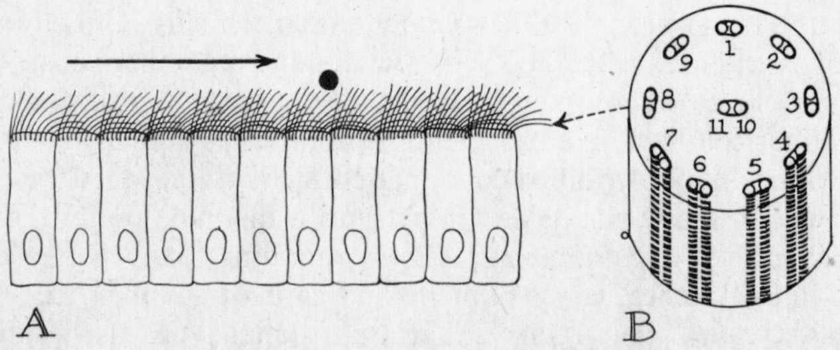

Fig. 3-2. Cilia. (*A*) Moving a particle. (*B*) Cross section, to show structure. (Reprinted with permission of W. B. Saunders Co., Philadelphia, from B. G. King and M. J. Showers, *Human Anatomy and Physiology,* 6th ed., 1969.)

and nine *peripheral fibrils,* which appear as doublets. The movement of cilia occurs as a rapid forward bending (*effective stroke*) in one direction and a subsequent slower, straightening (*recovery stroke*). The beat of each cilium occurs slightly after the beat of the one preceding it; the result is a wavelike movement in one direction that propels the surface layer of mucus or other secretory material. A ciliated border is characteristic of respiratory and reproductive epithelia. In the former, the action clears the mucous membrane of bacteria and other particulate matter.

Flagella. Flagella occur infrequently in the human body, the only cells bearing a typical flagellum being male spermatozoa. The mode of origin and structure of a flagellum are identical to that of a cilium.

Attachments and Binding Structures of Epithelial Cells. The cells of epithelial tissues are so firmly bound to each other and to the underlying basement membrane (basal lamina) that considerable mechanical force is required to separate them. The means by which they are held in place are as follows:

ATTACHMENT TO BASAL LAMINA. This is by adhesion plates, or *hemidesmosomes* (half-desmosomes), basal cell processes of the plasma membrane that extend into the basal lamina.

ATTACHMENT OF CELLS TO ADJOINING CELLS. This is accomplished through *cellular interdigitations,* in which fingerlike processes of a cell project into invaginations of the adjacent cell, and *cell junctions.* A junction between two columnar epithelial cells may be one of four types: tight junctions, gap junctions, intermediate junctions, and desmosomes (Fig. 3-3). In *tight* or *occluding junctions* (*zonula occludens*), the apposing cells' membranes are in close contact; in *gap junctions,* a narrow intercellular space separates the apposing cells; in *intermediate junctions,* the cytoplasmic surfaces of adjoining cells are thicker than usual, and embedded in the dense areas are minute tonofilaments; in a *desmosome* (*macula adherens*), from a dense region or *plaque* in each of the adherent cells, U-shaped *tonofilaments* extend into the cytoplasm. These tonofibrils were formerly thought to be "intercellular fibers," but

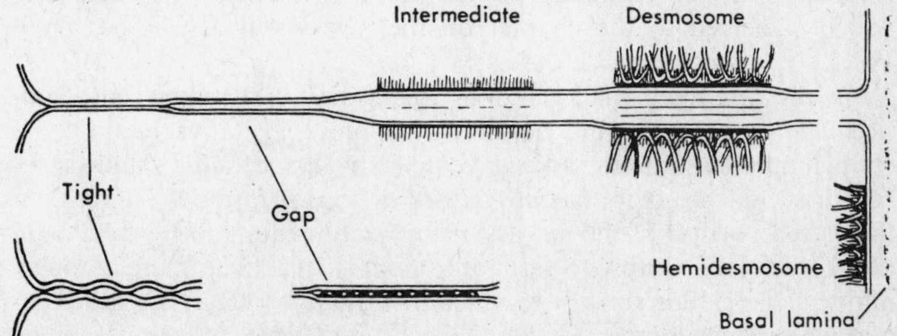

Fig. 3-3. Types of cell junctions. (Reprinted with permission of W. B. Saunders Co., Philadelphia, from C. R. Leeson and T. S. Leeson, *Histology,* 3rd ed., 1976.)

electromyographic studies have shown that there is no continuity of the cytoplasm of adjoining cells.

Regeneration of Epithelium. The epithelium that covers the surface of the body, lines the intestine and female genital tract, and is present in some glands, especially holocrine glands, possesses the ability to replace cells lost through functional activity, trauma, or disease processes. In the stratified epithelium of the epidermis, esophagus, and vagina, cells of the basal or germinative layer multiply by mitosis and move to the outer surface, replacing those that are lost. In the intestine, cells at the base of a villus migrate along the surface of the villus to the apex, where they are shed; replacement occurs in two to four days. In a holocrine gland, such as a sebaceous gland in which both cells and their contents comprise the secretory product, cells are being renewed constantly. When the epithelium is destroyed, as from a burn, cells from the margin of the wound migrate over the connective tissue to form a single sheet of continuous cells to which additional cells are added. In general, epithelium regenerates rapidly, even in such specialized organs as the liver.

Glands. Because they arise from an epithelial surface, glands are included in this discussion of epithelial tissues. The specialized glands associated with various body systems are described in greater detail in later chapters.

Functionally, a gland is a cell or a group of cells that elaborates or manufactures a specific substance that is discharged from the cell onto a surface or into the blood or lymph. The product thus formed does not become a part of the body tissues but is either absorbed and used by them or discharged from the body.

From the definition just given, some neurons can be considered to be gland cells. They produce substances that enter the bloodstream and affect distant organs and all liberate transmitter substances that cross synaptic clefts and act on membranes of postsynaptic cells. Effector neurons produce substances at their axon terminals that mediate responses initiated. But since neurons are primarily concerned with the conduction of nerve impulses, a discussion of their secretory activities will be deferred to the chapter on the nervous system (Chapter 3, Vol. 2).

Structurally, a gland is a cell or an aggregation of secreting cells. Multicellular glands arise and develop by a process of invagination; the epithelium grows into the adjacent connective tissue, and a simple tubelike or saclike structure (an *alveolus* or *acinus*) is formed. The cells in the closed portion of the sac assume a *secretory* function, while those near the surface narrow to form an *excretory* duct. The more complex multicellular glands arise by repeated invaginations; the result is a complicated, branched organ.

SECRETION. Secretion is the process whereby a cell obtains materials from the blood and lymph and transforms them into products that are then passed from the cell.

The *mechanism of secretion* is of this nature. Amino acids are synthesized and combined with others to form proteins on the surface of ribosomes on the endoplasmic reticulum (ER). The complete protein breaks away from the ER in the form of membrane-bound vesicles, which accumulate in the region of the Golgi apparatus. Here the product is concentrated, and, in some cases, additional material, such as glycoproteins, may be added. The final product leaves the Golgi apparatus as a *secretory granule.* These granules accumulate in the free end of the cell from which they are released. The mechanism of release varies in different types of gland cells.

Gland cells are not active at all times but pass through alternating active and resting phases (Fig. 3-4). In a *resting cell,* secretory granules accumulate in the apex of the cell. In an *active cell* (one discharging its secretion), the number of granules decreases, the number of water vacuoles increases, and the Golgi apparatus hypertrophies. In an *exhausted cell* there are few or no secretory granules, and the amount of cytoplasm is reduced.

Usually all the cells of a single alveolus or acinus are in the same state of activity. Cells may repeat the secretory cycle many times, but eventually they wear out and die. They are then replaced by a new generation of cells, or a new alveolus develops.

The functioning of a gland is under *neural* or *chemical control.* Most exocrine glands, for example, salivary glands and sweat glands, are under neural control. Glands are innervated by fibers of the automatic nervous system; hence, they function involuntarily. Secretory activity

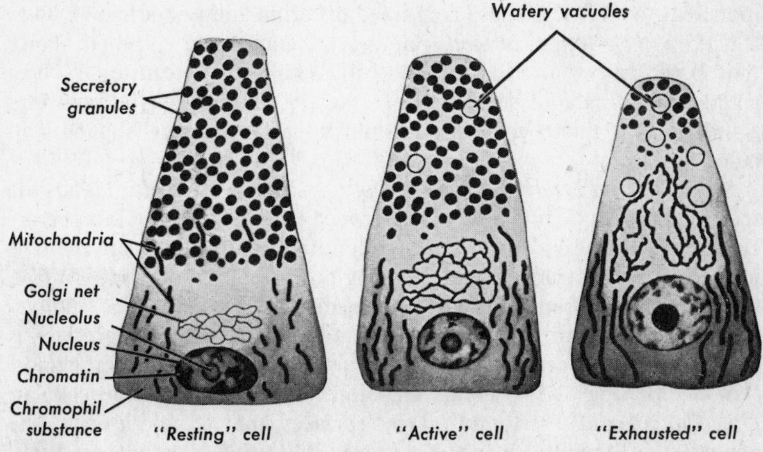

Fig. 3-4. Diagram showing stages of secretion in a serous cell. (Courtesy of I. Gersh. Reprinted with permission of W. B. Saunders Co., Philadelphia, from W. Bloom and D. W. Fawcett, *Histology,* 9th ed., 1968.)

may be stimulated or inhibited, depending on which division (sympathetic or parasympathetic) of the autonomic nervous system is involved. Secretion is usually brought about by release of a chemical, acetylcholine, at the nerve endings; secretion is reduced by constriction of arterioles and capillaries that supply blood to the gland.

All endocrine glands except the adrenal medulla are under chemical control; thus, an increase of glucose in the blood stimulates the islets of Langerhans in the pancreas and insulin output is increased. Most endocrine glands, however, are controlled by hormones or releasing factors produced by other endocrine glands. For example, the thyroid, parathyroid, adrenal cortex, and gonads are regulated by hormones secreted by the anterior lobe of the pituitary gland, which, in turn, is regulated by releasing factors produced by the hypothalamus.

POLARIZATION. Gland cells (and, for that matter, most epithelial cells) are said to be *polarized;* that is, their *proximal* or *basal end* (that nearest to the basement membrane) differs in structure from the *distal* or *free end.* The nucleus lies at the proximal end, whereas, in the active phase of secretion, the secretory products (granules, mucigen, or other substances) fill the distal end, from which they are discharged.

CLASSIFICATION OF GLANDS. Glands may be classified on the basis of (1) the presence or absence of excretory ducts, (2) the nature of the secretion, (3) whether the secretion is a product of the cell or a part of the cell itself, and (4) their structure.

Presence of Excretory Ducts. Exocrine (external-secreting) glands possess ducts and empty their products onto a free surface (e.g., salivary glands). *Endocrine* (internal-secreting) glands are ductless; their secretions are absorbed into blood or lymph (thyroid gland). Some glands (pancreas, ovary, testes) are combined exocrine and endocrine glands.

Nature of Secretion. Mucous glands secrete mucus, a viscid substance that principally contains mucin (goblet cells of intestine, tracheal glands). *Serous* glands secrete a clear, watery albuminous fluid (parotid gland). *Mixed* glands contain both mucous and serous cells (submaxillary gland).

Secretion as a Product of or a Part of the Cell. In *merocrine* glands the secretory product is the only part extruded (salivary gland). In *apocrine* glands the apical end of the cell, containing accumulated products, is broken off and extruded; the remaining portion of the cell is left intact, the cell re-forms, and the process is repeated (mammary gland). In *holocrine* glands the entire cell, along with its contained secretory product, is extruded, and the cell is replaced by a new cell (sebaceous or oil gland).

Structure (Fig. 3-5). Glands are either unicellular or multicellular. *Unicellular* glands are found on free surfaces; they secrete mucus. The *goblet cells* of the intestines are examples. *Multicellular* glands are tubular, alveolar (acinous), or tubuloalveolar (tubuloacinous).The *tubular* type have their secreting portion in the form of a blind, narrow tube.

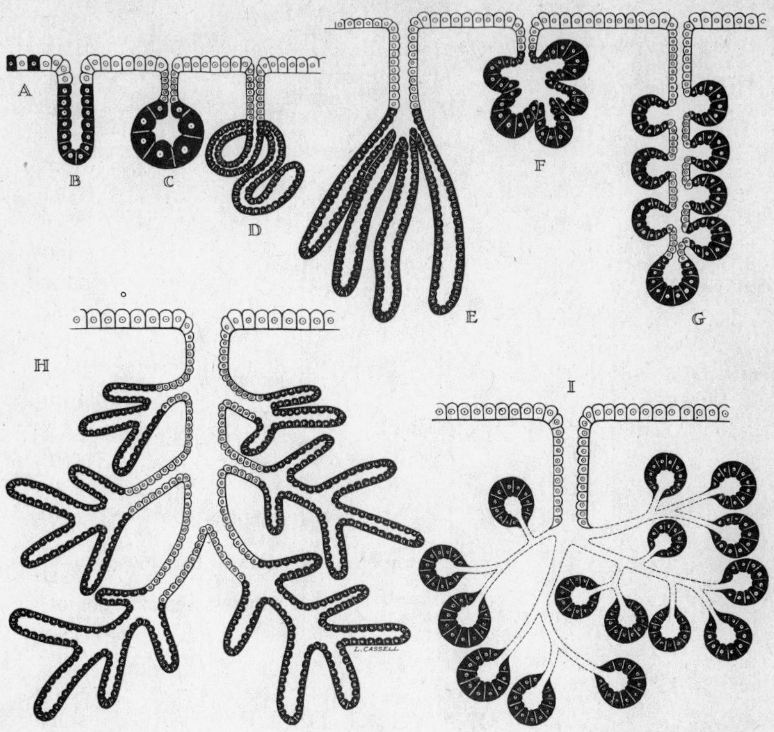

Fig. 3-5. Types of glands, diagrammatically shown as invaginations from the epithelium sheet. Shaded portions represent secreting cells. (*A*) Surface cells differentiated as unicellular glands. (*B*) Simple tubular gland. (*C*) Simple acinous or alveolar gland, formed from a group of glandular cells. (*D*) Simple coiled tubular gland and (*E*) Simple branched tubular gland. (*F, G*) Simple branched alveolar glands. As complexity increases, each tubule or acinus has its own duct, and all ducts communicate with a common duct that leads to a single outlet. (*H*) Compound tubular gland. (*I*) Compound alveolar gland. (Reprinted with permission of W. B. Saunders Co., Philadelphia, from B. G. King and M. J. Showers, *Human Anatomy and Physiology,* 6th ed., 1969.)

They may be *simple-straight* (in the large intestine), *simple-coiled* (sweat glands), or *simple-branched* (gastric glands); or they may be *compound* (in the kidney, testis, liver), with a large number of tubes branching repeatedly. The *alveolar* (acinous) type has a flask-shaped secreting portion called an *alveolus* or *acinus;* they are *simple-branched* (sebaceous glands) or *compound* (mammary gland). The compound *tubulo-alveolar* (tubulo-acinous) type has a secreting portion that consists of irregularly branched tubules and saccular outgrowths; examples are the majority of exocrine glands (salivary glands, pancreas).

TYPES OF GLANDS

Basis of Classification	*Types of Glands*		
1. Presence or absence of ducts	Exocrine		
	Endocrine		
2. Nature of product	Noncellular	Serous	
		Mucous	
		Mixed	
	Cellular	Cytogenic	
3. Mode of secretion	Merocrine		
	Apocrine		
	Holocrine		
4. Structure	Unicellular		
	Multicellular	Simple	Tubular: Straight, Coiled, Branched
			Alveolar (Acinous)
		Compound	Tubular
			Alveolar (Acinous)
			Tubulo-alveolar (Tubulo-acinous)

CONNECTIVE TISSUE

Connective tissue forms the supporting and connecting structures of the body. With one or two exceptions, connective tissue arises from the mesoderm (the middle germ layer of the embryo); more specifically, from the mesenchyme of this layer.

Mesenchyme is an embryonic tissue consisting of a network of branching, stellate cells whose slender, protoplasmic processes touch but do not anastomose. The intercellular spaces are filled with an amorphous, homogenous ground substance. Mesenchymal cells are ameboid and migrate freely. As development of the embryo proceeds, the nature of this intercellular substance changes. Fibers begin to appear, first the collagenous, then the elastic. Differentiation occurs with development, and the specialized adult types arise.

General Characteristics of Connective Tissue. The cells of connective tissue are relatively few in number, the bulk of the tissue consisting of intercellular material or *matrix.* Connective tissues are highly vascular; that is, they are well supplied with blood vessels. Rarely do they occur on free surfaces.

Functions. Connective tissue functions in the *support* of the body as a whole and in the support of individual organs and tissues. It forms *protective structures* such as bony cases that enclose organs. It provides a *storage place* for reserve food, and the loose connective tissue serves as a *pathway* of nutrients and oxygen from the capillaries to the tissues. Further, its *defensive cells* help prevent the spread of infective organisms and are involved in the production of antibodies.

Classification of Connective Tissue (Fig. 3-6). Connective tissue is of two types, embryonic and adult.

Embryonic tissues include mesenchyme and mucous tissue. *Mesenchyme* was described earlier. *Mucous connective tissue* is a jellylike type of tissue found under the skin and in various parts of a developing fetus. It consists of stellate fibroblasts and other cells in a matrix of mucoid intercellular material. In the umbilical cord it is called *Wharton's jelly.*

Adult connective tissues include connective tissue proper, cartilage, and bone. The various types are differentiated on the basis of (1) kinds of fibers and relative abundance of each, (2) arrangement of fibers and degree of compactness, (3) nature of the matrix, and (4) types of cells and their arrangement in the matrix. The principal types are *connective tissue proper,* in which the intercellular substance is of a fibrous nature and the typical cell is a *fibroblast; cartilage,* in which the matrix is dense, opaque, and nonvascular and the typical cell is a *chondrocyte;* and *bone,* in which the matrix is dense, hard, and calcareous and the typical cell is an *osteocyte.*

Sometimes the *vascular tissues* (blood and lymph) are classified as

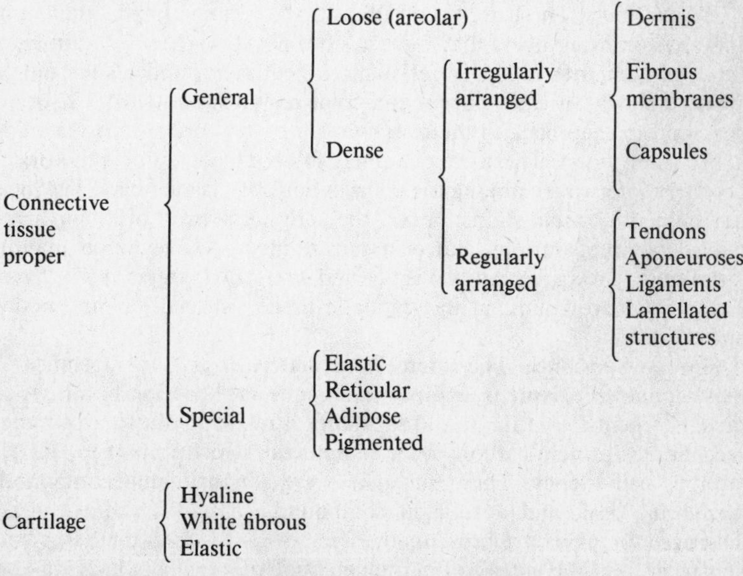

connective tissue since they have the same embryonic origin and resemble connective tissue in structure in that they consist of cells with much intercellular material. But since they are *fluid tissues* and do not serve a supportive or connective role, they will be discussed as part of the cardiovascular system in Chapter 10.

CONNECTIVE TISSUES PROPER. The connective tissue that is most generalized in structure and widely distributed is *loose connective tissue.* Its structure is as follows:

Loose or Areolar Connective Tissue. This soft, pliable tissue consists of *cells* embedded in an intercellular *matrix* of *fibers* and an *amorphous ground substance.* It forms the interstitial tissue of most organs, comprises the submucosa of most tubular organs, surrounds blood vessels and nerves, and constitutes much of the subcutaneous tissue and deep fascia.

Fibers. Fibers of three kinds occur in loose connective tissue. They are collagenic, elastic, and reticular.

1. *Collagenic fibers:* These are *white* fibers occurring in bundles, usually arranged as interlacing *wavy* strands. Each fiber is composed of *fibrils,* which under the electron microscope have a characteristic crossbanding. The fibers contain *collagen,* which upon boiling is hydrated to form *gelatin.* Collagen is the source of glue. White fibers are not elastic and yet are flexible; in bundles they possess great strength. In weak acids or alkalis, the fibers swell; they are digested by pepsin in an acid solution, but they resist trypsin digestion. Collagenous fibers are thought to be formed extracellularly from *tropocollagen* under the influence of fibroblasts.
2. *Elastic fibers:* These are *yellow* fibers that appear as single, branching and anastomosing fibers that form a loose network. They contain *elastin,* a protein that is highly resistant to heat, acid, and alkalis but is digested slowly by both pepsin and trypsin. When stretched, elastic fibers tend to snap back to their original state when pressure is released.
3. *Reticular fibers:* These are extremely fine, branching fibers that form networks interwoven among the collagenous and elastic fibers. They are not stained by usual stains; hence, they are not seen in ordinary light-microscope preparations. But in tissues stained by silver impregnation techniques, they are readily visible and are called *argyrophilic fibers.* These fibers are abundant in lymphatic tissue, especially lymph nodes and the spleen.

Ground Substance. The intercellular matrix or ground substance of loose connective tissue is an amorphous substance containing spaces or *areolae.* Chemically this ground substance is made up of two mucopolysaccharides (protein-carbohydrate complexes), which impart to it a gelatinous consistency. These substances are a nonsulfated compound, *hyaluronic acid,* and a sulfated compound, *chondroitinsulfuric acid.* Their gelatinous nature prevents the free flow of fluids in the tissue and acts to prevent the spread of particulate matter, especially bacteria and

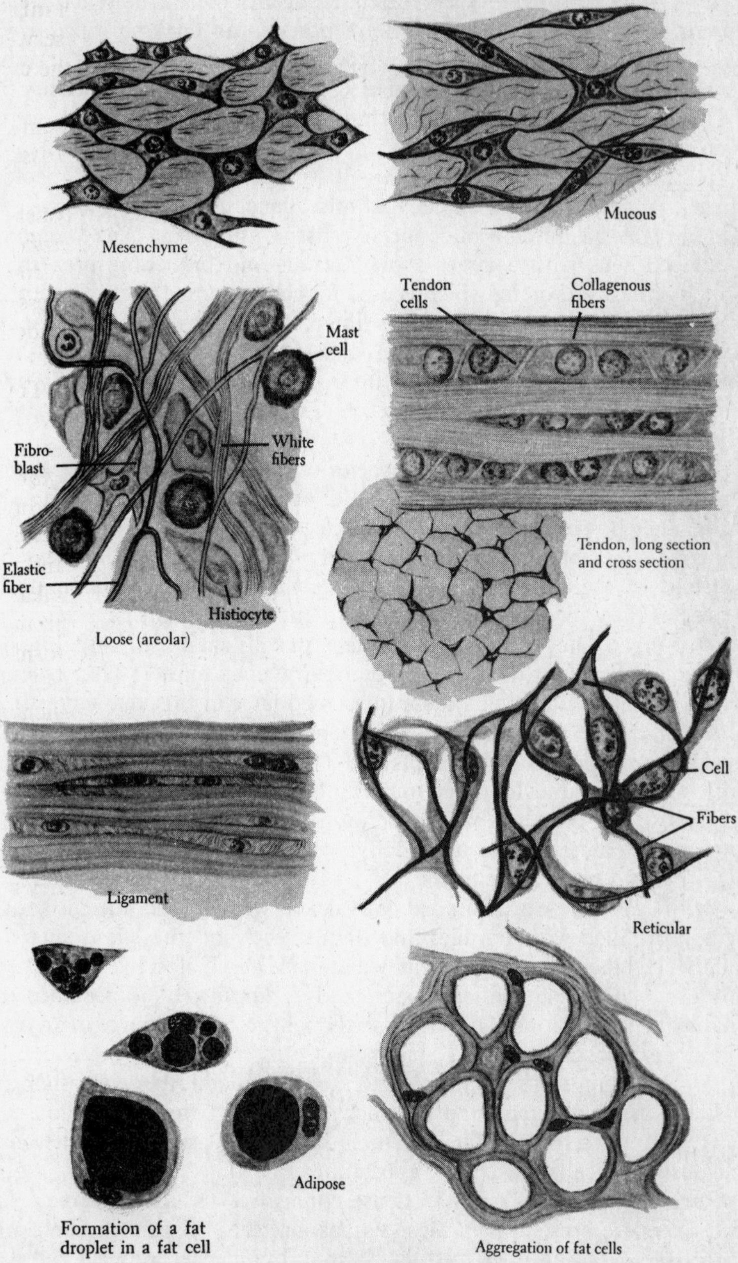

Fig. 3-6. Types of connective tissue proper.

viruses, thus tending to keep infectious agents localized. It has been found, however, that certain bacteria produce an enzyme, *hyaluronidase,* that depolymerizes hyaluronic acid. This factor and similar substances are called *spreading factors* because they reduce the viscosity of the ground substance and thus permit the rapid spread of infection or such substances as snake venom.

Cells. The cells of loose connective tissue include fibroblasts, macrophages, plasma cells, mast cells, and miscellaneous cells.

1. *Fibroblasts:* Common in connective tissues, fibroblasts are irregularly shaped cells with a large, ovoid nucleus and branching processes. They synthesize protein, which is used for cellular growth or, discharged from the cell, in the formation of intercellular substances, especially fibers. They remain relatively fixed in position, but in inflamed tissues they exhibit motility by moving with a gliding movement. In fully differentiated tissues, the fibroblasts are less active and are designated *fibrocytes.*
2. *Macrophages:* Also called *histiocytes* or *clasmatocytes,* macrophages are of variable shape: sometimes ovoid, at other times stellate or fusiform. They possess to a marked degree the capacity to ingest particulate matter, such as bacteria, a process called *phagocytosis.* Usually they are constant in their position (*fixed macrophages*), but in inflammatory processes they become actively ameboid and move about as *wandering macrophages.* They are important defensive cells against disease, ingesting bacteria, cellular debris, and other extraneous matter. They belong to the reticuloendothelial system (discussed later in this chapter).
3. *Plasma cells:* These are small flattened or oval cells with a small nucleus, usually eccentrically placed. They are the source of various antibodies (immunoglobulins) important in the body's immunologic responses to the presence of foreign antigens. They are especially abundant in the connective tissue underlying the mucous membranes of the digestive and respiratory tracts.
4. *Mast cells:* These are rounded or oval cells with a small nucleus and a cytoplasm filled with granules that stain selectively with stains such as methylene blue. They are especially abundant along blood vessels. They produce *heparin,* an anticoagulant, and *histamine,* of importance in anaphylactic reactions. They are also thought to be the source of *serotonin,* a vasoconstrictor.
5. *Miscellaneous cells:* In addition to the cells listed, other cells, such as eosinophils, undifferentiated mesenchymal cells, lymphoid wandering cells, pigment cells, and fat cells, may be present. In inflamed tissues, various blood cells make their appearance.

Dense Connective Tissue. In dense connective tissue, the same cells and fibers are present as in loose tissues, but the collagenous fibers are more abundant and interlaced tightly to form a dense, compact, feltlike network. Reticular and elastic fibers are also present. These tissues fall into two groups, according to the arrangement of fibers:

1. *Irregularly arranged tissues:* Tissues in which fibers are irregularly arranged include the *dermis* of the skin, *capsules* enclosing organs, and *fibrous membranes.* Important fibrous membranes are (a) *fascia,* sheets of which lie beneath the skin and form investing tissues of muscles; (b) *perichondrium,* the membrane surrounding cartilage; and (c) *periosteum,* the membrane surrounding bone.

2. *Regularly arranged tissues:* In tissues in which fibers are regularly arranged, the collagenous fibers run parallel, in closely packed bundles. They form structures of great tensile strength, such as tendons, aponeuroses, and ligaments. A *tendon* is a flat or cordlike band that serves to attach muscle to bone. Its collagenous fibers are arranged in parallel bundles. Fibroblasts, the only cells present, are arranged in parallel rows between the bundles. An *aponeurosis* is a flat sheet of connective tissue that attaches muscles to bones or other tissue; it is similar in structure to a tendon. A *ligament* also is similar to a tendon in structure except that its fibers are less regularly arranged and some elastic fibers may be present. Ligaments connect bone to bone, and organs to organs.

Other structures in which dense connective tissue is found are lamellated structures such as the corpuscle of Pacini or the perineurium surrounding nerves fibers.

Special Connective Tissues. Connective tissues with special properties include elastic, reticular, adipose, and pigmented tissues.

1. *Elastic tissue:* In this type of connective tissue, elastic fibers predominate. It is found in or between organs that can be stretched or deformed but regain their original position or shape when tension is relieved. It is present in certain ligaments (the ligamentum nuchae) and in the walls of certain tubular structures (aorta, larger arteries, trachea, and the bronchi).

2. *Reticular tissue:* This type of connective tissue consists of a syncytial network of cells with many *argyrophilic* intercellular fibers running in all directions and forming a *reticulum.* These fibers are so named because of their property of staining intensely in certain silver histiologic techniques. This type of tissue forms the framework of lymphatic organs such as the spleen, lymph nodes, and bone marrow. It is also found in certain endocrine glands, in the walls of blood vessels, and in the digestive and respiratory tracts underlying mucous membranes.

3. *Adipose tissue:* Adipose tissue is made up principally of cells that have the capacity for taking in fat and storing it. Mature fat cells contain a large droplet of neutral fat occupying the major portion of the cell. The cytoplasm is reduced to a thin layer surrounding the droplet. The nucleus is flattened and pushed to one side, giving the fat cell the appearance of a signet ring. Groups of fat cells are separated by areolar tissue. Adipose tissue is found in the superficial fascia under the skin; around such organs as the kidney, urinary bladder, and heart; in mesenteries and the

greater omentum; and as individual cells or in small groups in any loose connective tissue, especially that along the blood vessels and nerves. As to its *functions,* adipose tissue (a) serves as a reservoir of reserve food, (b) protects the organs it surrounds against cold (or heat loss), (c) protects against mechanical injury, (d) helps to support and hold organs in place, and (e) fills in the angular areas of the body.

4. *Pigmented tissue:* This is tissue containing cells, the cytoplasm of which is filled with a pigment, usually *melanin.* It is found in the iris, ciliary body, and choroid of the eye.

CARTILAGE (Fig. 3-7). This is a specialized form of connective tissue that, along with bone, functions primarily in support. It consists of cells, fibers, and ground substance, the latter two comprising the intercellular substance or *matrix.* The matrix contains a glycoprotein that gives it a firm but resilient structure. The cells, called *chondrocytes,* lie in cavities called *lacunae.* There are three types of cartilage: hyaline, white fibrous (fibrocartilage), and elastic.

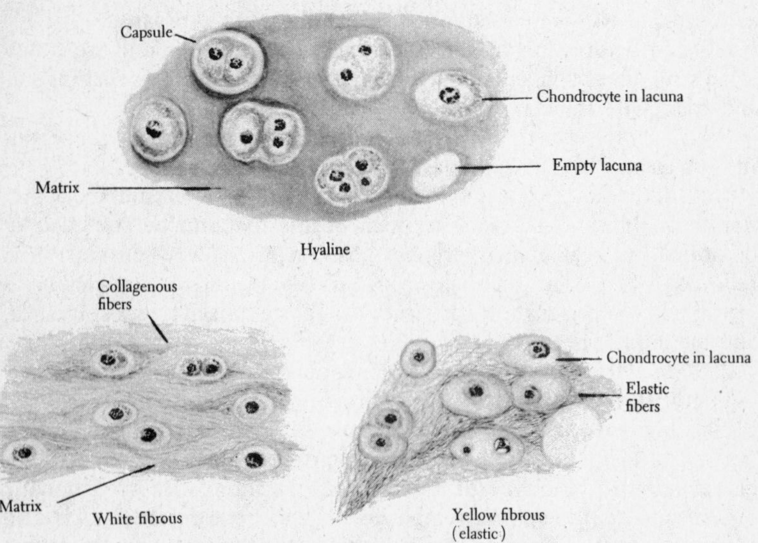

Fig. 3-7. Cartilage.

Hyaline cartilage. When fresh, hyaline cartilage has a bluish white, translucent appearance. The chondrocytes, which occur singly, in pairs, or in small groups, lie in lacunae surrounded by a basophilic *territorial matrix* called a *capsule.* The dense *interterritorial matrix* occupies the space between the lacunae. This matrix contains *chondromucin*, which, upon hydrolysis, yields a sulfonated polysaccharide, *chondroitin sulfate.* Hyaline cartilage is flexible, slightly elastic, and avascular (lacks blood and lymph vessels). It is enclosed by a dense connective tissue membrane, the *perichondrium,* except over the articular surfaces of bone.

Hyaline cartilage is found covering the ends of the bones at joints, in costal cartilages between the ribs and sternum, in the septum of the nose, and in the cartilages of the larynx. It also forms the early skeleton of a fetus and comprises the cartilage model within which bone formation takes place in endochondral ossification.

White Fibrous Cartilage. This cartilage, also called fibrocartilage, consists of a matrix not quite as dense as that of hyaline cartilage. It contains many collagenous fibers that are arranged more or less in rows and possesses great strength and flexibility. The cells, ovoid and enclosed in capsules, are relatively few in number and usually widely separated. This type of cartilage is found in intervertebral discs between the bodies of vertebrae and in the pubic symphysis.

Elastic Cartilage. This consists of a matrix with many elastic fibers that form an interlacing network. Collagenous fibers are present but not readily seen. This type of cartilage is flexible and elastic. It is found in the cartilage of the external ear, the wall of the auditory tube, the epiglottis, and certain laryngeal cartilages.

Cartilage in General.

1. *Growth:* Cartilage increases in amount by two methods: interstitial growth and appositional growth. In *interstitial growth,* cartilage cells (chondrocytes) divide and produce new intercellular substance. It occurs especially in young cartilage. In *appositional growth,* new cartilage is formed by cells from the inner layer of the perichondrium. These *chondrogenic cells* separate from the perichondrium, become chondrocytes, and add more cartilage to the surface layers.

2. *Nutrition:* Cartilage lacks a direct blood supply in that no capillaries are present. Nutrients for cartilage cells come from blood vessels outside the perichondrium by way of diffusion through the perichondrium and the cartilage matrix. Because of this, regeneration, which is essential in the repair of injured cartilage, occurs slowly, new cartilage cells being provided by the perichondrium. It occurs more readily in young than in old individuals.

3. *Calcification:* This results from the deposition of hydroxyapatite salts in the matrix of cartilage. It occurs as a normal process in the formation of bone (see *endochondral ossification,* p. 72) and as a regressive change in old age. Some cartilage calcifies readily, but calcification may never occur, as in the cartilage of the nose and the ear.

Bone or Osseous Tissue (Fig. 3-8). Bone is a rigid, calcified tissue consisting of cells (*osteocytes*) imbedded in a matrix of calcium and phosphorus salts. The structural unit of bone tissue is an *osteon* (formerly *haversian system*), which comprises concentric cylinders of matrix and bone cells surrounding a central cavity, the *haversian canal.* The canal contains blood vessels, lymphatics, and some connective tissue.

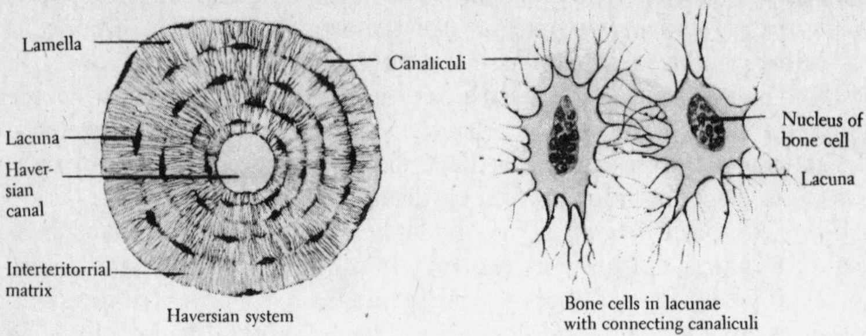

Fig. 3-8. Bone.

About each haversian canal are concentric layers of matrix called *lamellae.* Between the lamellae are *lacunae,* minute cavities within which lie the bone cells or *osteocytes.* Extending from each lacuna are minute canals called *canaliculi,* which radiate through the matrix and connect the lacunae. These provide channels by which nutrients from the haversian canal reach all bone cells.

THE RETICULOENDOTHELIAL SYSTEM

Associated with the connective tissues are the cells and fibers of the *reticuloendothelial system* (RES) or *macrophage system.* This is not a system of organs but a "system" of cells that includes, collectively, all the macrophages of the body. *Macrophages* are cells that have the ability to ingest particulate matter and also to phagocytose certain colloidal dyes, such as trypan blue and other vital stains. These cells ingest and digest bacteria, tissue debris from worn-out cells or injured tissue, and foreign particulate matter that may enter the circulation. Included in this system are the *macrophages* of loose connective tissue, the *Kupffer cells* of the liver, the *reticular cells* of lymphatic tissue, myeloid tissue, and spleen, the *dust cells* of the lungs, the *lining cells* of sinusoids of the adrenal gland and hypophysis, and the *microglia* of the central nervous system.

This system of cells plays an essential role in the destruction of red blood cells and subsequent bile pigment formation and iron metabolism. It is also an important part of the defense mechanisms of the body in preventing the growth and multiplication of pathogenic organisms.

MUSCLE TISSUE

Muscle tissue is that specialized for the production of movement, both for the body as a whole and in its various parts. This is accomplished by contraction or the shortening of its cells. Muscle tissue is classified on the basis of structure (presence of cross bands or striations) and function, whether voluntary (under the control of the will) or involuntary. The three types of muscle are *smooth, involuntary muscle,* found principally in the walls of hollow organs; *skeletal* or *striated voluntary muscle,* which is attached to bones of the skeleton or fascia; and *cardiac* or *heart muscle,* striated and involuntary, which is found in the walls of the heart. The three types are shown in Figs. 6-1 to 6-4. For details of their structure and functioning, see Chapter 6.

NERVOUS TISSUE

Nervous tissue is that specialized for the transmission of impulses from one part of the body to another. Its cells (*neurons* and *neuroglia*) are found in the brain, spinal cord, ganglia, nerves, and the sensory portion of sense organs. *Neurons* are irritable, conducting cells; *neuroglial cells* form the interstitial tissue that supports and protects the neurons. For a detailed discussion of nervous tissue and illustrations of representative neurons and a neuroglial cell, see Chapter 3, Vol. 2.

4: THE INTEGUMENT (The Skin, Its Derivatives and Appendages)

The integumentary system, consisting of the skin and its derivatives, is the largest and one of the most complex systems of the body. The surface area of the skin covers about 1.8 m^2 (19.4 ft^2) of the body of the average male adult and about 1.6 m^2 (17.2 ft^2) in the average adult female.

FUNCTIONS OF THE INTEGUMENT

It may be difficult to think of the integument as a system, but it is a complex of organs (sweat glands, oil glands, etc.) and it serves a number of functions, the principal ones being (1) to protect the underlying tissues from injury which may be physical, chemical, electrical, thermal, or biological; (2) to prevent excessive loss of water or drying of tissues; (3) to act as a temperature regulator, preventing loss of heat in a cold environment and facilitating loss of heat in a hot environment; (4) to serve as a reservoir for food and water; (5) to assist in the processes of excretion (through sweating), eliminating water, salts, and, to a limited extent, urea; (6) to serve as a sense organ for the cutaneous senses; (7) to prevent, in large measure, the entrance of foreign substance, especially microorganisms; and (8) to serve as the seat of origin of the antirachitic vitamin (vitamin D); (9) to serve as a metabolic organ involved in the metabolism and storage of fat; and (10) to serve as a self-maintaining and self-cleansing organ through its sebaceous and other secretions.

STRUCTURE OF THE INTEGUMENT

The integument comprises the skin (epidermis and dermis) and the derivatives of the skin (Figs. 4-1 and 4-2). The table on the facing page shows the distribution of the various elements.

Epidermis. The epidermis (outermost layer of the skin) is composed of stratified squamous epithelium. It arises from the ectoderm of the embryo. It varies in thickness from 0.07–0.12 mm over most of the body to 0.8–1.4 mm on the palms of the hands and the soles of the feet. The epidermis is devoid of blood vessels. It has five layers as follows:

1. *Stratum corneum:* the outermost layer. It consists of nonliving keratinized, flattened cells; the outer surface (*stratum disjunctum*) is shed continuously by the process called *desquamation.*

2. *Stratum lucidum:* a clear, translucent layer that lies beneath the

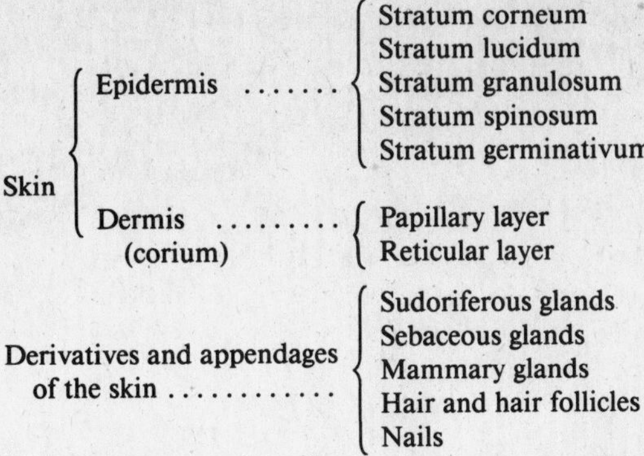

- Skin
  - Epidermis
    - Stratum corneum
    - Stratum lucidum
    - Stratum granulosum
    - Stratum spinosum
    - Stratum germinativum
  - Dermis (corium)
    - Papillary layer
    - Reticular layer
- Derivatives and appendages of the skin
  - Sudoriferous glands
  - Sebaceous glands
  - Mammary glands
  - Hair and hair follicles
  - Nails

stratum corneum. It consists of three to five layers of flattened, closely packed cells. It is clearly seen only in thick skin.

3. *Stratum granulosum:* a thin layer, two to four cells in thickness, consisting of flattened cells whose cytoplasm contains granules of *kera-*

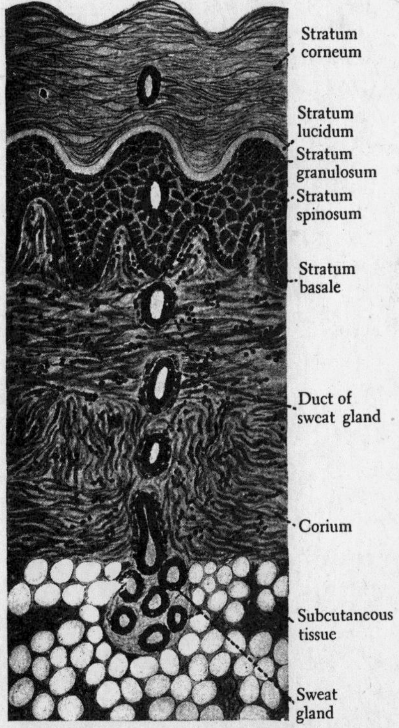

Fig. 4-1. General view of the skin. (Reprinted with permission of Macmillan, London and Basingstoke, from W. J. Hamilton, *Textbook of Human Anatomy,* 1976.)

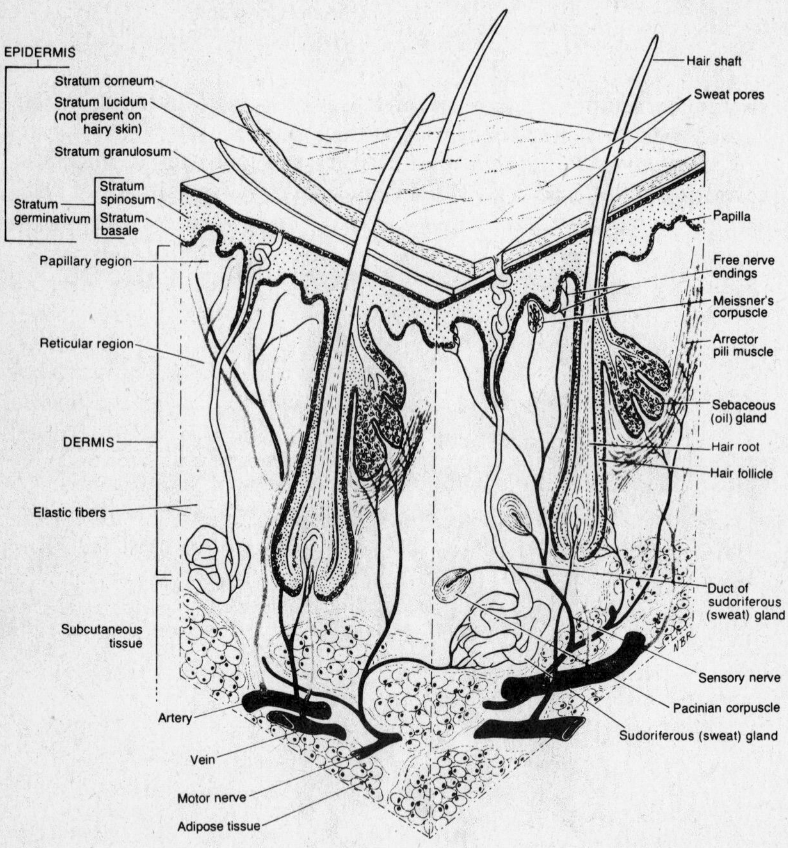

Fig. 4-2. Structure of the skin and underlying subcutaneous layer. (Reprinted with permission of Harper & Row, Publishers, Inc., from G. J. Tortora and N. P. Anagnostakos, *Principles of Anatomy and Physiology,* 3rd ed., © 1981.)

tohyaline. It is in this layer that cornification begins.

4. *Stratum spinosum:* a layer, also called the *prickle-cell layer,* that is made up of several layers of polyhedral cells, flattened in the outermost layers. The surface of these cells is covered with short, cytoplasmic projections that were called *tonofibrils* and erroneously interpreted as protoplasmic connections (intercellular bridges) between cells. Electron micrographs, however, have revealed that the points of contact between the cells are desmosomes and that no cytoplasmic continuity exists.

5. *Stratum germinativum:* also called *stratum basale* and *stratum cylindricum;* this layer, the deepest of the epidermis, consists of a single row of columnar cells which rest on a basal lamina adjacent to the dermis. Cells in this layer are constantly dividing, displacing cells into the layer above. Mitotic figures are often seen in these cells.

Keratinization. Cells of the skin are formed in the stratum germinativum and move outward to the stratum corneum, where they are shed. During *keratinization,* the cells change from physiologically active cells containing much water to a layer of dry, tough, resistant material composed of *keratin.* Keratin is a protein, an albuminoid, and is the principal material comprising hair, nails, claws, horns, beaks, and similar structures in various animals. Keratin is formed from *keratohyalin,* a substance present in cells of the stratum granulosum. In the surface layer of skin, tough, resistant keratin protects the underlying tissues. Where the skin is subject to wear, as on the soles of the feet or the palms of the hands, an extra-thick layer of keratin acts to protect the exposed surfaces.

Dermis or Corium. The dermis is a tough, flexible, highly elastic layer of connective tissue underlying the epidermis. It varies in thickness, being 2–4 mm thick in the palms of the hands and soles of the feet and very thin in the eyelids, scrotum, and penis. An *outer papillary layer* overlies a *reticular layer.*

The outer papillary layer lies next to the epidermis, into which numerous small processes, the papillae, project. The papillae are highly vascular and contain numerous nerve endings. The layer contains collagenous and elastic fibers closely matted together.

The reticular layer consists principally of bundles of collagenous fibers that interlace, forming a feltlike network. Interspersed among the fibers are numerous elastic fibers. This layer is not sharply delimited from the papillary layer above or the subcutaneous tissue beneath.

The dermis is well supplied with blood and lymphatic vessels and with nerves. Cutaneous sense organs are present in this layer. Afferent nerve fibers carry impulses from receptors located here, and efferent nerve fibers carry impulses to blood vessels, smooth muscles, and glands. The dermis is the portion of the skin of an animal that, when tanned, yields leather.

In the skin of the scrotum, the dermis contains, in addition to the foregoing structures, scattered bundles of smooth muscle fibers, which are also found in the skin of the prepuce, the perianal region, and the nipples. Contraction of these muscles produces a wrinkled appearance.

SKIN COLOR

The color of the skin depends upon three factors. Its basic color is yellow due to the presence of a pigment, *carotene*, in the dermis and epidermis and the underlying layer of fat. This is modified by the presence of blood vessels that impart a reddish or bluish hue and the presence of *melanin*, a pigment that is responsible for various shades of brown.

When blood flow to the skin is increased, as in vasodilation, the skin assumes a pink or reddish color; when blood flow is reduced, as in vasoconstriction, the skin turns pale. The degree of redness (erythema) or blueness (cyanosis) depends largely on the degree to which oxyhemoglobin is reduced in its flow through cutaneous vessels. Cyanosis indicates lack of oxygen.

The pigmentation of the skin in colored peoples, or in white peoples following excessive exposure to the sun's rays or ultraviolet rays, is due to the presence of *melanin* produced by special cells called *melanocytes.* Melanin occurs in the form of fine granules that, when abundant, clump together. The degree of color of the skin (white, yellow, red, brown, black) depends upon the amount of this pigment and its dispersal in the epidermis. The color of hair also depends upon this pigment.

Melanin is produced by melanocytes that develop from *melanoblasts,* derived from neural crest cells. Melanocytes lie in the basal layer (stratum germinativum) of the epidermis, but their branching processes extend for a distance from the main cell body. Through these processes melanin granules are dispersed throughout the epidermis and may be taken up by other epidermal cells. The degree of pigmentation depends upon the amount of melanin produced and the number of cells that take up the pigment. The production of melanin depends upon the presence of an enzyme, *tyrosinase.*

Melanin may be present in the dermis in cells called *chromatophores.* These cells phagocytose melanin granules produced by cells of the epidermis. True melanocytes are not present in the dermis except in infants of Mongolian races, where they appear in *blue spots* in the sacral region.

BLOOD AND LYMPHATIC VESSELS

Arteries in the subcutaneous layer form a network underlying the dermis from which branches pass inwardly to supply such structures as sweat glands and hair bulbs in the subcutaneous tissue; other branches pass outwardly and enter the dermis. At the junction of the articular and papillary layers, the vessels form a *subpapillary network* from which single capillary loops pass to the dermal papillae. No blood vessels enter the epidermis. Veins that drain the skin follow a course similar to that of the arteries. There are numerous arteriovenous anastomoses through which blood can bypass the capillaries.

Lymphatic vessels, numerous in the skin, begin as blind capillaries in the dermal papillae and form a dense network in the papillary layer. They lead to larger vessels, which follow the course of cutaneous veins.

NERVES AND SENSE ORGANS OF THE SKIN

Since the skin is the part of the body in contact with the external environment, it is richly supplied with sensory receptors (see Chapter 5, Vol. 2). These receptors respond to the stimuli of pressure, injury, cold, heat, and hair movement. Sensory or afferent fibers carry impulses arising from receptors to the spinal cord and brain. Efferent fibers of the sympathetic division of the autonomic nervous system supply the smooth muscles of blood vessels and the arrector pili muscles of hair follicles.

ABSORPTION THROUGH THE SKIN

Generally speaking, there is little absorption through the skin. Water and aqueous solutions of such substances as salts and sugars are not absorbed. Fatty substances, such as oils and ointments, may be absorbed, but to a limited extent. A few substances, such as lead, mercury, and certain aniline dyes, are taken up readily. Microorganisms, as a rule, cannot pass through unbroken skin but can enter the pores (the openings of sweat glands and hair follicles), as is indicated by the frequent infection of these structures.

DERIVATIVES AND APPENDAGES OF THE SKIN

These include the various glands (sweat, sebaceous, ceruminous, mammary), the hair, and the nails. All these structures are derived from the epidermis; only the hair and nails are, strictly speaking, appendages (Fig. 4-2).

Sweat Glands. These glands, also called *sudoriferous glands,* occur over the entire body surface, with the exception of the prepuce, glans penis, the margin of the lips, and the deeper portion of the auditory meatus. Each consists of a coiled *secretory portion,* located deep in the dermis or in the subcutaneous layer, and an *excretory duct,* which carries the secretion to the surface of the skin.

The secretory portion is a much-coiled tubule that, in section, appears as a compact nest of cut tubules. The cells are of two types: *myoepithelial cells,* elongated, flattened cells that, by contraction, aid in the discharge of the secretion; and *gland cells,* cuboidal or low columnar cells, whose bases pass between the myoepithelial cells to rest on the basement membrane. These cells contain fat droplets, pigment granules, and glycogen.

The *excretory duct* is narrow; its wall consists of two layers of cells, enclosing a narrow lumen. Myoepithelial cells are lacking. In the dermis the duct is straight; in the epidermis, where the stratum corneum is thick and cells are lacking in its wall, it is spiral.

The *number* of sweat glands varies from 60 to 80 per cm^2 on the back to 400 per cm^2 on the palms of the hands. The total number in the body is estimated at from 2 to 5 million. Most sweat glands (eccrine glands) are of the *merocrine* type; sweat glands in the axillae, around mammary papillae, on the major labia, and around the anus are considered to be of the *apocrine* type. They are larger and produce a thicker secretion than those in other regions, and their ducts have a wider lumen and usually open into hair follicles. They begin functioning at puberty; they are supplied by adrenergic nerve fibers.

Sweat or *perspiration* consists of water, salts, principally sodium chloride—(0.2 to 0.5 percent), urea (0.08 percent), and traces of other substances. The amount secreted in 24 hr under average conditions is 500 to 1000 ml. During strenuous exercise, 1000 ml or more may be lost in 1 hr.

In moderate temperatures, sweat is produced continuously at a low rate. Since it evaporates as rapidly as it is formed, it is not apparent, hence it is called *insensible perspiration.* This amounts to about 500 ml per day. The cooling effect resulting from evaporation of sweat plays an important role in temperature regulation. When excessive quantities of sweat are produced, the loss of water and various ions, especially sodium and chloride, may seriously alter water and electrolyte balance, with heat cramps or heat prostration a possible result.

Sweat glands are innervated by efferent fibers of the sympathetic nervous system, which produce acetylcholine at their endings. A center in the hypothalamus of the brain controls sweating. It is stimulated directly by a rise in body temperature or indirectly by stimulation of heat receptors in the skin. Sweating usually is associated with vasodilation of cutaneous vessels, but it may occur when vessels are constricted, as in the production of *cold sweat,* most noticeable on the forehead, palms of the hands, and soles of the feet. This may result from psychic influences such as fright, embarrassment, or nervousness. Sweating may also accompany nausea, asphyxiation, severe pain, or the ingestion of highly spiced foods.

Sebaceous Glands. These are simple or branched alveolar glands that secrete an oily product, *sebum,* consisting of various lipids and cellular debris. The exact role of sebum is uncertain, although it probably acts as a water-repellant and bacteriostatic agent on the surface of the skin. Sebaceous glands usually open into a hair follicle, although they may open directly on the surface of the skin. They are holocrine glands, their secretion being the product of the disintegration of fat-infiltrated cells. New cells replace those lost in the secretory process. The exact control of secretory activity is uncertain, but it is influenced by temperature, sex hormones (especially androgens), and age.

Meibomian (tarsal) Glands. These are specialized sebaceous glands of the upper and lower eyelids. They are elongated, branched alveolar glands whose ducts open on the edge of each eyelid. Their oily secretion seals the margins of the lids when they are closed and acts to prevent the overflow of tears when the lids are open.

Ceruminous Glands. These are modified sweat glands located in the external auditory meatus. Their ducts open on the surface directly or with the duct of a sebaceous gland into a hair follicle. They are apocrine glands, and their secretion, plus sebum and desquamated epithelial cells, form *cerumen,* or ear wax.

Mammary Glands. These modified sweat glands are specialized for the production of milk. Since milk is used in the nourishment of the young, these glands are described in the discussion of the reproductive system, Chapter 7, Vol. 2.

Hair and Hair Follicles. A *hair* is an elastic, horny, threadlike structure that develops from the epidermis. It consists of a *shaft,* the part that projects from the epidermis and may extend slightly below the surface of the epidermis, and a *root,* the part that is embedded within the dermis, with its lower portion expanded to form the *hair bulb.* Within the undersurface of the hair bulb is a projection of the dermis, the *papilla.*

STRUCTURE OF A HAIR. A single hair consists of three layers of cells: *Cuticle,* the outermost layer, consists of a single layer of thin non-nucleated cells; the cells overlay one another, like scales.

The *cortex,* the main horny portion of a hair, is composed of several layers of compact cells; the color of hair is due to the presence of pigment in this layer.

The *medulla,* the central axis, is lacking in fine hairs; it consists of two or three rows of polyhedral cells and may not extend the entire length of the hair.

Intercellular spaces may contain air, which accounts for the gray or silvery color of hair when pigment is lost.

THE HAIR FOLLICLE. The *hair follicle* is the structure enclosing the hair root. It consists of an epithelial sheath and a connective-tissue sheath. The former has an *inner* root sheath consisting of three layers, the cuticle (a single layer of overlapping horny scales), Huxley's layer (a double layer of flattened nucleated cells), and Henle's layer (a single layer of cuboidal cells); and an *outer* root sheath, a layer that is continuous with the stratum germinativum of the epidermis. The connective tissue sheath (outer or *dermic* coat) is that part of the dermis that encloses the epithelial portion.

The hair follicle usually occupies an oblique position in the skin. Attached to it, also obliquely, is the *arrector pili muscle,* a bundle of smooth muscle fibers, which arises in the superficial portion of the dermis. Contraction causes the hair follicle to assume a more vertical position. Impulses through the sympathetic nervous system bring about contraction, as from cold or fright. When the arrector pili muscle contracts, the epidermis is depressed and the region immediately around the hair is lifted, producing the condition called "goose flesh."

GROWTH AND REPLACEMENT OF HAIR. The bulb of the hair root consists of a mass of growing and multiplying cells. As these increase in number, they move outward and become massed closely together. Chemical changes occur, and the cells are transformed into the horny cells of the hair and the inner root sheath. The *dermal papilla,* which projects into the bulb, contains capillaries through which nourishment is received.

Hairs in each part of the body have definite periods of growth, after which they are shed and then replaced. In some animals, this is a periodic process; in man, growth and loss take place continuously. The life duration of scalp hairs is two to five years; of eyebrows and eyelashes, three to five months. The replacing hairs of the eyebrows, eyelashes, and scalp of children are progressively larger and coarser than the preceding set. Hormones, especially the sex hormones, influence the growth of hair at puberty in the axillary and pubic regions of both sexes and on the face and trunk of males. There is no hair on palms and soles.

Hair grows at an average rate of 1.5 to 3.0 mm/wk.

LANUGO. This is the layer of fine, silklike hairs that almost completely covers the embryo. It is prominent at about the seventh month of development but is shed before or very shortly after birth.

Nails. The nail, a flat, horny, scalelike modification of the epidermis, corresponds to the hoof or claw of lower animals. It serves to protect the end of the digit. A nail consists of a *body,* the attached, uncovered portion; a *free edge,* the distal unattached portion; and a *nail root,* the portion embedded in a groove of the skin.

The skin that covers the nail root and the lateral edges of the nail is called the *nail wall;* the skin under the nail, the *nail bed.* The *lunula* is the semicircular white portion at the base of the nail.

A nail grows in length and thickness through cellular proliferation at the proximal end of the nail bed, in the region of the root. Nails lost or torn away will regenerate if the *stratum germinativum* is not damaged.

THE SKIN AS AN INDICATOR OF DISEASE

The skin is of importance to the physician in the observation of changes in the functioning of the body during health and illness. The appearance of the skin is frequently a reflection of internal disorder. Examples are:

Increased redness (erythema), which results from dilation of capillaries. It may indicate hyperemia, hypertension, infection, fever, allergy, or an emotional disturbance.

Pallor, which may be due to constriction of capillaries (as in cases of fright or exposure to cold) or to a reduced number of red blood cells (as in anemia).

Cyanosis, a bluish appearance, which results from incomplete oxygenation of the blood, as in pneumonia, asphyxiation, heart failure, and heart anomalies ("blue baby").

Yellowness, or a "jaundiced" appearance, which results from a liver disorder in which bile pigments are absorbed by blood and circulated throughout the body.

Increased pigmentation, as in Addison's disease.

Extreme dryness, as in vitamin deficiency or heat stroke.

Extreme moisture, a sign of hyperthyroidism.

SOME DISEASES AND DISORDERS INVOLVING THE SKIN

Abscess. A localized collection of pus in any part of the body. In the skin, commonly called a *pimple.*

Acne. An inflammatory condition involving sebaceous glands, resulting in formation of comedones, papules, pustules, and sometimes infected cysts. The cause is unknown, but hereditary factors, diet, microorganisms, and the activity of androgenic hormones may be involved.

Albinism. Complete absence of pigment.

Alopecia. Baldness; loss of hair. May be due to hereditary factors, pathological conditions, or irradiation.

Bedsore. A pressure sore resulting from tissues being subjected to prolonged pressure from an external object; also called *decubitus.*

Blackhead. A comedo; blackened mass consisting of sebaceous material and epithelial debris that fills a hair follicle.

Boil. A furuncle; a localized, inflamed nodule with a central core of pus.

Burn. Tissue injury caused by thermal, electrical, radioactive, or chemical agents. In *first-degree burns,* damage is limited to the epidermis; in *second-degree burns,* both the epidermis and dermis are damaged and blisters are formed; in *third-degree burns,* both the epidermis and dermis are destroyed, with charring or coagulation of underlying tissues.

Callosity, callus. A thickening of the epidermis of the skin that results from pressure or friction.

Cancer. A malignant tumor, usually an epithelioma.

Cold sore. See *Herpes simplex.*

Comedo. A blackhead.

Corn. A clavus; a hard, cone-shaped thickening of the horny layer of the epidermis, usually occurring over a toe joint or between the toes. Corns are caused by pressure or friction.

Cyst. A closed sac containing a fluid or other material.

Decubitus. A bedsore.

Dermatomycosis. A fungus infection of the skin or its appendages, as ringworm, athlete's foot, barber's itch, or moniliasis.

Dermatitis. A bacterial infection of the skin, as impetigo or erysipelas.

Drug eruption. An eruption of the skin or mucous membranes following parenteral administration of a drug. It may be due to the toxicity of the drug or an allergic response to it.

Eczema. A noncontagious, itching, inflammatory condition of the skin.

Furuncle. See *Boil.*

Herpes simplex. A virus infection characterized by the appearance of single or multiple fluid-filled vesicles on the skin or mucous membranes; a cold sore or fever blister. There are two types. Type I, the labial type, occurs above the waist, commonly on the lips; type II, the genital type, occurs below the waist, commonly involving the genital organs, especially the vagina in females.

Herpes zoster. A nervous disease involving sensory neurons but characterized by inflammation of the skin at the terminations of the neurons involved. Commonly called *shingles.*

Hyperhidrosis. Excessive perspiration.

Hyperkeratosis. Hypertrophy of the horny layer of the skin.

Hypertrichosis. Excessive hairiness; growth of hair in areas ordinarily not hairy.

Impetigo. A bacterial skin infection.

Nevus. A birthmark.

Pediculosis. Infection of the skin or hair by lice (head lice, body lice, or pubic lice, also called *crabs*).

Pruritus. Itching.

Psoriasis. A chronic, inflammatory skin disorder characterized by reddish patches covered by silvery scales.

Pustule. A small elevation of the skin, containing pus.

Scabies. The itch, a parasitic skin disorder caused by the itch mite, *Sarcoptes scabeii,* which bores into the epidermis, forming burrows.

Shingles. See *Herpes zoster.*

Urticaria. Hives, nettle rash; an inflammation of the skin characterized by redness, development of wheals and welts, and extreme itching.

Verruca. A wart.

Vertiligo. Piebald skin; leukoderma. Loss of pigment in localized areas of the skin.

Wart. A circumscribed, benign elevation of the skin resulting from hypertrophy and hyperplasia of dermal papillae and the overlying epidermis. It is caused by a virus. Also called *verruca.*

5: THE SKELETAL SYSTEM

The skeletal system lends form to the body. Specifically, its functions are (1) to provide a framework for all the body systems; (2) to provide attachments for muscles, ligaments, tendons, and fascia; (3) to enclose and protect vital organs, such as the heart, lungs, brain, and sense organs; (4) to serve as a site for manufacture of blood cells (the "hemopoietic" function); and (5) to serve as a storehouse for calcium.

The components of the skeletal system are (1) *bones,* which comprise the hard framework of the body; (2) *cartilage,* which forms the connecting and supporting structures; and (3) *ligaments,* which bind the bones together. The microscopic structure of bone and cartilage has been discussed in Chapter 3. Ligaments bear an intimate relationship to the functions of the skeletal system; consequently, they are discussed in this chapter.

BONES

Bone consists of organic and inorganic materials. The *organic substance* includes the living cells and about one-third of the interstitial substance, or *matrix.* The *inorganic substance* comprises the remaining two-thirds of the matrix and is made up of inorganic salts, which are complex compounds of calcium and phosphorus.

If a bone is placed in weak hydrochloric acid, the mineral matter will dissolve and the organic framework will remain. Such a decalcified bone can be easily bent or even tied into a knot. If a bone is heated, the organic matter will be destroyed and only the mineral matter or ash will remain. Such a bone is extremely brittle, but it retains its external form and, to some extent, its microscopic structure.

Classification of Bones. Bones are classified in three ways:

On the basis of shape, bones are *long* (most of the bones of the extremities), *short* (bones of the wrist and ankle and sesamoid bones), or *flat* (bones of the cranium, the scapula, and the ribs). Certain bones, such as the vertebrae and some skull bones, are classified as *irregular.*

On the basis of embryonic origin, bones are either *membranous* (developing directly from connective tissue membrane) or *cartilaginous* (developing from cartilage). Postnatally, *sesamoid* bones may develop in tendons or in the capsules of joints. A sesamoid bone occurs in the tendon of the short flexor muscle of the thumb as it passes over the head of the metacarpal bone; two sesamoid bones are situated in the tendon of the corresponding muscle of the big toe (the flexor hallucis brevis on the plantar surface of the head of the first metatarsal). The knee cap, or *patella,* situated in the tendon of the quadriceps femoris muscle, is the largest sesamoid bone.

On the basis of structure, bone tissue is either *compact* (the hard, dense, outside layer of all bones) or *spongy* (also called *cancellous,* containing many small cavities filled with marrow and generally enclosed by compact bone).

Structure of a Typical Long Bone (Fig. 5-1). Using the femur as an

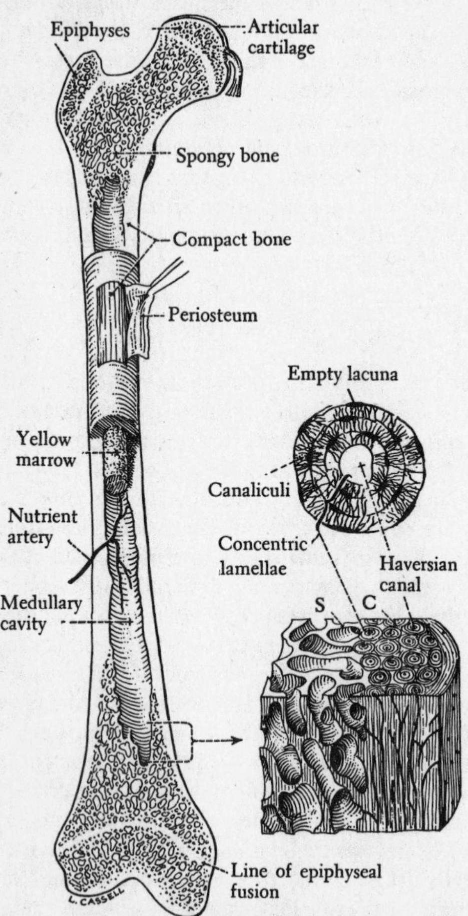

Fig. 5-1. Structure of a long bone. *S*, on inset, indicates spongy bone; *C*, compact bone. (Inset adapted from Toldt.) (Reprinted with permission of W. B. Saunders Co., Philadelphia, from B. G. King and M. J. Showers, *Human Anatomy and Physiology,* 6th ed., 1969.)

example, the following components of a long bone are found: the *diaphysis,* the shaft or main central portion (principally compact bone); the *epiphysis,* the portion at each end (spongy bone covered with a thin layer of compact bone); the *medullary* or *marrow cavity,* containing yellow bone marrow, within the diaphysis; *periosteum,* the membrane covering the bone; and *endosteum,* the fibrous membrane lining the marrow cavity.

There are three kinds of epiphyses: *pressure* epiphyses, at the articular ends of bones; *traction* epiphyses, associated with the insertions of muscles, and *atavistic* epiphyses, which represent bones that were separate in earlier stages of evolution but have lost their function and have disappeared in man.

The *periosteum* is a tough, fibrous, vascular membrane that envelops bone except over surfaces covered with articular cartilage. It consists of two layers: An inner *osteogenic layer* contains *osteoblasts,* which are involved in the formation of new bone; it may also contain *osteoclasts,* which are involved in the destruction of bone. The outer *fibrous layer* consists of dense connective tissue fibers continuous with the attachments of muscles, tendons, and ligaments. Fibers also extend into the bone substance to form a firm attachment. These are called *Sharpey's fibers.* Branches of periosteal blood vessels, called *Volkmann's canals,* enter the bone and connect with vessels of the haversian canals. In addition to these blood vessels, each bone is supplied by one or two *nutrient arteries,* each of which passes by way of a *nutrient foramen* and *canal* to the marrow cavity, there branching into distal and proximal branches that supply the extremities. *Veins* parallel the nutrient arteries. Many smaller veins leave by the foramina of the extremities. *Lymph vessels* and *nerves* are present in the periosteum and haversian canals.

Nerve Supply of Bone. Myelinated and unmyelinated nerve fibers enter the bone with the blood vessels. Some nerve fibers can be traced to the Haversian canals, others in close relation to the bone cells, while many end blindly in the bone matrix.

Bone Marrow. Bone marrow is the substance occupying the medullary cavity and the spaces in cancellous bone. *Yellow bone marrow* is found in the marrow cavities of long bones. It consists principally of fat cells. In pathological conditions such as anemia, it may be replaced by red marrow. *Red bone marrow* is found in spongy bone at the ends of most long bones and in the flat bones of the cranium, sternum, ribs, and bodies of vertebrae. It consists of a *stroma* (framework) and free cells within the mesh of the stroma. Red bone marrow is concerned with the formation of blood cells; the process is described in detail in Chapter 9.

Formation of Bone (Ossification). Bone develops from the mesoderm and begins to make its appearance at about the eighth week of embryonic development. There are two modes of formation: intramembranous and endochondral (intracartilaginous).

INTRAMEMBRANOUS OSSIFICATION. By this mode, bone is formed within fibrous membranes. Certain bones of the face (mandible, maxilla, nasal, lacrimal), the flat bones of the cranium (frontal, parietal), and the clavicle develop in this way. Each bone is preceded by a membrane of mesenchymal connective tissue. Within this membrane one or more points of ossification arise, where *osteoblasts* or bone-forming cells appear and begin to deposit a *bone matrix* consisting of calcium salts (carbonates and phosphates) in the form of *spicules.* As the matrix increases in mass, strands called *trabeculae* are formed; these extend radially in all directions. The osteoblasts are arranged in a single layer on the surface of the spicules, and, as development proceeds, the spicules increase in width and thickness to produce more matrix. In the process some osteoblasts become entrapped within small cavities of the matrix, where they become mature *bone cells* called *osteocytes.* Ultimately, they lose their capacity to form bone, but they are essential to the maintenance of bone that is already formed.

When trabeculae are first formed, they are isolated; as more matrix is produced, they meet and coalesce, eventually comprising a latticelike structure known as *cancellous bone.* Meanwhile, the mesenchymal cells on the outer surface of the membrane give rise to a fibrous membrane, the *periosteum.* At about the same time, on the inner surface, the osteoblasts begin to deposit parallel plates of compact bone; this process is known as *periosteal ossification.* As a result of these two processes, two thin layers of compact bone (the *inner* and *outer tables*) are formed. They are separated by a layer of spongy bone called *diploe.* In the development of bone, much bone tissue that is formed is resorbed and replaced by new bone. It is thought that resorption is accomplished by the action of large multinucleated cells called *osteoclasts* that are found in hollow cavities (*Howship's lacunae*) in eroded bone.

ENDOCHONDRAL OSSIFICATION. Endochondral bones are formed by replacement of cartilage. A temporary cartilage model of the bone precedes the formation of the actual bone. In cartilage that is about to be replaced with bone, the following changes take place: Cartilage cells that have been producing cartilage matrix begin to destroy that matrix. Lacunae enlarge, and the cells become arranged in rows. Some of the matrix begins to calcify, and the cartilage cells disintegrate and disappear. Spaces then appear; these are gradually filled with *primary marrow tissue.* The perichondrium over the area becomes active. Cells proliferate, new cells invade the cartilage in budlike masses, and the perichondrium becomes transformed into periosteum. Now blood vessels invade the honeycombed cartilage, and osteoblasts deposit bone matrix until the entire cartilage is replaced with spongy bone. In the last stage of the process, bone marrow develops in the spaces of the bony tissue.

It should be noted that *cartilage is not transformed into bone.* The cartilage is first destroyed; then bone is formed in the place that was occupied by the cartilage. For this reason, this type of bone is called *replacement* or *substitution bone.*

Growth of a Long Bone (Fig. 5-2). As previously stated, the shape of the cartilage model is in general the shape of the adult long bone. There are usually several ossification centers. The first to appear is that in the diaphysis (shaft), where bone starts to form in a ring-shaped area at the center. As soon as osteoblasts appear and bone formation commences, the membrane covering the cartilage model (perichondrium) becomes the periosteum. Bone formed by this process is called *periosteal bone.* It forms a band or collar about the central region of the diaphysis.

In the center of the diaphysis, cartilage is eroded away and the *primary marrow cavity* is formed. Bone continues to be added progressively toward each end until the entire shaft is ossified. As the diameter of the bone increases, the diameter of the marrow cavity also increases through the action of osteoclasts, which destroy the bone already formed in that area. While ossification is proceeding in the diaphysis, ossification centers develop in the epiphyses. The cartilage in the region between the bone being formed in the diaphysis and the bone being formed in the epiphysis is called the *cartilage plate.* This plate remains relatively constant in thickness inasmuch as it replaces its cells as rapidly as the cartilage at its surface is replaced with bone. When proliferation of cartilage cells ceases and cartilage matrix is replaced with bone, the epiphysis unites with the diaphysis, and the longitudinal growth of the bone is terminated.

The first *ossific* or *ossification center* to appear in a bone is called the *primary center.* The one in the center of each long bone is present by the eighth week of fetal life. *Second centers* are those of the epiphyses, only one of which (the distal epiphysis of the femur) is present at birth. The development of epiphyses and their fusion with the shafts of various bones takes place progressively up to the age of about 25 years for a male. Most secondary centers appear earlier in females than do corresponding centers in the male, and fusion with the diaphyses also occurs earlier in females. The age of a bone (and an entire skeleton) can often be determined by noting which epiphyses have or have not united with the diaphyses. The last epiphysis to appear and unite is that at the distal end of the clavicle, which, incidentally, is the first bone to begin to ossify.

Factors in Formation of Bone. Many factors are involved in the formation of bone. The principal mineral of bone is *hydroxyapatite* $[3Ca_3(PO_4)_2 \cdot Ca(OH)_2]$, which accounts for the calcium and phosphate, but other salts are present also.

Bone *calcium* and blood calcium are in equilibrium. If blood calcium is maintained at a stable level, there is little exchange between blood and bone, but if the blood level of calcium drops, calcium is withdrawn from bone (decalcification) through the action of a *hormone* from the

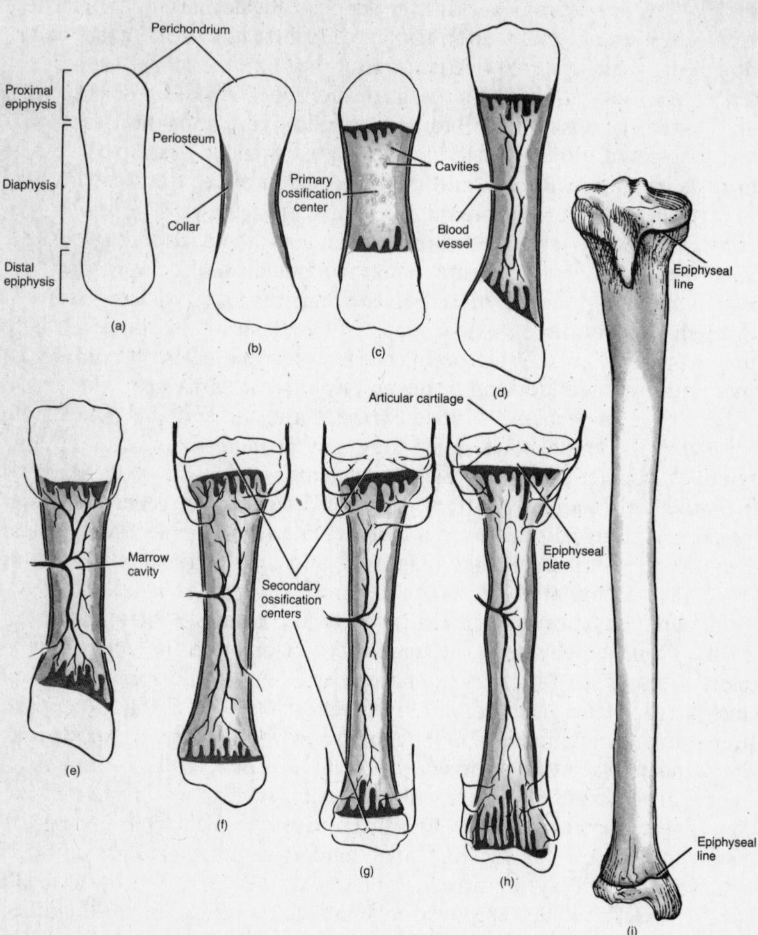

Fig. 5-2. Stages in the growth of a long bone. (*a*) Cartilage model. (*b*) Collar formation. (*c*) Development of primary ossification center. (*d*) Entrance of blood vessels. (*e*) Marrow-cavity formation. (*f*) Thickening and lengthening of the collar. (*g*) Formation of secondary ossification centers. (*h*) Remains of cartilage as the articular cartilage and epiphyseal plate. (*i*) Formation of the epiphyseal lines. (Reprinted with permission of Harper & Row, Publishers, Inc., from G. J. Tortora and N. P. Anagnostakos, *Principles of Anatomy and Physiology,* 3rd ed., © 1981.)

parathyroid gland. Dysfunction of these glands can cause such serious bone disorders as osteitis fibrosa and osteomalacia (see Disorders and Diseases of the Bones and Joints, p. 120). Another hormone, *calcitonin,* produced by the thyroid gland, stimulates osteogenic cells, inducing the formation of bone. The growth hormone (*somatotrophin*) of the anterior pituitary also plays a role in the formation of bone.

Vitamin D is essential for the formation of bone in children. In its absence, calcium and phosphorus are not absorbed by the intestine. Osseous tissue, although it continues to grow, does not calcify. As a result, bone fails to harden and *rickets* develops. In adults, a deficiency of vitamin D may cause *osteomalacia.*

Phosphorus deficiency and an imbalance between phosphorus and calcium are also known to influence the growth of bone unfavorably.

Bone as a Living, Adaptable Structure. Every organ of the body is a complex of tissues arranged in a definite pattern, having the ability to grow, develop, and repair itself. Because a bone has these properties, it must be regarded as a living structure or an *organ of the body.* The more or less definite *form* of each bone is closely correlated with the function or functions it serves.

Although bone consists largely of inorganic matter and appears relatively fixed in structure, it is nevertheless remarkably *adaptable to environmental influences.* Bone is continually undergoing change through the processes of modeling and remodeling. Studies with radioactive isotopes indicate that the inorganic salts are undergoing constant use and continual replacement. Disuse as occurs in debilitating diseases or physical inactivity results in atrophy of bone with loss of substance; processes become less pronounced. With increased use, as occurs in increased physical activity, bone tends to hypertrophy and the bone is strengthened. Processes where increased stress is applied increase in size. Bone modification is apparent following loss of teeth, for example. Bone in the alveolar processes of the mandible or maxilla, which give them support, is resorbed. Habitual squatting with feet crossed produces "squatter's facets" on the external malleoli of the tibiae; the cradling practice in which the infant is made to rest its head on a hard board usually results in a flat ("Armenoid") occiput. Emotional disturbances in the mother during pregnancy, and such disturbances in the child during the growing period may affect bone growth deleteriously.

Descriptive Terms Applied to Bones. All bones have irregular surfaces: *Elevations* or *projections* serve for attachment of muscles, for articulation with other bones, or for protection of vital parts; *depressions, grooves,* or *openings,* for the passage of blood vessels, nerves, tendons, and ducts of glands; *concavities,* for receiving the articulating surface of other bones or providing space for organs. To facilitate description and recognition of these irregularities, the following terms are used.

Elevations, Projections

Condyle a rounded process, usually smooth, for articulation
Crest a ridge or projecting structure
Epicondyle an elevation on a bone above a condyle
Head the expanded end of a bone
Line a mark or narrow ridge
Neck a constricted region by which a head is connected to the body of a bone
Process a general term for any elevation or projection
Spine a more or less pointed process
Trochanter a very large process
Trochlea a process shaped like a pulley
Tubercle a small, rounded projection
Tuberosity a larger, roughened projection

Depressions, Grooves, Openings, Concavities

Aditus an entrance into a cavity
Alveolus a deep pit or socket
Antrum a cavity or hollow space
Facet a small, flat surface
Fissure a slitlike opening
Foramen an opening through which blood vessels or other structures pass
Fossa a depression or concavity
Fovea a slight depression
Hamulus a hook-shaped process
Hiatus a slit or gap
Incisura a notch or incision
Meatus a short canal
Sinus a cavity or hollow in the interior of a bone, lined with mucous membrane and filled with air
Sulcus a groove or narrow, elongated concavity

Other Terms

Fontanel a membranous space between the cranial bones in fetal life and infancy
Ramus a part of a bone that forms an angle with the body of the bone
Suture the line of junction between two adjacent bones of the skull

DIVISIONS OF THE SKELETAL SYSTEM

The skeletal system (Fig. 5-3), comprising 206 bones, has two general divisions, the *axial skeleton* (the basic framework of the body) and the *appendicular skeleton* (the extremities), subdivided as follows:

AXIAL SKELETON

		Number of Bones		
Skull	Cranium	8	29	80
	Face	14		
	Hyoid bone	1		
	Auditory ossicles	6		
Vertebral column	Vertebrae		26	
Thorax	Sternum	1	25	
	Ribs	24		

APPENDICULAR SKELETON

		Number of Bones	
Upper extremity	Pectoral girdle	4	126
	Arms, forearms, hands	60	
Lower extremity	Pelvic girdle	2	
	Thighs, legs, feet	60	

THE AXIAL SKELETON

The axial skeleton includes the bones of the skull, the vertebrae, and the bones of the thorax.

The Skull. The 29 bones of the skull comprise those in the cranium and face, the auditory ossicles, and the hyoid bone.

	Single Bones	*Paired Bones*
Cranium	Frontal Occipital Sphenoid Ethmoid	Parietal Temporal
Face	Mandible Vomer	Maxillae Zygomatic Lacrimal Nasal Inferior nasal conchae Palatine
Hyoid bone		
Auditory ossicles		Malleus Incus Stapes

THE SKULL AS A WHOLE. Examination of the skull from various aspects enables one to better understand the relationships of the cranium and the face and the various bones comprising them, which are described in detail later.

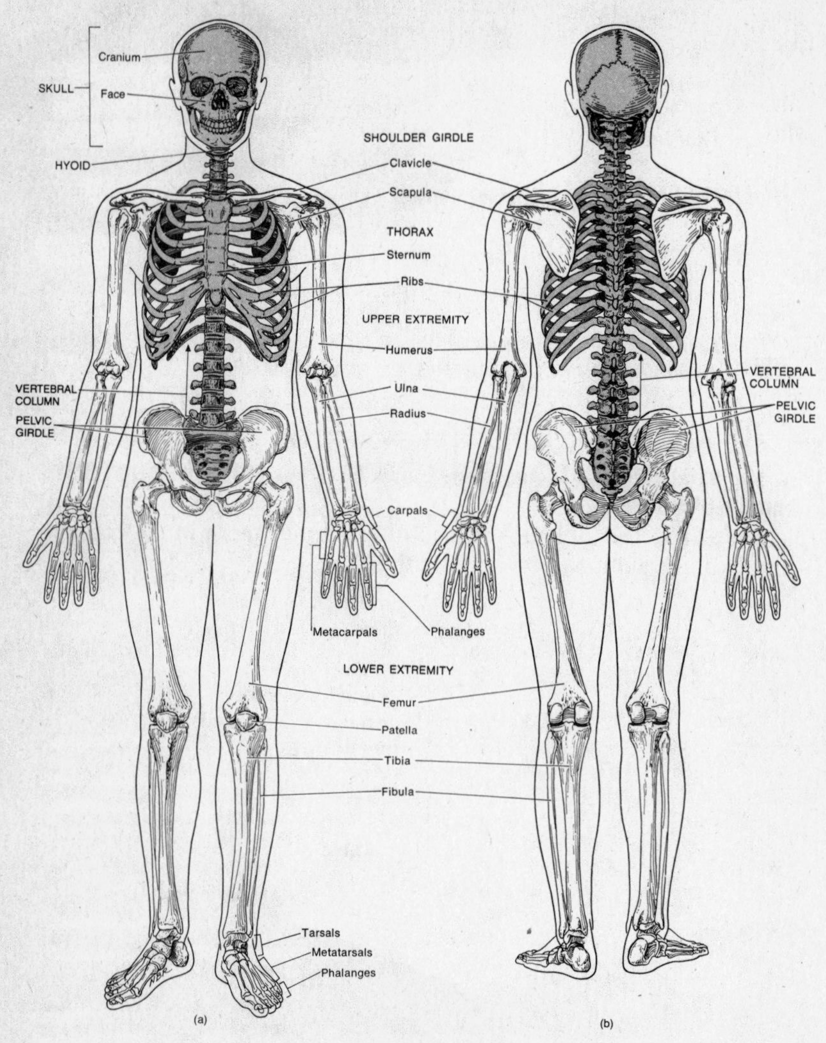

Fig. 5-3. The skeletal system. (*a*) Anterior view. (*b*) Posterior view. (Reprinted with permission of Harper & Row, Publishers, Inc., from G. J. Tortora and N. P. Anagnostakos, *Principles of Anatomy and Physiology,* 3rd ed., © 1981.)

Anterior or Frontal Aspect (Fig. 5-4). Viewed from the front, the skull presents the following features. It is ovoid in shape, wider above, the frontal bone forming the region of the forehead. Four cavities are noted: two *orbits* lodging the eyes; a pear-shaped *nasal opening* leading to the nasal cavity; and the large *buccal* or *mouth opening* between the rows of teeth.

The *nasal opening* is bordered laterally by the two *maxillae* and above by the two *nasal bones.* Looking into the nasal cavity, the median *nasal septum* (part of the ethmoid plus vomer) and *nasal conchae* (folds on lateral walls) can be seen.

The *mouth* or *oral cavity* is bordered below by the *mandible* and above by the two *maxillae.* Each of these bones bear teeth, the roots of which are lodged in their free arch-shaped, alveolar borders. The arch of the mandible bears a projecting *mental protuberance*, a distinctive human characteristic, and lateral to the protuberance are two *mental foramina.*

The *orbits* are bordered superiorly by the *frontal bone.* The thickened region underneath the eyebrow is the *superciliary arch.* The lateral and inferior borders are formed by the *zygoma*, or *cheekbone* and the *maxilla.* A process of the latter extends upward, articulating with the frontal bone; another extends laterally, articulating with the zygomatic bone. These act as buttresses to absorb the biting and chewing forces exerted on the maxilla by the mandible.

Lateral Aspect (Fig. 5-5). The skull, as seen from the side, shows clearly its two major divisions: the *cranium,* or *brain case,* and the *face bones.* Cranial bones seen are the frontal, parietal, occipital, temporal (with its mastoid process), and the sphenoid (its great wing). An opening, the *external acoustic meatus,* lies in the lower portion of the temporal bone. Inferior and medial to the meatus is a long *styloid process,* and extending anteriorly, a process of the temporal unites with a process of the zygomatic to form the *zygomatic arch.* A broad, shallow depression lying above the arch is the *temporal fossa* from which the fibers of a broad, fan-shaped muscle, the temporalis, arise. This muscle closes the jaw. The body of the mandible continues into an upward-projecting *ramus,* which articulates with the temporal bone immediately anterior to the acoustic meatus. The ramus bears a *condyloid process* to which the temporalis muscle is attached. The *lacrimal bone* within the orbit can be seen.

Posterior Aspect. From behind, only cranial bones can be seen. Two *parietal bones* form the roof of the cranium. A single *occipital bone* lies between the two *temporal bones,* which are located at the sides, inferior to the parietal bones. The *mastoid processes* of the temporal bones form prominent projections.

Base of the Skull, Inferior Aspect (Fig. 5-6). The base of the skull, viewed from the outside, reveals these features. If the mandible has

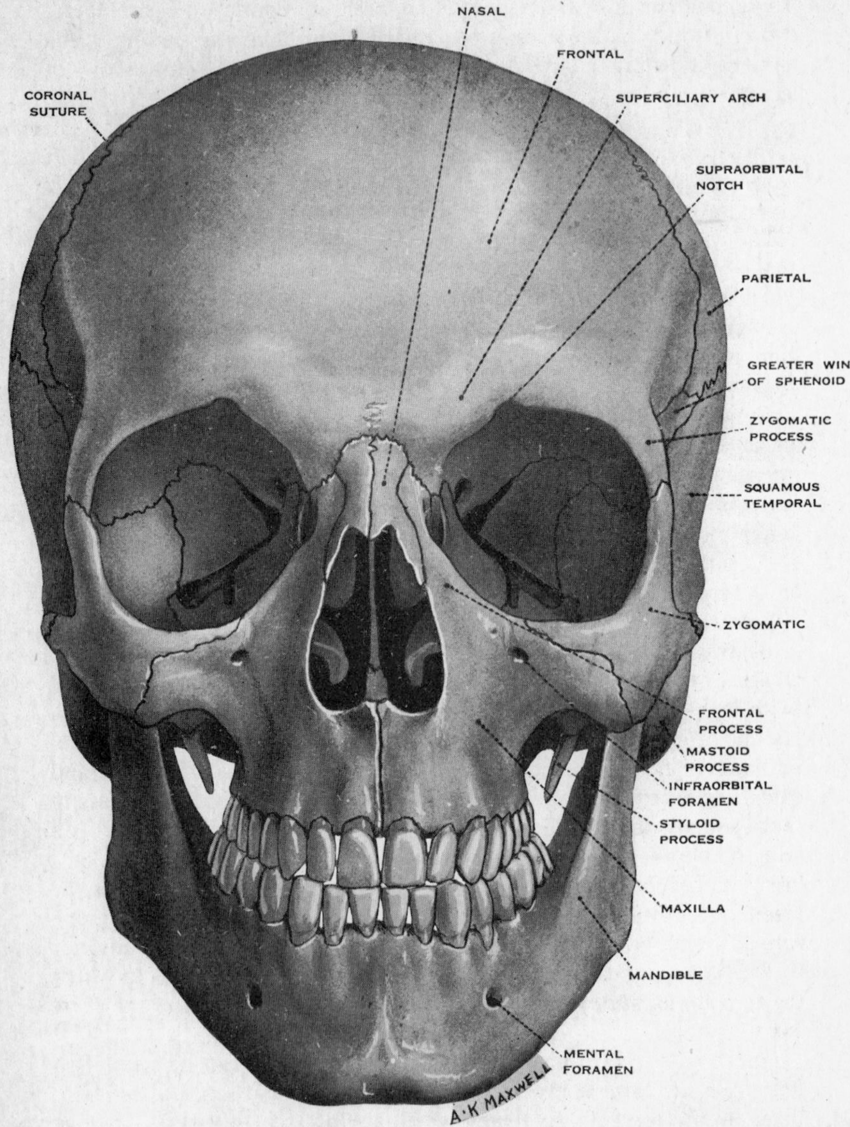

Fig. 5-4. The skull, anterior aspect. (Reprinted with permission of Macmillan, London and Basingstoke, from W. J. Hamilton, *Textbook of Human Anatomy,* 1976.)

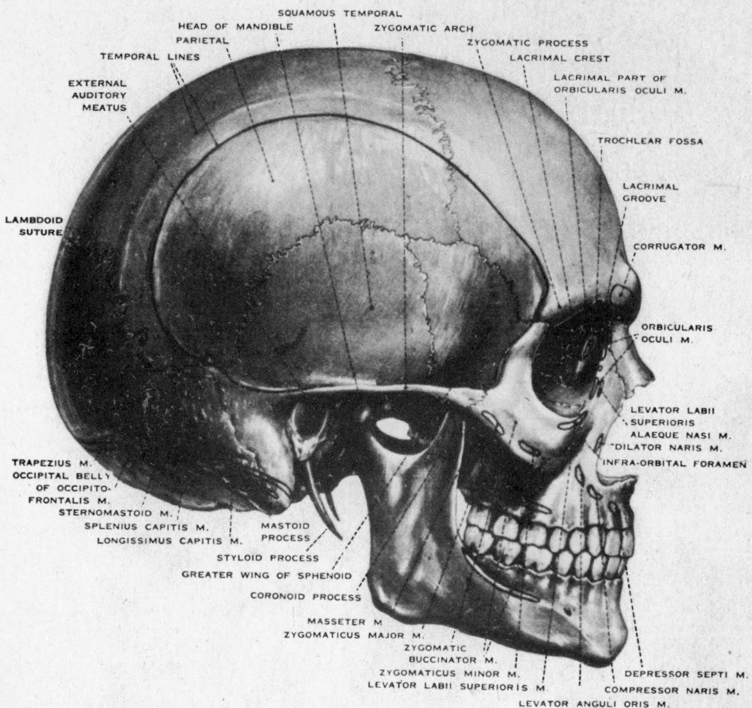

Fig. 5-5. The skull, lateral aspect. (Reprinted with permission of Macmillan, London and Basingstoke, from W. J. Hamilton, *Textbook of Human Anatomy,* 1976.)

been removed, three regions, anterior, middle, and posterior, can be identified. A description of each follows.

1. *Anterior region:* Anteriorly, the roof of the mouth formed by the hard palate and bordered by the maxillary teeth can be noted. The *hard palate* is formed by the *palatine processes* of the *maxilla* and the horizontal processes of the *palatine bones.* Laterally, the inferior surfaces of the *zygomatic arches* can be seen.

2. *Middle region:* Two large openings leading anteriorly can be noted. These are the posterior openings *(choanae)* of the nasal cavity. They are separated by the *vomer bone,* forming the posterior portion of the *nasal septum.* Lateral to the choanae are two pairs of thin plates of bone, the *lateral* and *medial pterygoid processes* of the sphenoid bone. Conspicu-

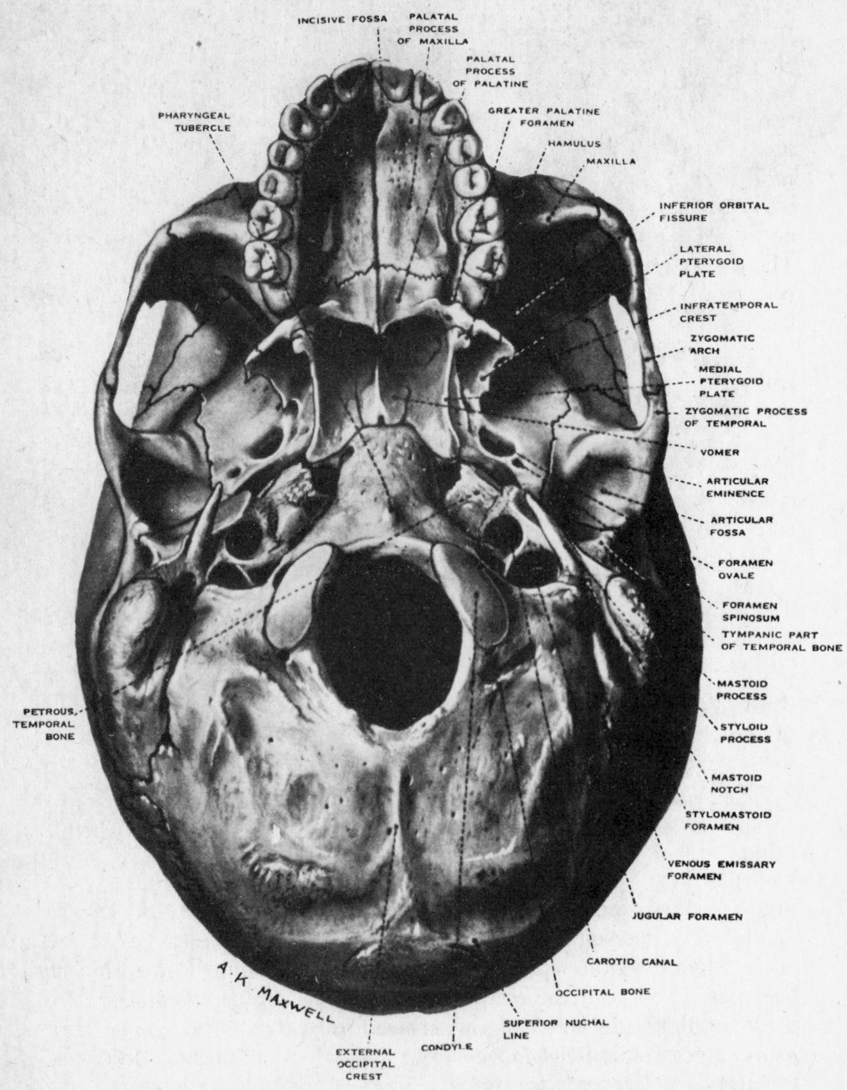

Fig. 5-6. Base of skull, inferior aspect (external surface). (Reprinted with permission of Macmillan, London and Basingstoke, from W. J. Hamilton, *Textbook of Human Anatomy,* 1976.)

ous openings are the paired *foramen lacerum, foramen ovale, carotid canal* and *jugular foramen.*

3. *Posterior region:* This is occupied principally by the *occipital bone.* In its center is a large opening, the *foramen magnum,* through which the spinal cord and blood vessels enter the cranium. Lateral to the foramen magnum are two rounded processes, the *occipital condyles,* by which the skull articulates with the first vertebra, the atlas. Lateral to the condyles are the *mastoid* and *styloid processes* of the temporal bone and the *articular fossae* for the articulation of the mandible.

Base of the Skull, Superior Aspect (Fig. 5-7). This surface can be divided into three regions: the *anterior, middle,* and *posterior cranial fossae.* These fossae are occupied, respectively, by the frontal lobes, anterior portions of the temporal lobes, and the cerebellar hemispheres and brain stem.

1. *Anterior fossa:* This fossa is formed by the orbital plates of the frontal bone and lies over the orbits and nasal cavity. In the midline of the fossa and lying directly over the nose is a depressed area perforated by many small openings. This is the *cribriform plate* of the *ethmoid bone.* Extending upward in the midline of the cribriform plate is a pointed process, the *crista galli,* to which a membrane of the brain, the *falx cerebri,* is attached.

2. *Middle fossa:* This fossa lies over the sphenoid bone, which it roughly parallels in shape, usually described as resembling a butterfly. The midportion is constricted and lies over the *body of the sphenoid.* On its dorsal surface is the *hypophyseal fossa* or *sella turcica,* a depression that lodges the pituitary gland. The lateral portions of the fossa lie over the great wings of the sphenoid bone and tympanic and petrous portions of the temporal bone, the latter marking the posterior border of the fossa. Conspicuous processes in the anterior region of the fossa are the *lesser wings of the sphenoid* and the *anterior clinoid processes.*

Several openings can be seen in this fossa. Important ones are the *optic canal, superior orbital fissure, foramen rotundum, carotid canal, foramen ovale, foramen spinosum,* and *foramen lacerum.* These openings, all of which are paired, transmit various nerves and blood vessels.

3. *Posterior fossa:* This fossa is the largest and deepest of the cranial fossae. Most of its floor is formed by the occipital bone, in the center of which is a large opening, the *foramen magnum.* The *petrous portion* of the temporal bone forms a conspicuous ridge that forms the anterior border of the fossa on each side. On its posterior surface is the *internal acoustic meatus,* which transmits the optic nerve. On the posterior and lateral walls on each side is a deep groove that lodges the *transverse sinus,* a large vein that leaves the skull through the *jugular foramen.* The *hypoglossal canal,* which lies lateral to the foramen magnum, transmits the hypoglossal nerve.

CRANIAL BONES (Fig. 5-8). The cranial bones comprise the *cranium,*

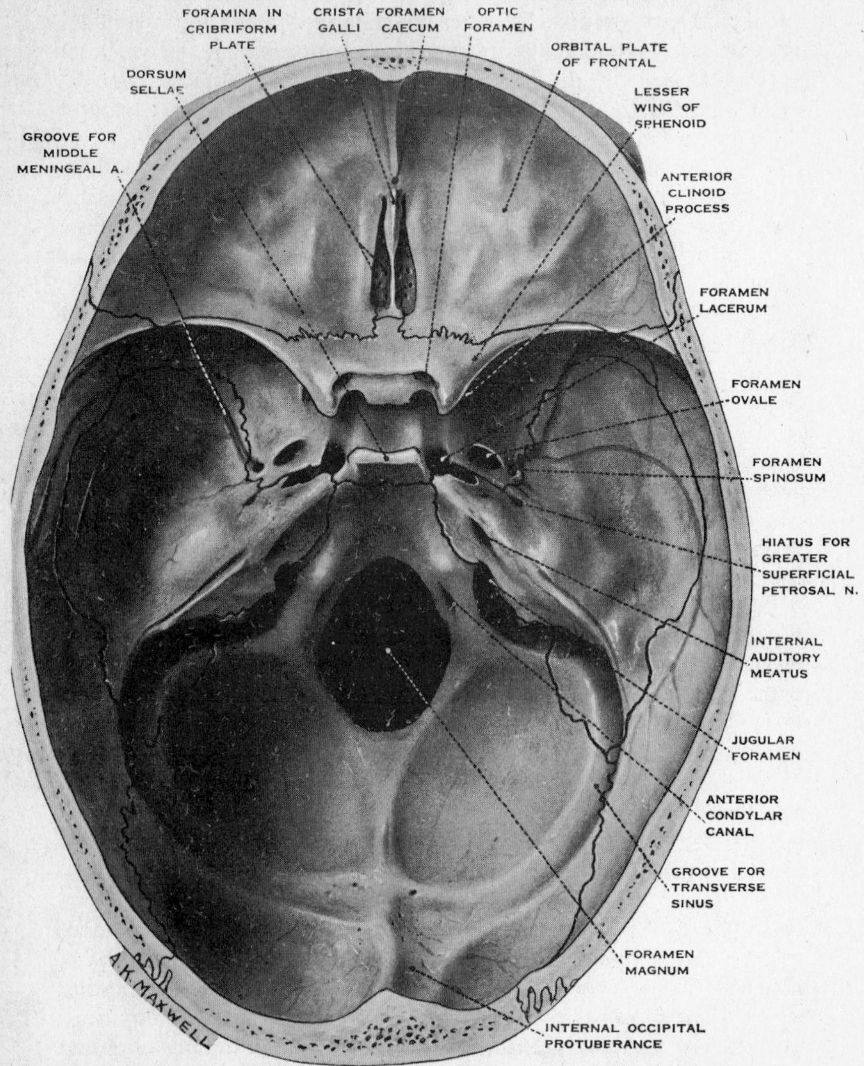

Fig. 5-7. Base of skull, superior aspect (internal surface). (Reprinted with permission of Macmillan, London and Basingstoke, from W. J. Hamilton, *Textbook of Human Anatomy,* 1976.)

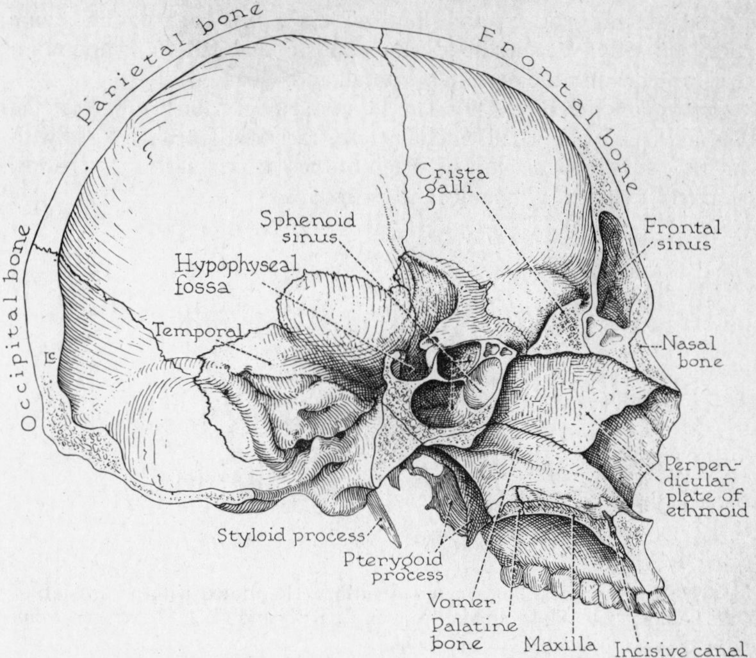

Fig. 5-8. The skull, medial aspect of left half. (Reprinted with permission of W. B. Saunders Co., Philadelphia, from B. G. King and M. J. Showers, *Human Anatomy and Physiology,* 6th ed., 1969.)

the part of the skull that encloses the brain. They include the single *frontal, occipital, sphenoid,* and *ethmoid bones* and the paired *parietal* and *temporal bones.* The *vault* or roof of the cranium is called the *calvaria* or skull cap. A description of each of the bones follows.

Frontal Bone. Early in its formation the frontal bone develops in two halves, but by the end of the second year of life these halves are fused in 91.5 percent of whites. This bone forms the anterior portion of the cranial vault and the major portion of the orbit. Its parts are the *glabella,* an area in the midsagittal plane, situated between the superciliary arches; the *superciliary arches,* ridges superior to the orbits beneath the eyebrows; the *supraorbital foramen* (sometimes a notch rather than a foramen), which transmits the supraorbital nerve and blood vessels; and the *frontal sinuses,* air cavities enclosed in bone directly over the orbit. *Articulations:* The frontal bone is joined with 12 other bones (the sphenoid, the ethmoid, and the paired parietals, nasals, maxillae, lacrimals, and zygomatics).

Parietal Bones. These bones form the roof and part of the sides of the cranium. The *sagittal suture* lies at the junction of the two bones in the midline. The *sagittal sulcus* is a groove on the inner surface at the junction for the superior sagittal sinus. *Articulations:* Each parietal bone articulates with five other bones (the opposite parietal, the temporal on the same side, and the occipital, frontal, and sphenoid).

Temporal Bones (Fig. 5-9). The temporal bones form a part of the sides and base of the skull. Each encloses an ear and bears a fossa with which the lower jaw articulates. Each of these bones has four portions: squamous, mastoid, tympanic, and petrous.

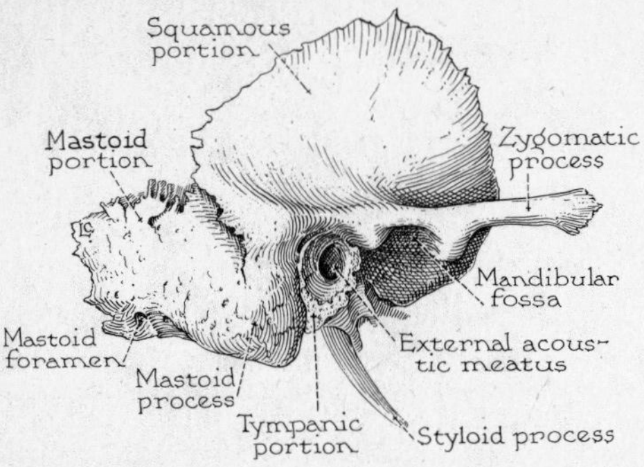

Fig. 5-9. Right temporal bone, external surface. (Reprinted with permission of W. B. Saunders Co., Philadelphia, from B. G. King and M. J. Showers, *Human Anatomy and Physiology,* 6th ed., 1969.)

The *squamous portion* includes a thin, flat plate of bone called *squama,* which forms a part of the side of the skull above the ear. It bears a *zygomatic process,* which extends forward and unites with a process of the zygomatic bone to form the *zygomatic arch.* Its inferior surface bears the *mandibular fossa,* which receives the condyle of the mandible.

The *mastoid portion* lies posterior and inferior to the external acoustic meatus. It bears the prominent *mastoid process,* containing *air cells,* which communicate with the cavity of the middle ear by way of the mastoid antrum. On the inner surface of the mastoid region is a *groove* for the *transverse sinus,* which is separated from the mastoid cells by a very thin plate of bone.

The *tympanic portion* consists of a curved plate of bone that surrounds the *external acoustic meatus,* a canal leading to the middle ear.

The *petrous portion* projects medially and forms a prominent ridge in the floor of the cranial cavity. It bears a sharp, pointed *styloid process,* which extends downward and serves for attachment of the tongue and the hyoid muscles. On the posterior surface of the petrous portion is the *internal acoustic meatus,* which transmits the statoacoustic and facial nerves. A large foramen, the opening to the *carotid canal,* lies medial to the styloid process; posterior to it is the *jugular foramen,* which lies between the temporal and occipital bones. *Articulations:* Each temporal bone articulates with five bones (occipital, parietal, sphenoid, zygomatic, and mandible).

Occipital Bone. This bone forms the posterior and inferior portions of the cranial cavity. Its *foramen magnum,* a large opening, transmits the spinal cord. *Occipital condyles,* rounded prominences on each side of the foramen magnum, articulate with the first vertebrae (the atlas). Two prominences (the *external occipital protuberance* and the *external occipital crest*) lie in the middle on the posterior surface. Extending laterally from these are the *superior* and *inferior nuchal lines.* Two *transverse grooves* on the inner surface are for transmission of large blood vessels, the right and left *transverse sinuses,* their junction being called the *confluence of sinuses. Articulations:* The occipital bone articulates with six other bones (the sphenoid, the atlas, and the paired parietals and temporals).

Sphenoid Bone (Fig. 5-10). The sphenoid bone (its form resembles that of a bird in flight) comprises the anterior portion of the base of the cranium. Its *body* is a cube-shaped median portion containing two *sphenoidal sinuses* separated by a septum. Lateral extensions from the body (two *great wings* and two *lesser wings*) form part of the walls of the cranial cavity and the orbits. A concavity, the *hypophysial fossa* (sella turcica) lodges the hypophysis (pituitary gland). Its posterior border, the *dorsum sellae,* bears two *posterior clinoid processes.* Two *anterior clinoid processes* extend backward from the lesser wings. The inferior surface of the sphenoid bears two processes, the *lateral* and *medial pterygoid plates.* Prominent *openings* in this bone are the foramen rotundum, foramen ovale, foramen spinosum, optic foramen or canal, and supraorbital fissure. *Articulations:* The sphenoid bone articulates with 12 other

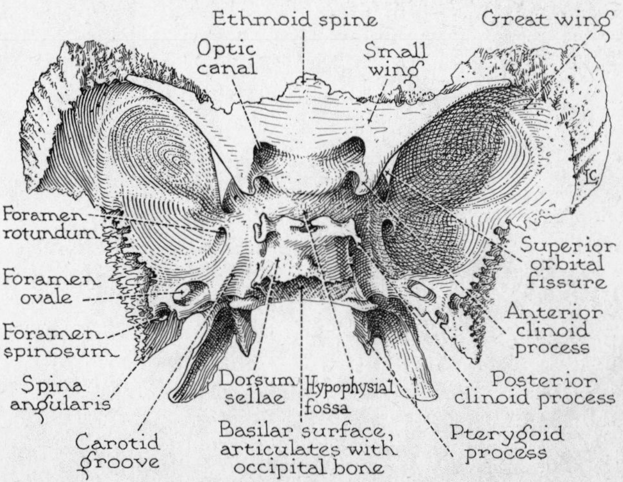

Fig. 5-10. Sphenoid bone, upper surface. (Reprinted with permission of W. B. Saunders Co., Philadelphia, from B. G. King and M. J. Showers, *Human Anatomy and Physiology,* 6th ed., 1969.)

bones (vomer, ethmoid, frontal, occipital, and the paired parietals, temporals, zygomatics, and palatines).

Ethmoid Bone (Fig. 5-11). The ethmoid bone lies between the orbits at the base of the cranium. It forms the principal supporting bone of the nasal cavity. The ethmoid has four parts: *cribriform plate,* a horizontal plate forming the base of the cranium and the roof of the nasal cavity, perforated by numerous openings for olfactory nerves and bearing a median projection, the *cristi galli;* two *lateral masses* forming the lateral walls of the nasal cavity, each bearing two scroll-like folds, the *superior* and *middle conchae* (turbinate bones) on their median surfaces and containing the *ethmoidal air cells* or sinuses; and the *perpendicular plate,* a thin quadrangular median plate that forms the upper portion of the nasal septum. *Articulations:* The ethmoid bone articulates with 11 other bones (vomer, frontal, sphenoid, and the paired lacrimals, palatines, inferior conchae, and nasals).

Other Cranial Bones. These include the small, irregular bones located in the sutures, called *wormian* or *sutural bones. Epipteric* bones are wormian bones found in the region of the sphenoidal fontanel, where the parietal, sphenoid, frontal, and temporal bones meet, *the pterion.*

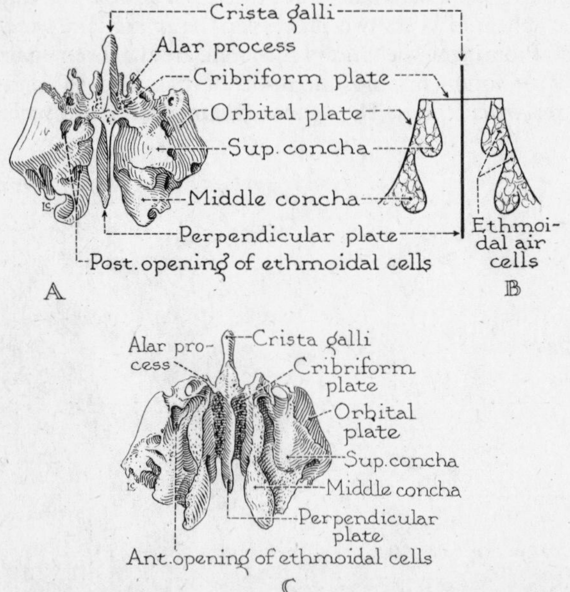

Fig. 5-11. Ethmoid bone. (*A*) From behind. (*B*) Diagram of the parts of the ethmoid bone. (*C*) Ethmoid bone from in front and below. (Reprinted with permission of W. B. Saunders Co., Philadelphia, from B. G. King and M. J. Showers, *Human Anatomy and Physiology,* 6th ed., 1969.)

Bones of the Face. The parts and interrelationships of the 14 bones of the face are discussed in the following paragraphs. Details can be seen in Figs. 5-4, 5-5, 5-6, and 5-8.

Maxillae. The two maxillae constitute the upper jaw, most of the roof of the mouth and floor of the nasal cavity, part of the median surface of each orbit, and the lateral wall of the nasal cavity. The *zygomatic* and *frontal processes* extend upwardly and laterally. On the inferior border of each maxilla is an *alveolar process,* which bears sockets holding the upper teeth. Its *horizontal* (*palatine*) *processes* form the hard palate. The *infraorbital foramen* lies beneath the orbit; the *incisive foramen* is located anteriorly between the palatine processes near the incisor teeth. The *lacrimal groove* lodges the lacrimal sac, the expanded portion of the nasolacrimal duct, which opens into the nasal cavity. The *maxillary sinus,* an air sinus, lies in the body of a bone beneath each orbit. *Articulations:* Each maxilla articulates with nine other bones (frontal, ethmoid, nasal, zygomatic, lacrimal, inferior nasal conchae, vomer, and the opposite maxilla).

Zygomatic (*Malar*) *Bones.* These bones form the prominences of the cheeks and part of the lateral and inferior walls of the orbits. Each bears a *temporal process* that projects posteriorly and articulates with the zygomatic process of the temporal bone to form the *zygomatic arch* or *zygoma.* A *frontal process* projects superiorly and articulates with the frontal bone. *Articulations:* Each zygomatic bone articulates with four other bones (frontal, temporal, sphenoid, and maxilla).

Lacrimal Bones. These thin, flat bones form part of the median wall of each orbit. Each has a *lacrimal groove* for the lacrimal canal. *Articulations:* Each lacrimal bone articulates with four other bones (the frontal, ethmoid, maxilla, and inferior nasal concha).

Nasal Bones. These small, flat bones form the bridge of the nose. *Articulations:* The nasal bones articulate with four other bones (frontal, ethmoid, maxilla, and the opposite nasal bone).

Inferior Nasal Conchae. These small, curved bones extend horizontally along the lateral walls of the nasal cavity. Each is curved on itself, in the manner of a scroll, and projects medially into the nasal cavity. *Articulations:* Each inferior nasal concha articulates with four other bones (ethmoid, maxilla, lacrimal, and palatine).

Palatine Bones (Fig. 5-12). These bones form part of the lateral wall and floor of the nasal cavity and part of the roof of the mouth and the floor of the orbit. Their paired *horizontal plates* unite to form the posterior portion of the hard palate. Extending upward from these plates are the *vertical plates.* A *pyramidal process* extends backward and laterally from the junction of the vertical and horizontal plates. An *orbital process,* extending upward and laterally from the vertical plate, forms part of the posterior floor of the orbit, and usually contains a small sinus that communicates with the sphenoidal and ethmoidal sinuses. A *sphenoid*

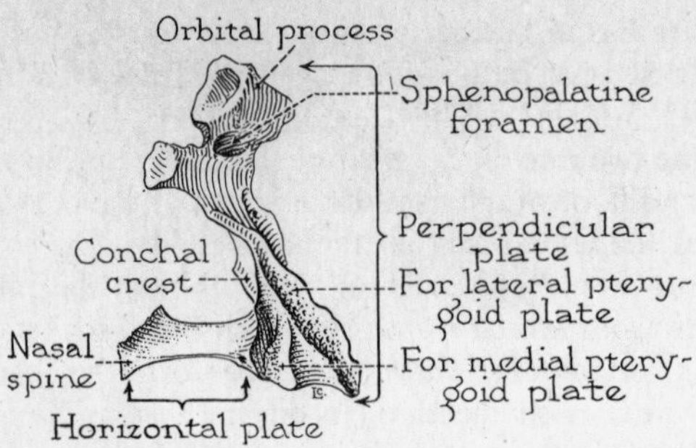

Fig. 5-12. Right palatine bone, from behind. (Reprinted with permission of W. B. Saunders Co., Philadelphia, from B. G. King and M. J. Showers, *Human Anatomy and Physiology,* 6th ed., 1969.)

process extends upward and medially from the vertical plate. *Articulations:* Each palatine bone articulates with six other bones (sphenoid, ethmoid, maxilla, inferior concha, vomer, and opposite palatine).

Vomer (Fig. 5-8). This thin, flat bone lies in the median plane and forms the posterior and inferior portion of the nasal septum. *Articulations:* The vomer articulates with six other bones (sphenoid, ethmoid, and the paired maxillae and palatines) as well as with the septal cartilage.

Mandible. This is a U-shaped bone comprising the lower jaw. It consists of a horizontal *body* and two upward-projecting *rami.* The two halves of the body are united at the *symphysis,* the lower portion of which forms the *mental protuberance* (chin prominence). The upper portion of each half of the body forms the *alveolar process.* This process contains eight sockets (*alveoli*) in which are found the roots of the lower teeth. On the lateral surface is the *mental foramen* for transmitting the mental nerve and vessels. The halves of the mandible unite at the symphysis during the second year.

Projecting upward from the posterior ends of the body are the two *rami.* Each ramus bears two processes: an anterior *coronoid process,* which serves for attachment of the temporal muscles, and a *condyloid process,* consisting of a *capitulum* and a *neck.* This latter process articulates with the mandibular fossa of the temporal bone. Where the posterior border of the ramus meets with the inferior border of the body, there is a prominence, the *angle of the jaw.* On the inner surface of each ramus is a *mandibular foramen,* which leads to the *mandibular canal.* These foramina transmit the inferior alveolar vessels and nerve. *Articulations:* The mandible articulates with the two temporal bones.

HYOID BONE. This horseshoe-shaped bone (see Fig. 6-17) lies in the neck, suspended from the tips of the styloid processes of the temporal bone by two *stylohyoid ligaments.* It has a *body* from which two hornlike processes, the *greater cornua,* project posteriorly from its lateral surface. Two *lesser cornua,* small, conical eminences, project upward at the junction of the body and the greater cornua. The hyoid bone has no articulations.

AUDITORY OSSICLES. In the tympanic cavity on each side of the skull lie the three ear bones, or auditory ossicles, the *malleus, incus,* and *stapes.* They comprise a part of the middle ear, and function in the transmission of sound vibrations from the tympanic membrane to the vestibule of the inner ear. They are described in Chapter 5, Vol. 2.

SPECIAL FEATURES OF THE SKULL. Certain characteristics of the bony structure of the skull are not encountered elsewhere in the skeleton and consequently merit special explanation. These are sutures, fontanels, air sinuses and mastoid cells, and cavities of the skull.

Sutures. A *suture* is a line of junction or fusion between bones, especially the cranial bones. The principal cranial sutures are the *sagittal,* between the parietal bones; the *coronal,* between the parietal and frontal bones; the *lambdoidal,* between the parietal and occipital bones; the *squamosal,* between the parietal and the temporal bones; and the *metopic* or *interfrontal,* between the halves of the frontal bone before fusion.

After adolescence, bones begin to unite across the sutures, which gradually become obliterated, the inner surfaces or *tables* usually fusing earlier than the other surfaces. The degree of obliteration of the sutures, plus wear evidenced by the teeth, gives a very rough indication of the age of the skull. The sagittal suture begins to close at the age of 22 and is usually well advanced by age 35; the coronal suture begins to close at age 24 and is almost entirely closed by the age of 38, although complete fusion may be delayed for several years. Other sutures begin to close between the ages of 26 and 37, but some of these sutures may not become completely obliterated until late in life.

One suture, the metopic (or interfrontal), closes early in life, commencing during the second half of the first year and being completed by the end of the second year. Its obliteration, however, as with all other sutures, is variable and in unusual cases may not occur at all. This suture remains unclosed in 8.5 percent of white and in 1.2 percent of black adults; in a radiographic view, it has occasionally been mistaken for a fracture.

Fontanels (Fig. 5-13). These are membrane-covered spaces in the skull of a fetus or an infant, located at junctions of the parietals and bones adjacent to them. The fontanels represent the areas at the angles of the cranial bones that are last to undergo ossification; the membranes in these areas remain unossified until after birth. They constitute the "soft spots" (of which the *bregmatic fontanel* is the most prominent) in the head of the infant. The names, location, and times of closure of the fontanels are noted in the accompanying table.

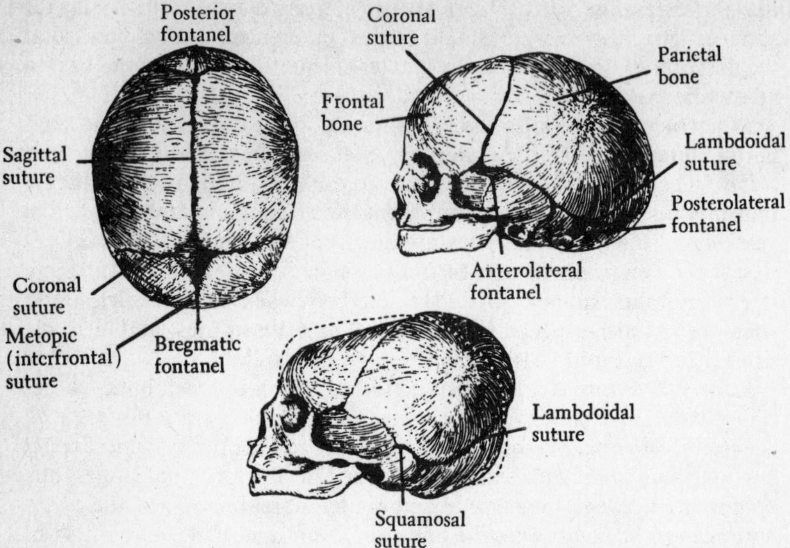

Fig. 5-13. Three views of the skull of a newborn infant, showing sutures and fontanels. In the lower drawing, the molded head is shown; dotted lines indicate the alteration that takes place during birth. (From *Atlas of Human Anatomy,* Barnes & Noble, Inc., 1961.)

LOCATIONS AND TIMES OF CLOSURE OF FONTANELS

Fontanel	*Juncture (Sutures)*	*Time of Closure*
Anterior (bregmatic)	Coronal and sagittal	During 2nd year
Posterior	Lambdoidal and sagittal	About 2 months after birth
Anterolateral	Coronal and squamosal	About 3 months after birth
Posterolateral	Lambdoidal and squamosal	About end of 1st year

Air Sinuses. The air sinuses are cavities in certain of the skull bones. Those lying adjacent to and communicating with the nasal cavities are the *paranasal sinuses.* The small, irregular spaces in the mastoid processes of the temporal bones are *mastoid air cells.* These communicate with each other and with the middle ear through the *tympanic antrum.* Not infrequently the mastoid air cells invade adjacent bones. The orbital process of the palatine bone contains a sinus which communicates with the sphenoidal sinus; its inflammation may result in pain behind the eyeball.

The paranasal sinuses include the *frontal sinus* (usually paired), lying superior to the orbits; the *maxillary sinuses* lying inferior to the orbits and lateral to the nasal cavity; *ethmoid air cells* (3 to 18 in number), lateral to the nasal cavity, honeycombing the lateral masses of the ethmoid bone and sometimes extending into the nasal conchae or adjacent skull bones; and the *sphenoid sinus,* lying superior and posterior to the nasal cavity. These sinuses are lined with a ciliated mucous membrane which is continuous with that of the nasal cavity. The lining, made up of pseudostratified or simple columnar ciliated epithelium, is thinner and has fewer glands and goblet cells than the nasal epithelium. The lamina propria is thin and continuous with the periosteum of the bones. Drainage is into the nasal cavity; the ducts open into the meatuses between the conchae. The ducts draining the maxillary and sphenoidal sinuses are disadvantageously placed for efficient drainage because their nasal openings lie above the floor of the sinus drained.

The paranasal sinuses lighten the skull, play a secondary role as resonating chambers in voice production, and, through their mucous secretions, aid in moistening the nasal cavity.

Cavities of the Skull. These cavities are the *orbits,* the *nasal cavity,* the *oral cavity,* and the *cranial cavity.*

The *orbits* contain the eyes and associated structures. Each is formed by parts of the frontal, zygomatic, ethmoid, sphenoid, lacrimal, and palatine bones and the maxillae, and each communicates posteriorly with the cranial cavity by means of the supraorbital fissure and the optic canal. Medially each communicates with the nasal cavity through the nasolacrimal canal, laterally and inferiorly with the infratemporal and pterygopalatine fossa by the inferior orbital fissure.

The *nasal cavity* consists of two *nasal fossae* (separated by the *nasal septum*), which communicate anteriorly with the outside through the *anterior nares* and posteriorly with the nasopharynx through the *posterior nares,* or *choanae.* The bones enclosing and forming the supporting structures for the nasal cavity are the ethmoid, palatine, vomer, sphenoid, and maxillae.

The *oral cavity* or *mouth* is enclosed by the teeth-bearing mandible and maxillae and by the palatine and sphenoid bones.

The *cranial cavity* contains the brain. The bones that enclose it, comprising the *cranium,* have been described earlier in this chapter.

CRANIOMETRY. Craniometry is the science that deals with techniques involved in making measurements of skulls in order to establish exact and comparable records for use in the comparative study of skulls of individuals or groups. A large number of points on the skull are utilized for reference. A few of the more important *craniometric points* are as follows.

Glabella. In the midsagittal plane on a bony prominence between the supraorbital ridges.

Basion. In the midsagittal plane on the anterior margin of the foramen magnum.

Bregma. At the junction of the coronal and sagittal sutures.

Nasion. At the intersection of the midsagittal plane with the nasofrontal suture.

Pterion. The region on the lateral surface of skull where the sphenoid, frontal, parietal, and temporal bones come together.

The Vertebral Column. The vertebral column serves as the main axis of the body, providing general rigidity yet permitting flexibility of movement. It is an enclosing protective case for the spinal cord and the roots of the spinal nerves. It provides surfaces for the articulation of the skull, ribs, and pelvic girdle and for the attachment of muscles and ligaments. The outline of the vertebral column, as seen from the side, shows four *curves:* the cervical, thoracic, lumbar, and sacral. The thoracic and sacral curves are *primary* (present at birth); the cervical and lumbar are *secondary* (developing after birth). The 26 bones of the vertebral column comprise 24 individual vertebrae (7 cervical, 12 thoracic, 5 lumbar), the sacrum (5 fused vertebrae), and the coccyx (3 to 5 rudimentary vertebrae).

Although the vertebrae in different regions of the column vary in size and in details of structure, they exhibit a marked uniformity in their fundamental plan of structure. The 6th thoracic vertebra (Fig. 5-14), for example, has the following parts:

Body. A solid, cylindrical part forming the major portion of the vertebra.

Vertebral or neural arch. The remaining portion that, with the body, encloses the vertebral foramen, an opening in which lies the spinal cord. The neural arch has two *pedicles,* which form the sides of the arch; two *laminae,* which form the roof; and seven processes (two inferior articular, two superior articular, two transverse, and one spinous). A pronounced notch on the posterior surface of each pedicle is called the *intervertebral notch.* When vertebrae are in normal position, this notch forms a foramen through which a spinal nerve makes its exit from the spinal cord.

Between the bodies of successive vertebrae are *intervertebral discs,*

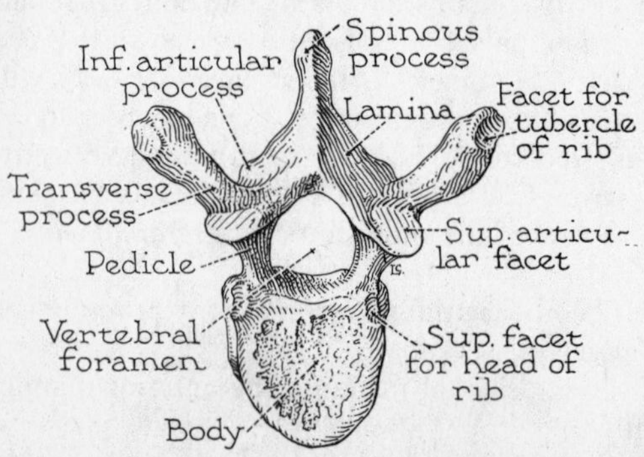

Fig. 5-14. A typical vertebra (6th thoracic), viewed from above. (Reprinted with permission of W. B. Saunders Co., Philadelphia, from B. G. King and M. J. Showers, *Human Anatomy and Physiology,* 6th ed., 1969.)

flexible elastic connections each consisting of a disc of fibrous cartilage enclosing a soft, resilient central mass, the *nucleus pulposus.* These discs hold the bodies of the vertebrae firmly together yet permit a limited degree of movement.

Also acting to hold the vertebrae in position are a number of strong ligaments. Among them are the *anterior* and *posterior longitudinal ligaments,* which extend longitudinally along the bodies; the *supraspinous ligament,* which connects the tips of the spinous processes; the *ligamentum flava,* which connects the laminae of adjacent vertebrae; the *interspinal ligaments,* which connect adjoining spinous processes; and the *intertransverse ligaments,* which connect the transverse processes, especially in the lumbar region.

In the neck region, the supraspinous ligament is replaced by the *ligamentum nuchae,* a triangular, fibrous membrane extending from the spinous process of the 7th cervical vertebra upward and attaching to the external occipital crest and protuberance. In quadrupeds it is a strong elastic membrane that helps to support the head.

CERVICAL VERTEBRAE (Figs. 5-15 to 5-19). Lying in the neck region, the cervical vertebrae include the atlas, the axis, and five additional vertebrae. For convenience of reference, they are numbered C1 through C7. These vertebrae have relatively small bodies and large vertebral foramina. Each of their transverse processes possesses an opening, the *foramen transversarium,* which transmits the vertebral artery. A spine is lacking on C1 (the atlas). On C2 through C6 the spines are rather blunt and compressed and their free ends are bifid (forked). On C7 the spine is long and thickened at its free end, producing a prominence at the base of the neck.

The *atlas* (C1) is highly modified in that a body and a spinous process are lacking. It is made up of two *lateral masses* and two *arches* (anterior and posterior); each arch bears a *median tubercle.* On the upper surface are two *superior articular surfaces,* cup-shaped concavities that receive the occipital condyles of the skull; they permit nodding movements of the head. Posterior to each process is a groove for the vertebral artery. On the lower surface are two *inferior articular surfaces;* these are flattened and permit turning movements on the axis. The *transverse atlantal ligament* extends across the vertebral foramen, dividing it into two parts, the anterior of which encloses the *dens.*

The *axis* or *epistropheus* (C2) bears the dens on its upper surface. The dens is a prominent elongated *odontoid process* that forms a pivot for rotation of the atlas and its attached skull. On each side of the dens is a *superior articular facet.* The transverse processes are small, and each bears a transverse foramen. The spine is prominent and bifid; at its base are two *inferior articular facets.*

THORACIC VERTEBRAE. Lying in the chest region are the 12 thoracic vertebrae (T1 through T12). Each of their transverse processes bears on

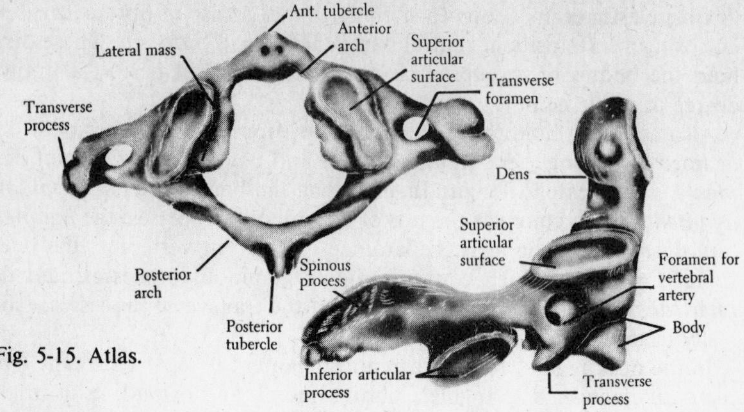

Fig. 5-15. Atlas.

Fig. 5-16. Axis.

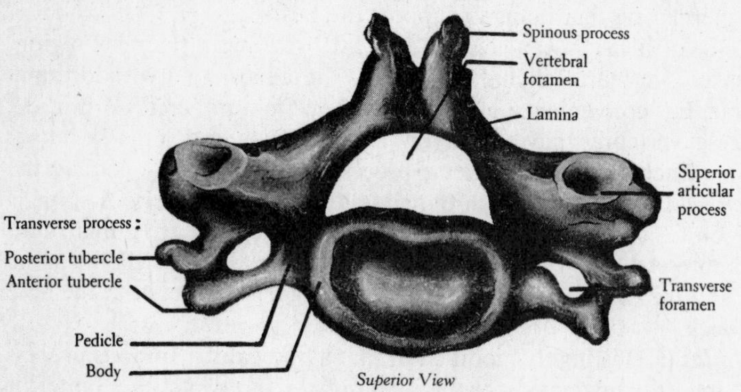

Fig. 5-17. 3rd cervical.

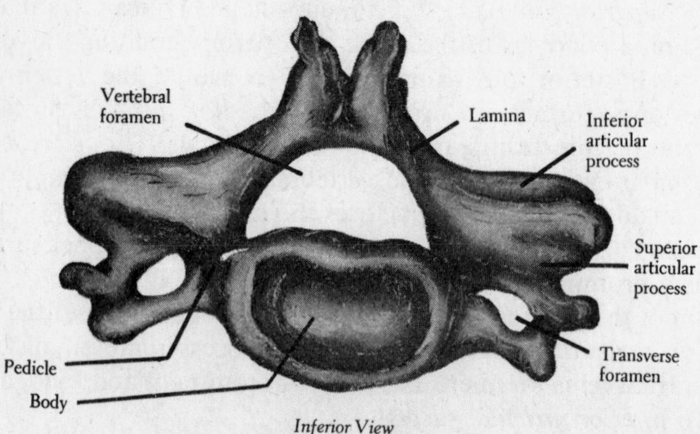

Fig. 5-18. 4th cervical.

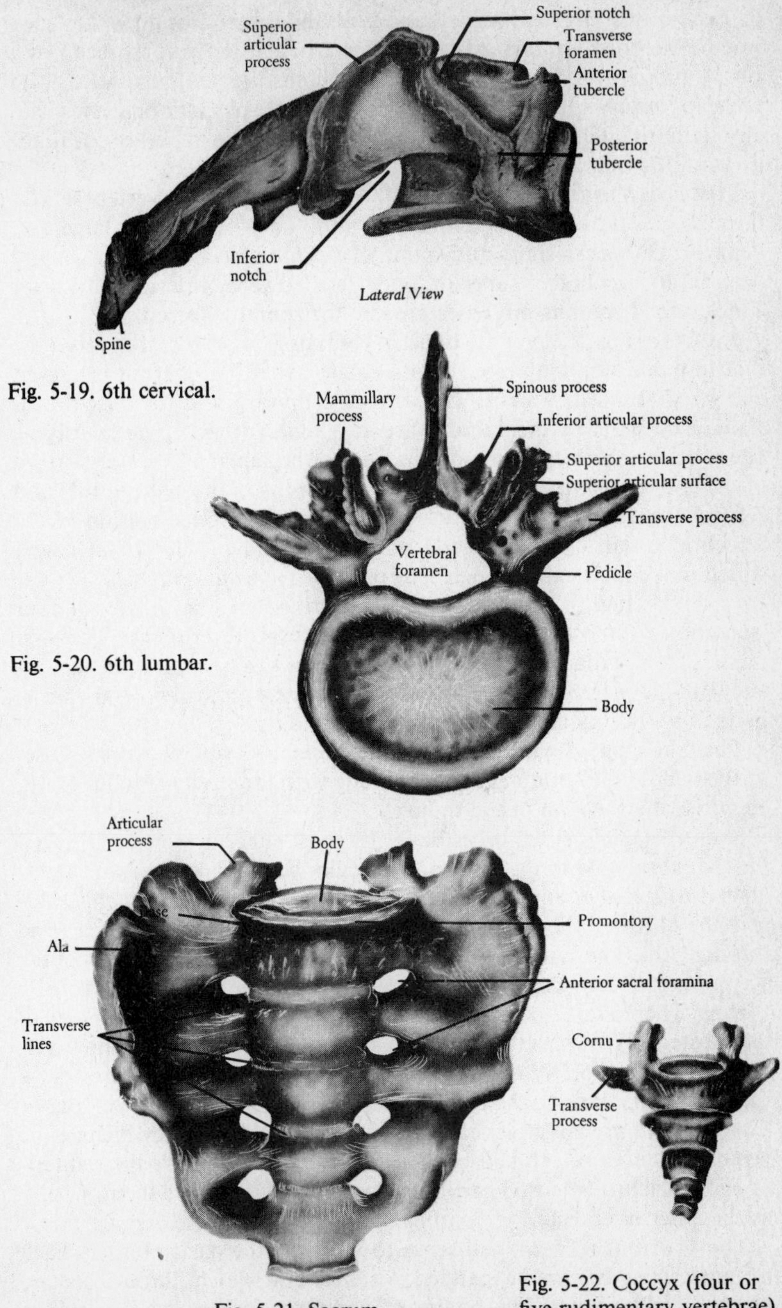

Fig. 5-19. 6th cervical.

Fig. 5-20. 6th lumbar.

Fig. 5-21. Sacrum.

Fig. 5-22. Coccyx (four or five rudimentary vertebrae).

its extremity a *facet* for articulation with the tubercle of a rib. On each side of the body are two *demifacets* for articulation with the head of a rib. A *spinous process* is well developed and directed downward. The *vertebral foramen* is generally circular in shape. Articular processes usually lie horizontal to the long axis of the body (the superior ones face upward, the inferior downward).

LUMBAR VERTEBRAE (Fig. 5-20). The five lumbar vertebrae (L1 through L5) lie in the region of the loin. Their *bodies* are large and heavy, their *spines* short and blunt. *Articular processes* are in general vertical to the body (superior ones face inward, inferior outward). Thick, broad spinous processes project horizontally dorsad.

SACRUM (Fig. 5-21). This bone is composed of five sacral vertebrae that fuse and become united into a single bone. The sacrum is curved and wedge-shaped, with its *base* directed upward and its *apex* downward. The parts of the sacrum are two *alae* (wings), processes lying lateral to the *promontory,* a prominence on the superior anterior portion of the body and two *articular surfaces* for the ilia of the pelvic girdle and two for the 5th lumbar vertebra. A *sacral canal,* a continuation of the vertebral canal, contains sacral nerves (instead of the spinal cord), which make their exit through four pairs of *sacral foramina.*

COCCYX (Fig. 5-22). The coccyx is formed by the fusion of four (sometimes three or five) rudimentary coccygeal vertebrae. The first piece is much broader than the others; it bears two *cornua,* which articulate with the sacrum, and rudimentary *transverse processes.* All the usual parts of a vertebra, except the body, are lacking.

The Skeleton of the Thorax. The thoracic vertebrae, already described under the vertebral column, along with the *sternum* and the ribs, comprise the skeleton of the thorax.

STERNUM. This is the breastbone. Its most superior portion is called the *manubrium,* with the following articular surfaces: a *clavicular notch,* a *1st costal notch,* and the superior half of the *2nd costal notch.* The middle portion, the *body* (or *gladiolus*) bears costal notches (2nd through 7th). The *xiphoid (ensiform) process* (the inferior portion of the sternum) is usually cartilaginous, though sometimes it is ossified.

RIBS. There are 12 pairs of ribs. (Figs. 6-22, 6-23) All of them articulate posteriorly with the vertebrae. Their anterior attachment, however, varies. The *true* or *vertebrosternal ribs* (the seven superior pairs) have their anterior ends attached directly to the sternum by *costal cartilages.* The *false ribs* are not attached directly to the sternum. They include the *vertebrochondral ribs* (8th, 9th, and 10th pairs), which have their anterior ends attached by costal cartilages to the cartilage above them, whereas the anterior ends of the *floating ribs* (11th and 12th pairs) are free.

The 4th rib (Fig. 5-23) will serve to illustrate the parts of a rib. Each rib possesses a *head,* an expanded posterior end which, through its two *facets,* articulates with the bodies of two thoracic vertebrae; a *neck,* a

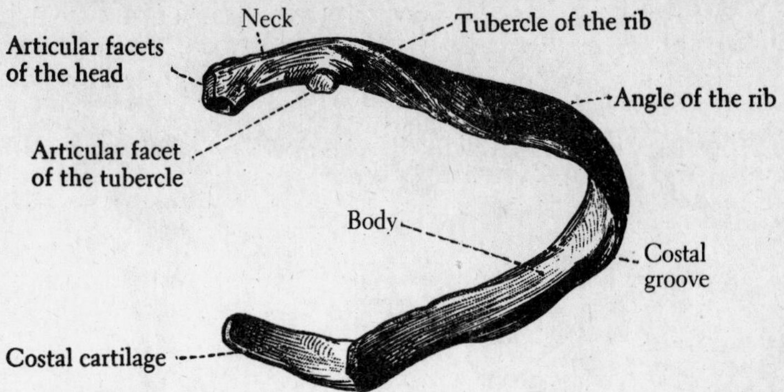

Fig. 5-23. A typical rib (4th right). (Reprinted with permission of The Macmillan Company from C. V. Toldt and A. D. Rosa, *Atlas of Human Anatomy for Students and Physicians,* 1948.)

slightly constricted region; and a *shaft,* the remainder of the rib. At the juncture of the shaft with the neck there is a *tubercle,* which has a facet for articulation with the end of the transverse process of a vertebra. The *angle* of the rib is a moderately sharp bend near the posterior end. The sternal end of the rib has an oval recess into which the costal cartilage fits. The *costal groove* is on the inferior border.

Some ribs have atypical features: The *1st rib* is very short and broad and markedly curved, with a prominent tubercle. It has two grooves on its superior surface for the subclavian artery and vein and the 1st thoracic nerve, and there is a scalene tubercle between the grooves. It lacks an angle and a costal groove.

The *2nd rib* is longer than the first. It has a pronounced eminence on its lateral surface and a short costal groove. *Ribs 10, 11,* and *12* have a single large articular facet on the head, and the *floating ribs,* 11 and 12, lack a neck, tubercle, angle, and costal groove.

THE APPENDICULAR SKELETON

The appendicular skeleton includes the bones of the upper and lower extremities. The upper extremities are comprised of the bones of the pectoral (shoulder) girdle, the arms, the forearms, and the hands. The lower extremities are comprised of the bones of the pelvic girdle, the thighs, the knees, the legs, and the feet.

The Upper Extremity. The upper extremity has 64 bones. They are described, along with their parts, in the succeeding paragraphs.

PECTORAL (SHOULDER) GIRDLE. Two scapulae and two clavicles make up the shoulder girdle.

Scapula (Fig. 5-24). The shoulder blade or *scapula* is a triangular bone

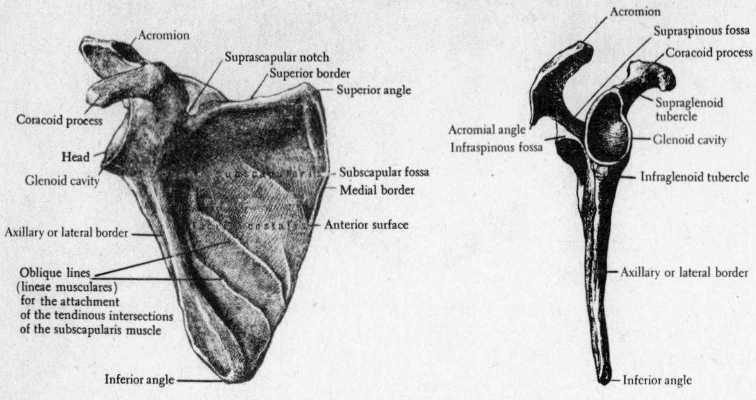

Fig. 5-24. Two views of right scapula. (Reprinted with permission of The Macmillan Company from C. V. Toldt and A. D. Rosa, *Atlas of Human Anatomy for Students and Physicians,* 1948.)

with *costal* (anterior) and *posterior surfaces* and three *borders* (*lateral* or *axillary, medial* or *vertebral,* and *superior*). Where the borders intersect they form the two upper *lateral* and *medial angles* and the lowermost *inferior angle.* Extending transversely across the posterior surface is a prominence, the *spine,* which separates two depressions, the *supraspinous* and *infraspinous fossae.* The spine extends laterally and terminates in a process, the *acromion,* which overhangs the *glenoid cavity,* an oval depression or socket that provides an articular surface or socket for the head of the humerus. At the apex of the glenoid cavity is the *supraglenoid tubercle;* at its lower margin, the *infraglenoid tubercle.* Extending forward from the upper border and turning laterally over the glenoid cavity is the *coracoid process.* Medial to this process, on the superior border, is a conspicuous *scapular notch.* The costal surface presents a broad, shallow concavity, the *subscapular fossa.*

Clavicle. The *clavicle,* or *collarbone,* is a somewhat S-shaped bone that connects with the scapula and the sternum. Its *medial* or *sternal end* articulates with the sternum, the *lateral* or *acromial end* with the scapula. On the inferior surface near the acromial end there is a roughened elevation, the *coracoid tuberosity,* which bears a *conoid tubercle;* on its inferior surface is a *subclavian groove.*

ARM (BRACHIUM). The *humerus,* or arm bone (Fig. 5-25) that extends from the shoulder to the elbow, consists of a shaft and two extremities.

The *shaft* has a *deltoid tuberosity,* an *intertubercular* or *bicipital groove,* and a groove for the radial nerve. The upper extremity of the humerus has a *head,* which articulates with the glenoid cavity; an *anatomical neck,* a shallow constriction; *greater* and *lesser tubercles,* separated by the bicipital groove; and a *surgical neck,* a region immediately distal to the tubercles, which is especially susceptible to fracture. The lower extremity of the humerus has a *trochlea,* an articular surface for the semilunar notch of the ulna; a *capitulum,* an articular surface for the head of the radius; a *coronoid fossa* to receive the coronoid process of the ulna; and a *radial fossa* to receive the head of the radius. It also bears *median* and *lateral epicondyles,* prominences on the medial and lateral surfaces that serve for attachment of muscles, and an *olecranon fossa,* a concavity on the posterior surface that receives the olecranon process of the ulna.

FOREARM (ANTEBRACHIUM). The *ulna* and the *radius* (Fig. 5-26) are the bones of the forearm.

Ulna. The ulna is slightly longer and lies medial to the radius. At its proximal end is the *semilunar notch,* which articulates with the trochlea of the humerus. Its *coronoid process* forms the anterior part of the semilunar notch; its *olecranon process* forms the posterior portion (the tip of the elbow). At the base of the coronoid process is the *radial notch,* which accommodates the head of the radius. An *interosseous border,* on the anterolateral surface, is sharp and prominent. The *head* of the ulna is a small, rounded process at the distal end. It bears an articular surface for the radius and a *styloid process,* the most distal portion of the bone, medially. Note that the ulna does not articulate with any of the carpal bones.

Radius. The radius lies lateral to the ulna. It consists of a *shaft* and two *extremities.* The *head,* the enlarged disclike end of the proximal extremity, has a depression for articulation with the capitulum of the humerus. Its rim, the *articular circumference,* articulates with the radial notch of the ulna. The *neck* is a constricted region distal to the head. A *radial tuberosity* provides for the insertion of the biceps muscle. On the medial surface of the shaft is a sharp *interosseous border* corresponding to a similar border on the ulna. An *interosseous membrane* binds the two bones together. The distal end bears a blunt *styloid process,* which lies lateral to a concave *carpal surface* with which the lunate and scaphoid bones of the wrist articulate. Medially, an *ulnar notch* provides an articular surface for the head of the ulna.

HAND. The bones of the hand (Fig. 5-27) comprise the *carpals* (wrist), *metacarpals* (palm), and *phalanges* (digits or fingers).

Carpals. The wrist, or *carpus,* consists of eight small bones arranged in two rows. The bones are named here in order, from the lateral (radial) side toward the medial (ulnar) side:

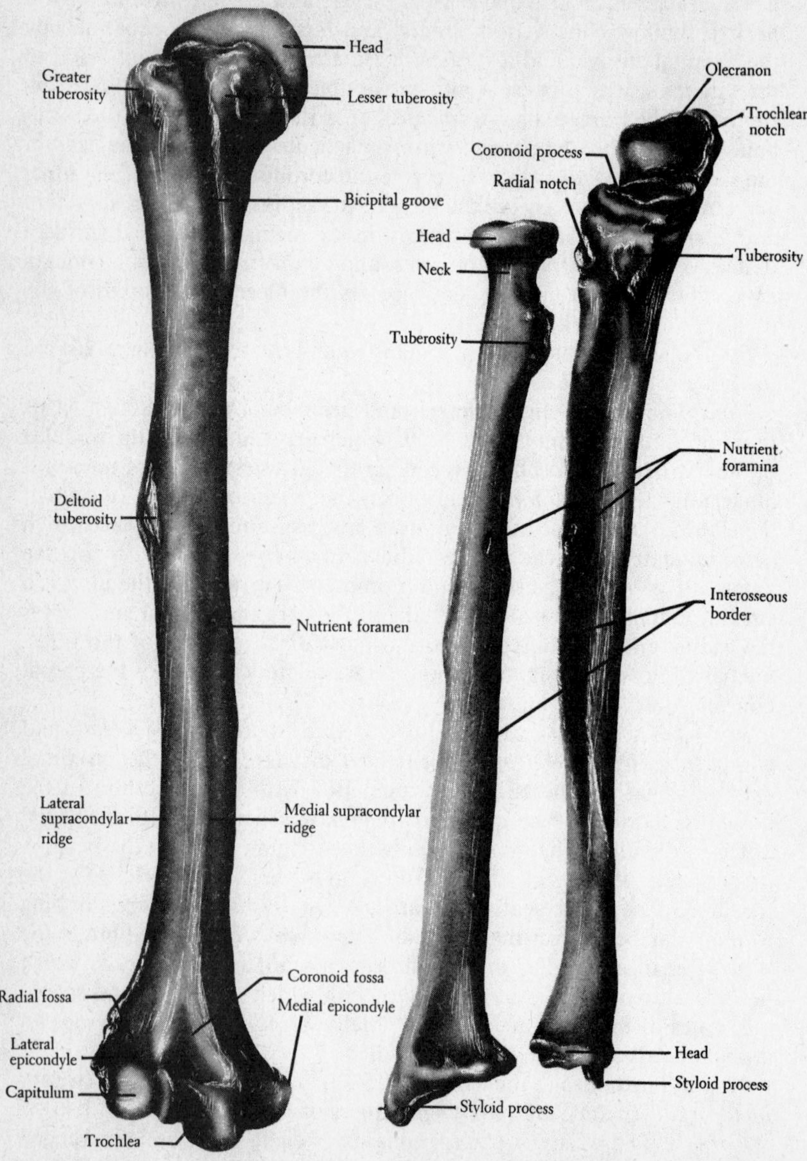

Fig. 5-25. Humerus, anterior view.

Fig. 5-26. Radius (left) and ulna, anterior views.

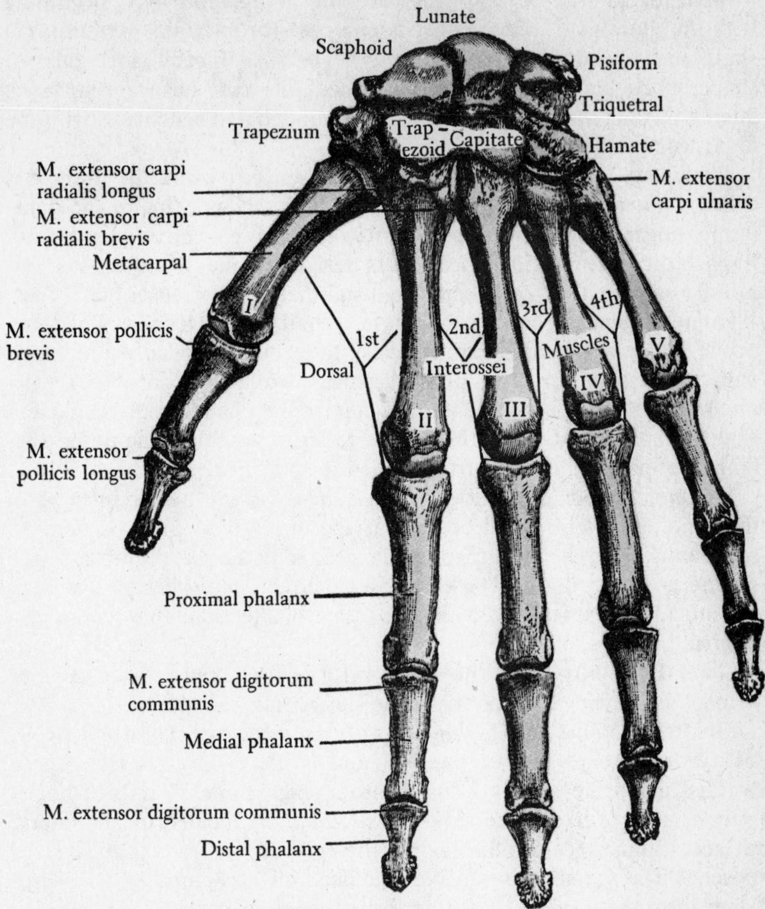

Fig. 5-27. Left hand, dorsal surface. (Reprinted with permission of Blakiston Division, McGraw-Hill Book Company, from *Morris' Human Anatomy,* 11th ed., edited by J. P. Schaeffer, 1953.)

Proximal row	Scaphoid (navicular) articulates with radius
	Lunate; articulates with radius
	Triangular (triquetral, cuneiform)
	Pisiform (smallest carpal bone)
	Trapezium (greater multangular) articulates with 1st metacarpal
Distal row	Trapezoid (lesser multangular)
	Capitate (largest carpal bone)
	Hamate (unciform) bears a *hamulus,* a hooklike process that projects toward the volar surface

Metacarpals. The bones of the palm are numbered I to V, beginning with the thumb side. Each consists of a base (or proximal extremity), a shaft, and a head (or distal extremity). The *base* of each is irregular in shape; it articulates with the carpal bones. The *shaft* has three surfaces: lateral, medial, and dorsal. The *head* is rounded and articulates with the first phalanx of the corresponding digit.

Phalanges. Each of the three bones in a finger (two in the thumb) is called a *phalanx.* It is therefore customary when speaking of the digits (either fingers or toes) to use the term *phalanges* collectively for a digit. In each digit the proximal phalanx is designated the "1st phalanx," the middle phalanx the "2nd," and the distal the "3rd" or "terminal." Each 1st and 2nd phalanx consists of a base, a shaft, and a trochlea. The *bases* of the proximal phalanges are concave, to receive the heads of metacarpals; the bases of the other phalanges bear two lateral concavities separated by a ridge. The distal end of each terminal phalanx is flattened and slightly expanded to form the *ungual tuberosity;* that of the other phalanges forms a pulleylike surface called the *trochlea.*

Sesamoid Bones. Embedded in the tendons of the flexor muscles of the hand are several small bones, varying in number. They occur most commonly over the metacarpophalangeal joint of the thumb and the second and fifth digits. These bones, called *sesamoid* because of their resemblance to sesame seeds, occur in various tendons that are subject to great pressure.

Lower Extremity. The lower extremity has 62 bones. They are described, along with their parts, in the succeeding paragraphs.

THE PELVIC GIRDLE. The *hipbone* (*os coxae* or *innominate bone*) is in reality three bones (ilium, ischium, and pubis), which in a child are distinct but in the adult are fused into a single bone. The three bones unite at a socketlike depression, the *acetabulum,* located on the lateral surface. (Sometimes a fourth bone, the *cotyloid* or *acetabular bone,* is present.) The acetabulum receives the head of the femur. The two hipbones form the pelvic girdle: anteriorly, they unite at the *pubic symphysis;* posteriorly, they articulate with the *sacrum.* They form the anterior and lateral walls of the pelvis. A large aperture, the *obturator foramen,* lies between the pubis and ischium.

Ilium. This is the largest of the three bones forming the hipbone. Each ilium consists of a lower portion or *body,* which occupies about two-fifths of the acetabulum, and a flattened, expanded upper portion, the *ala,* whose superior border, the *crest,* forms the prominence of the hip. On the inner surface, the *arcuate line* separates these two divisions. Conspicuous processes are the *anterior* and *posterior superior iliac spines,* which mark the ends of the crest, and the *anterior* and *posterior inferior spines,* which lie below them. A deep *greater sciatic notch* transmits the sciatic nerve. On the lateral surface of the ala are the *anterior, posterior,* and *inferior gluteal lines;* on the medial surface are the *iliac fossa* (a shallow concavity), the *arcuate line* (a diagonal prominence), an

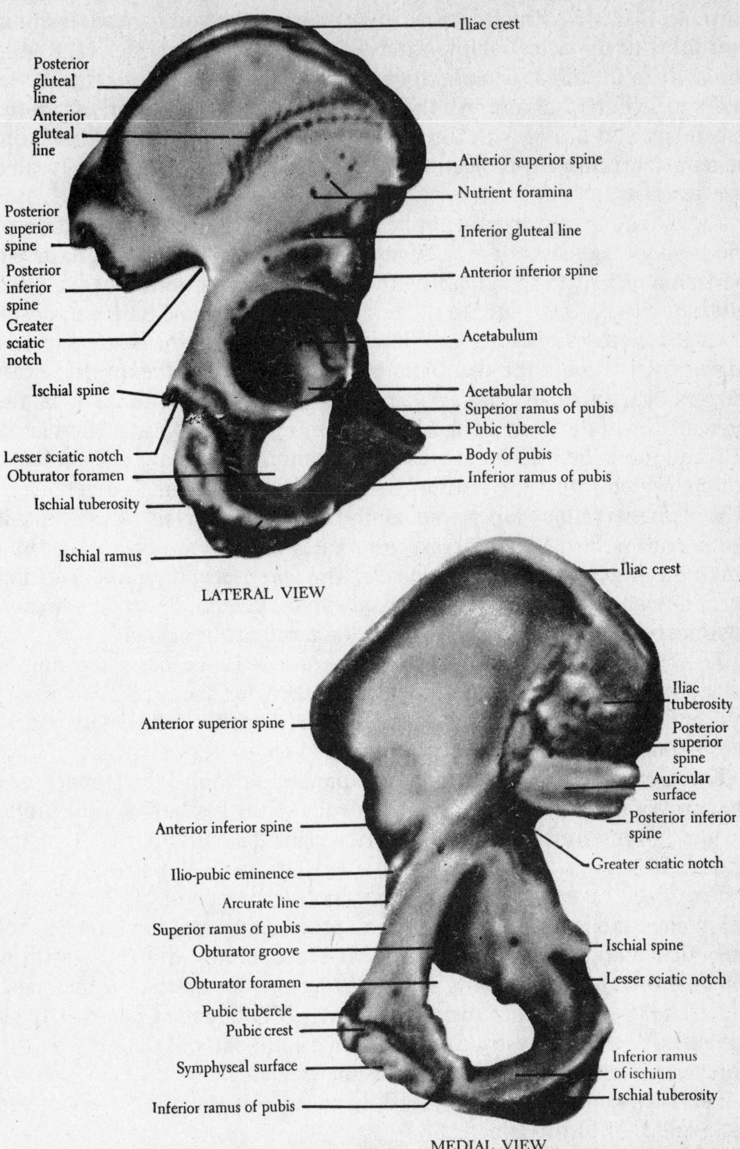

Fig. 5-28. Right innominate bone (os coxae).

auricular surface for articulation with the sacrum, and, above it, the *iliac tuberosity* for attachment of the sacroiliac ligaments.

Ischium. This bone forms the posterior and inferior portion of the innominate bone. It consists of a body and a ramus, which together form an L-shaped structure. The *body* or main portion comprises about two-fifths of the acetabulum. Extending downward and backward from the body is the *superior ramus,* which terminates in a large process, the *tuberosity* of the ischium. An *inferior ramus* extends anteriorly from the tuberosity and unites with the inferior ramus of the pubis. The ischium bears a sharp inward-projecting *ischial spine,* beneath which is located the *lesser sciatic notch.*

Pubis. The *pubis,* or pubic bone, which forms the anterior portion of the hipbone, consists of a body and two rami. The *body* is a flattened portion adjacent to the *pubic symphysis,* where it is united by an *interpubic disc* of *fibrocartilage* to the body of the pubic bone on the opposite side. Extending superiorly and laterally from the body is the *superior ramus,* which joins the iliac bone at the *iliopubic eminence.* Its upper surface bears a ridge, the *pectineal line,* which is continuous with the *arcuate line* of the ilium. The *inferior ramus* passes downward and laterally and joins the inferior ramus of the ischium to form the *ischiopubic ramus,* which forms the inferior border of the obturator foramen.

Where the pubic bones join at the pubic symphysis, a prominent *pubic crest* is formed. The crest terminates laterally on each bone in a small prominence, the *pubic tubercle.* The two descending inferior rami form a triangular *pubic arch.* The *subpubic angle* made by this arch is significantly different in the sexes, being more acute in males.

The Pelvis (Figs. 5-29 and 5-30). The pelvis is the basinlike ring of bones consisting of the two innominate bones, the sacrum, and the coccyx, and the ligaments that bind these bones together. The cavity within these bones is divided into the greater and lesser pelvic cavities.

The *greater* or *false pelvis* is the expanded portion lying superior to the brim of the true pelvis. It is bounded laterally by the ilia, posteriorly by the lumbar vertebrae, and anteriorly by the muscular wall of the abdomen.

The *lesser* or *true pelvis* is the portion of the pelvic cavity that lies below the margin formed by the sacral promontory and the iliopectineal lines that mark the *brim* of the pelvis. The opening within the brim is the *inlet* or *superior aperture.* The *outlet* or *inferior aperture* is the opening at the lower limits of the pelvic cavity. It is bounded posteriorly by the coccyx, anteriorly by the pubic arch, and laterally by the sciatic notches, spines, and inferior rami of the ischia.

The following are noteworthy differences between the pelvis of the male and that of the female:

Feature	*In the Male*	*In the Female*
General feature	Narrow, heavy, compact	Broad, light, capacious
Shape of inlet	Heart-shaped	Round or oval
Angle of pubic arch	Acute, narrow	Obtuse, broader
Ilia	Directed less vertically	More vertical

Feature	*In the Male*	*In the Female*
Iliac fossa	Deeper	More shallow
Acetabulum	Directed laterally	Directed slightly anteriorly
Ischial spine	Sharper, directed medially	Blunter, directed more posteriorly
Sacrum	Narrower, longer, set lower between ilia	Broader, shorter, set higher between ilia
Sacral curve	Pronounced	Less pronounced
Coccyx	Directed anteriorly	Directed inferiorly
Adaptation	For strength, speed	For childbearing

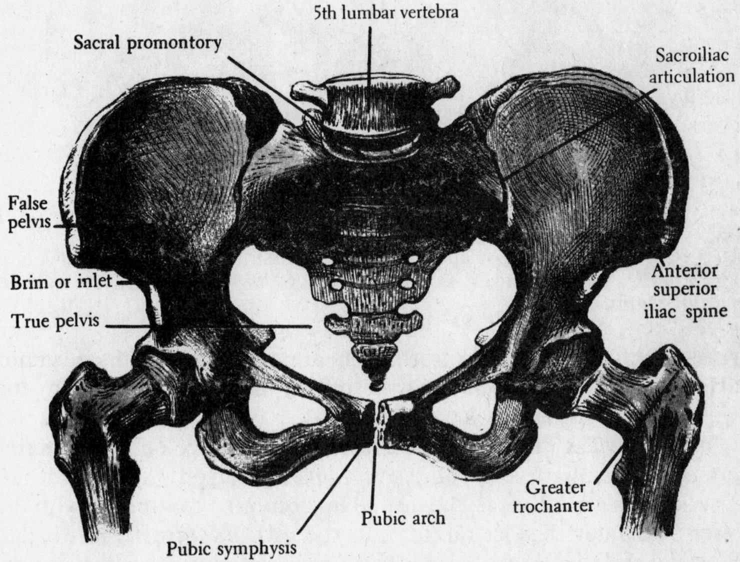

Fig. 5-29. The female pelvis. (Reprinted with permission of The Macmillan Company from C. V. Toldt and A. D. Rosa, *Atlas of Human Anatomy for Students and Physicians,* 1948.)

THE THIGH. The *femur* (thigh bone, Fig. 5-31) consists of a head, neck, shaft, and two extremities.

The *upper* or *proximal extremity* of the femur comprises the following parts: the *head,* the smooth, rounded articulating portion that fits into the acetabulum and has a small pit, the *fovea,* to which the *round ligament* (*ligamentum teres*) is attached; the *neck,* a constricted region; the *greater trochanter,* a large process extending laterally; the *trochanteric fossa,* a depression of the medial surface of the great trochanter; and the *lesser trochanter,* situated inferior, posterior, and medial to the

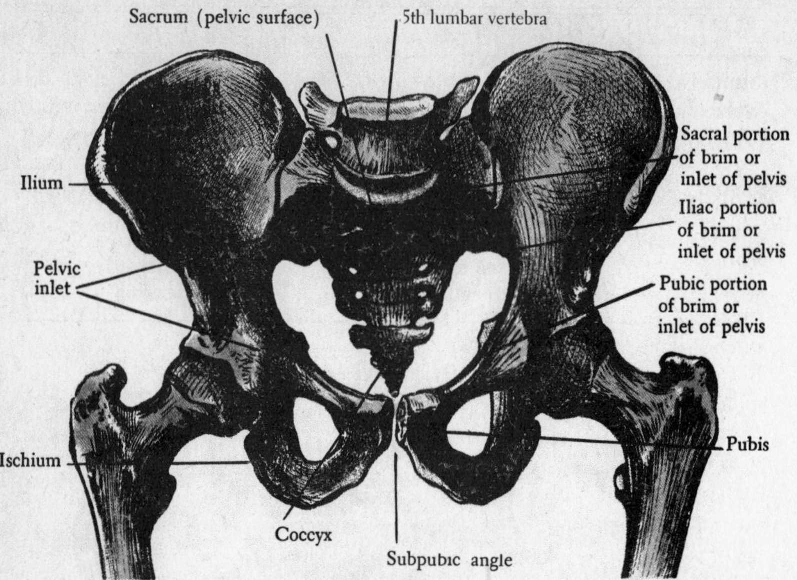

Fig. 5-30. The male pelvis. (Reprinted with permission of The Macmillan Company from C. V. Toldt and A. D. Rosa, *Atlas of Human Anatomy for Students and Physicians,* 1948.)

greater trochanter. The two trochanters are connected on the posterior surface by the *intertrochanteric crest* and on the anterior surface by the *intertrochanteric line.*

On the *shaft* is the *linea aspera,* a prominent ridge on the posterior surface. It possesses *medial* and *lateral lips* separated by a narrow band. Superiorly the lips diverge, the medial lip becoming continuous with the lesser trochanter. The lateral lip bears the *gluteal tuberosity* just before ending at the base of the greater trochanter. Inferiorly the lips also diverge and continue as the *medial* and *lateral supracondylar lines,* which continue to the epicondyles. These lines enclose a triangular area, the *popliteal surface,* which forms the upper portion of the floor of the *popliteal fossa,* the space posterior to the knee joint.

At the *lower* or *distal extremity* of the femur are *lateral* and *medial condyles,* large, rounded processes for articulation with the tibia; *lateral* and *medial epicondyles,* slight elevations lying superior to the condyles, the medial one bearing the *adductor tubercle;* the *intercondylar fossa,* a deep notch separating the condyles on the posterior surface; and a smooth *patellar surface* on the anterior surface between the condyles, for articulation with the patella.

The Patella. The *patella,* or kneecap (Fig. 5-32), the largest sesamoid bone in the body, develops on the anterior surface of the knee in the extensor tendon of the quadriceps muscle. It is triangular, with its base directed upward and its apex downward. Its anterior surface is rounded and rough; its posterior surface is smooth, bearing *medial* and *lateral articular facets.*

THE LEG. The tibia and the fibula (Fig. 5-33) are the bones of the leg.

Tibia. The *tibia,* or shinbone, consists of a shaft and two extremities. At its proximal extremity are *two condyles* (lateral and medial), each bearing articular surfaces for the femur; an *intercondyloid eminence,* a prominence between the condyles; a *tuberosity,* a prominent roughened area on the anterior surface inferior to the condyles; and a *fibular facet* (on the lateral condyle) for articulation of the fibula. The *shaft* has an *anterior crest,* the prominent ridge (or shin); a *lateral border,* bearing a thin interosseous crest; a *medial border,* a broad, rounded surface; and a *popliteal line,* a ridge on the upper posterior surface. The *distal extremity* bears a *medial malleolus,* a process that forms the inner prominence of the ankle; a *fibular notch,* an articular surface for reception of the distal end of the fibula; and an *inferior articular surface* for articulation with the talus at the ankle joint.

Fibula. The fibula lies parallel to the tibia on the lateral side. It consists of a shaft and two extremities. At the *proximal extremity* is the *head,* which articulates with the lateral aspect of the lateral condyle of the tibia. The *shaft* bears no distinct processes. Its anteromedial border, or *interosseous crest,* serves for attachment of the interosseous membrane. At the *lower extremity* is the *lateral malleolus,* a substantial process forming the outer prominence of the ankle. Its inner surface articulates with the fibular notch of the tibia and with the lateral surface of the talus.

Note that the ankle joint is the joint between the leg and the foot. The ankle bone is the talus.

THE FOOT. The bones of the foot (Fig. 5-34) are the tarsals, the metatarsals, and the phalanges of the toes.

Tarsus. The tarsus comprises seven bones: the calcaneus, the talus, the navicular bone, the cuboid bone, and the cuneiform bones (medial, intermediate, and lateral). The *calcaneus,* the largest and most posterior tarsal bone, forms the prominence of the heel. It articulates with the talus and navicular bones. The *talus* has a smooth, rounded superior surface, for articulation with the tibia; its lateral surfaces articulate with the lateral and medial malleoli of the fibula and tibia; inferiorly it articulates with the calcaneus and the navicular bone. The *navicular bone* is a slightly curved bone on the medial side of the foot, which articulates posteriorly with the talus, anteriorly with the three cuneiform bones, and sometimes laterally with the cuboid bone. Medially, it bears a prominent tuberosity. The *cuboid bone,* on the lateral side of the foot, articulates posteriorly with the calcaneus, anteriorly with the 4th and 5th metatarsals, and medially with the 3rd cuneiform bone and some-

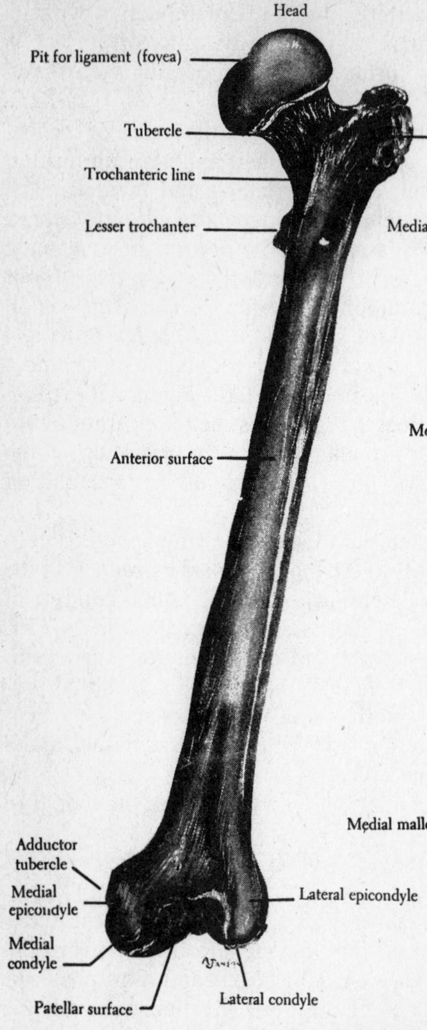

Fig. 5-31. Femur, anterior view.

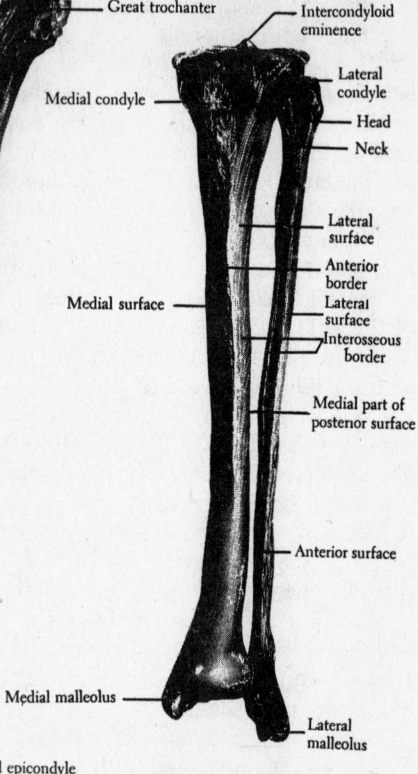

Fig. 5-33. Tibia (left) and fibula, anterior views.

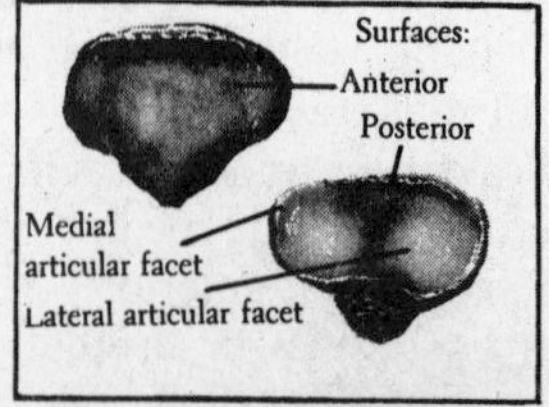

Fig. 5-32. Patella, anterior and posterior views.

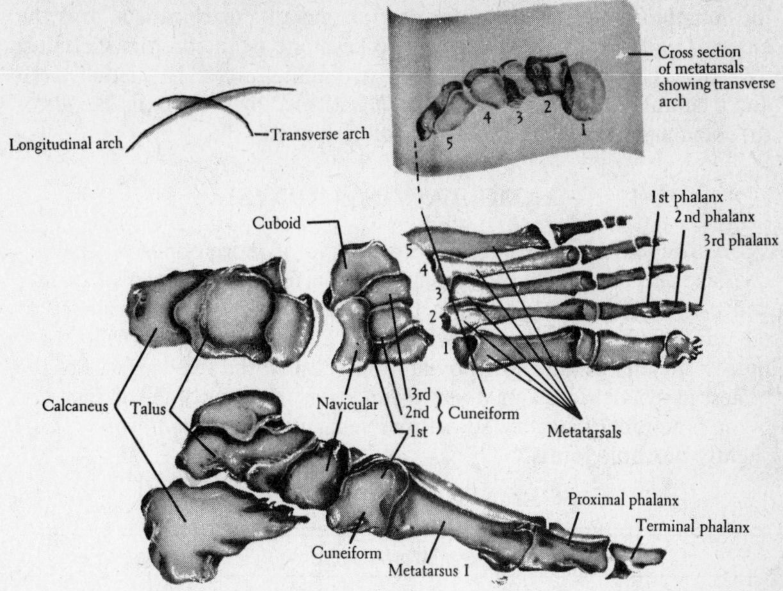

Fig. 5-34. Bones of the foot.

times with the navicular bone. The three *cuneiform bones* lie anterior to the navicular, medial to the cuboid, and posterior to the first three metatarsals.

Metatarsals. The five metatarsal bones of the foot are numbered I to V, from the medial side outward. Metatarsus I is on the side of the great toe (the hallux). The bases of the metatarsal bones articulate with the three cuneiform bones and the cuboid bone. The heads of the metatarsal bones articulate with the proximal phalanges of the digits. The rounded head of metatarsus I forms the "ball of the foot" at the base of the great toe.

Phalanges. As with the hand, there are two phalanges in the first digit (*hallux*) and three in each of the remaining toes. Each phalanx consists of a *base* or *proximal extremity;* a *shaft,* the middle portion; and a *head* or *distal extremity.* The distal extremities are expanded and flattened to form the horseshoe-shaped ungual tuberosities which bear the toenails.

Sesamoid Bones. Two sesamoid bones are commonly found under the head of the first metatarsal, in the tendon of the flexor hallucis brevis. They may also be present in other tendons of the foot.

The Arches of the Foot. Most effectively to perform its function of supporting the weight of the body, to cushion the effect of this weight being put upon the talus of a single foot at every step, and to provide some degree of resiliency, the bones of the foot are arranged in the form of two arches, a longitudinal and a transverse arch. The two pillars of the *longitudinal arch* are the posterior end of the calcaneus and the anterior ends of the metatarsals. The keystone of the arch is the talus. This arch is usually considered as two parallel arches, a median arch and a lateral arch. The *transverse arch* is formed by the proximal ends of the metatarsal bones and by the distal row of tarsals.

ARTICULATIONS (JOINTS)

An articulation is a junction between two or more bones or between cartilage and bone. The two main types of articulations are *synarthroses* and *diarthroses,* which differ in the degree of movement permitted at the joint, as shown in the following table. Diarthroses, or *synovial joints,* possess a cleft containing synovial fluid between the free surfaces of the bones; in synarthroses, or *nonsynovial joints,* the cavity is lacking. Another functional category sometimes recognized is *amphiarthroses,* or slightly movable joints.

CLASSIFICATION OF ARTICULATIONS

Type	*Degree of Movement*	*Example*
Synarthrosis (joint cavity absent)		
Fibrous		
Syndesmosis	Slight	Between distal ends of tibia and fibula
Suture	None	Between cranial bones
Cartilaginous		
Synchondrosis	Slight	Between epiphysis and shaft; between 1st rib and sternum
Symphysis	Slight	Between bodies of vertebrae; between pubic bones
Diarthrosis or synovial (joint cavity present)	Slightly or freely movable	Between limbs and body; between parts of limbs

Synarthrosis. This type of joint (Fig. 5-35) has a continuous binding substance, such as cartilage or fibrous tissue, between its components. It lacks an articular cavity, and the degree of movement is either limited or absent.

FIBROUS JOINTS. These include syndesmoses and sutures.

Syndesmosis. In this type of joint, the bones are held together by a tough, fibrous *interosseous membrane,* as in the union of the tibia and fibula and between the radius and ulna. Sometimes ligaments form a binding tissue, as in the connections between the laminae of successive vertebrae.

Sutures. These are joints between contiguous bones in which fibrous tissue is minimal. Their edges may be overlapping (*squamous*) or inter-

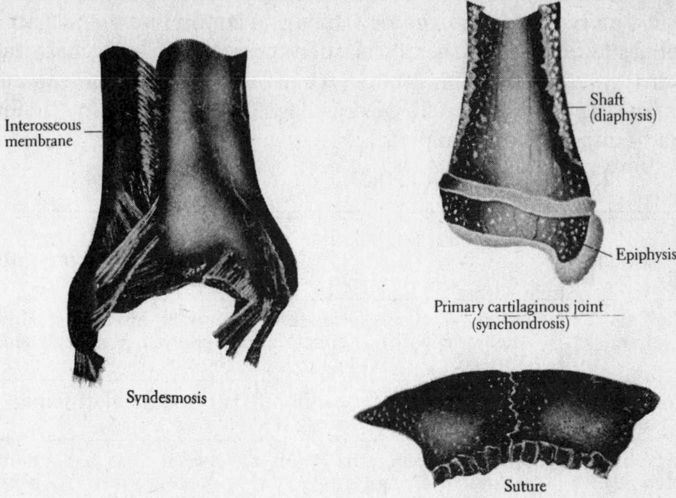

Fig. 5-35. Types of synarthrodial joints.

locking (*serrate* or *dentate*). They occur where rigidity of structure is essential, as between bones of the cranium. With increasing age, bones at sutures tend to fuse, obliterating the suture. When this occurs and the fibrous tissue is replaced with bone, the joint is called a *synostosis.*

Cartilaginous Joints. These include the synchondrosis and symphysis types.

Synchondrosis. In this type of joint, a plate or layer of hyaline cartilage separates the bones. Temporary joints of this nature form between the epiphysis and diaphysis of a long bone in development. Permanent joints of this type are those between the 1st rib and the sternum, where hyaline cartilage serves to join the two, or between the manubrium and the body of the sternum before fusion occurs.

Symphysis. In this type of joint, a disc of fibrocartilage firmly binds the bones together, as in the junction between the two pubic bones (*pubic symphysis*) or in the articulations between the bodies of successive vertebrae, where an intervertebral disc forms the connecting structure.

Diarthrosis. A diarthrosis or synovial joint is one in which the articular surfaces are covered with cartilage and separated by an articular cavity. The degree of movement ranges from slight to great.

Structure of a Diarthrodial Articulation. A typical diarthrodial or synovial joint consists of *articular cartilage,* which covers the articulating surfaces; an *articular capsule,* completely enveloping the joint and having two layers, an outer fibrous layer and an inner synovial layer; a *joint cavity,* a space within the capsule lined with the *synovial membrane* and containing a small quantity of *synovial fluid* or *synovia,* a clear, watery fluid; and *ligaments,* bands of connective tissue that bind the bones together, not generally elastic yet permitting movement and at the same time acting to limit the degree of movement. In some joints an *articular disc* composed of fibrous cartilage divides the cavity (the mandibular-temporal, for example).

TYPES OF DIARTHRODIAL ARTICULATIONS

Gliding (arthrodia)	Articular surfaces are flat.	Between carpal bones; between articular processes of vertebrae
Hinge (ginglymus)	Convex cylindrical surface articulates with a concave surface.	Between humerus and ulna (elbow); interphalangeal joints
Ball-and-socket (enarthrosis)	Head of one bone fits into concavity of another.	Hip and shoulder joints
Pivot (trochoid)	Pivotlike process turns within a ring, or a ring turns on a pivot.	Between atlas and odontoid process of axis; between head of radius and ulna
Ellipsoidal or *ovoid* (condyloid)	Ovoid or rounded surface of a bone fits into a shallow concavity of another.	Between occipital condyles and atlas; between carpals and radius
Saddle	The two articulating surfaces are saddle-shaped (concave), fitting reciprocally into each other.	Between carpal bone and 1st metacarpal of thumb

Diarthrodial articulations may also be classified as *plane* (same as gliding), *uniaxial* (same as hinge and pivot), *biaxial* (condyloid), and *multiaxial* (ball-and-socket and saddle). A ball-and-socket joint is shown in Fig. 5-36.

MOVEMENTS AT DIARTHRODIAL JOINTS. The types of movement (Fig. 5-37) that may occur at the various joints of the body and terms applied to these movements are discussed in the following paragraphs. The *anatomic position,* shown in Fig. 1-1, is the basic point of reference.

Flexion is movement in an anteroposterior plane, in which the angle between the bones is *decreased* (bending arm at elbow, bending wrist on forearm, bending head forward, bending body forward at hip joint). *Extension,* the opposite of flexion, is movement in an anteroposterior plane, in which the angle between the bones is *increased* (the opposites of the movements mentioned under flexion serve as examples).

Abduction is movement of the part *away from* the median plane of the body or the median plane of a part, as in lateral and upward movement of arm, lateral movement of leg, or movement of digits away from the third digit (in hand) or second digit (in foot). *Adduction* is the oppo-

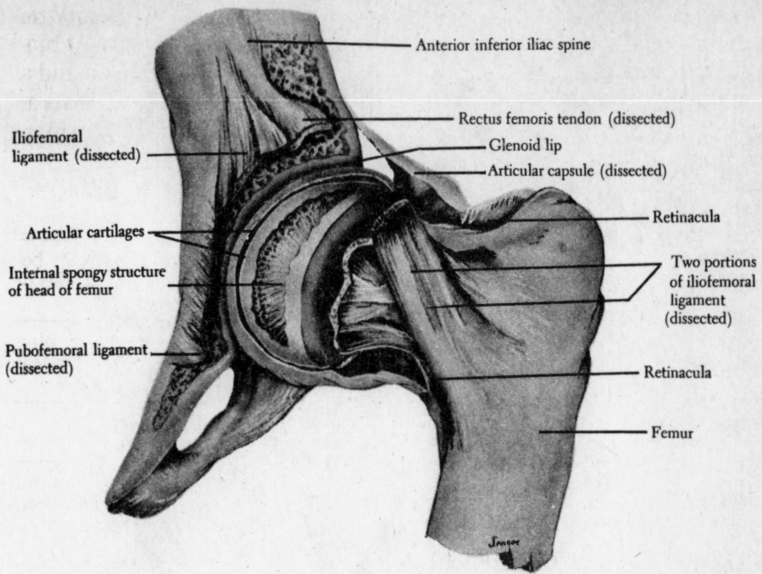

Fig. 5-36. The hip joint, a ball-and-socket type of diarthrosis.

site of abduction; movement is *toward* the median plane of the body or of a part.

Pronation is the medial rotation of the forearm so that the palm of the hand faces backward or downward. *Supination* is the outward rotation of the forearm so that the palm of the hand faces forward or upward.

Rotation is movement in which a part turns on its longitudinal axis, as in turning the head to left or right at the joint between the atlas and the axis or turning the arm or thigh outward (lateral rotation) or inward (medial rotation).

Circumduction is movement involving both angular and rotary movements; the proximal end of a limb remains fixed while the distal end describes an arc (swinging the arm in a circle).

Inversion is movement at an ankle joint, in which the sole of the foot is turned inward (medially). *Eversion* is the opposite movement of the same part, that is, turning it outward (laterally).

Elevation is movement in which a part is raised (movement of mandible in closing the jaw, raising of the scapula). *Depression* is movement in which the part is lowered (movement of mandible in opening the jaw, lowering of scapula).

Protraction is movement of a part forward (protrusion of mandible). *Retraction* is backward movement of a part (drawing back of mandible).

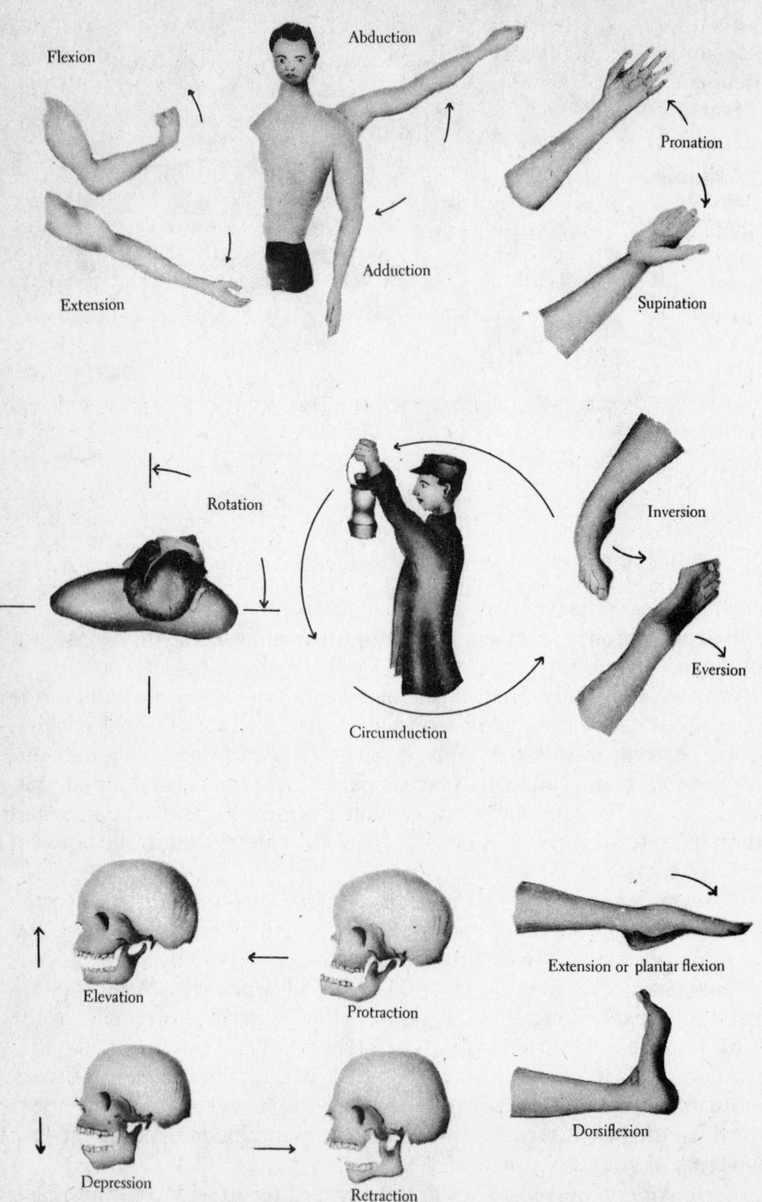

Fig. 5-37. Types of joint movements.

Hyperextension designates movements at the wrist, phalangeal joints, and metatarsophalangeal joints, in which the part is extended beyond the straight line formed by normal extension. *Plantar flexion* is extension of the foot. *Dorsiflexion,* the opposite of plantar flexion, is movement of the foot upward (that is, toward the anterior surface of the leg).

Some Important Articulations of the Body

JOINTS OF THE SKULL

Temporomandibular Joint. This is a compound joint between the mandible and the skull with the cavity divided into an upper and lower compartment by an *articular disc.* The upper compartment permits forward, backward, and sideways movements of the mandible (gliding movements), and the lower compartment permits the up and down movements. It is a gliding and hinge joint.

Atlanto-occipital Joint. This is the joint between the occipital bone and the atlas. On the superior surface of the atlas are two superior articular surfaces, concavities that receive the two occipital condyles of the skull. Rocking movements (flexion and extension) of the head take place. These two joints are condyloid.

Atlantoaxial Joints. These are the joints between the atlas and the axis. Three articulations are involved, one median and two lateral. The median articulation is between the dens of the axis and the anterior arch and transverse ligament of the atlas. It is a pivot type of joint, permitting rotation of the atlas and, with it, the head. The two lateral articulations are between the articular processes of the two bones. These are gliding joints.

JOINTS OF THE VERTEBRAL COLUMN AND THORAX

Joints Between the Vertebrae. These are principally of two types: joints between the bodies and joints between articular processes. The bodies of two successive vertebrae are separated by a disc of fibrous cartilage, the *intervertebral disc.* This is a symphysis type of joint. The joints between the articular processes are of the gliding type.

Joint Between the Vertebral Column and the Pelvis (Sacroiliac Joint). This joint is between the auricular surface of the ilium and the corresponding surface on the ala of the sacrum. Although the surfaces are irregular, presenting an interlocking arrangement, cartilage covers the surfaces and a synovial cavity is present. The joint is diarthrodial, with minimal gliding movement. In later life the cavity may disappear.

ARTICULATIONS OF THE RIBS

1. *Costovertebral Joints:* The head of a rib articulates with the bodies of two vertebrae, a gliding type of joint.

2. *Costotransverse Joints:* The tubercle of a rib articulates with the tip of a transverse process, a gliding type of joint.

3. *Costosternal Joints:* The 1st rib is joined directly to the sternum, a synchondrosis. The remaining true ribs (2–7) articulate with the distal extremity of a costal cartilage, forming a synchondrosis; the articulation

of a costal cartilage with the sternum is a diarthrosis with a gliding movement.

JOINTS OF THE APPENDICULAR SKELETON

The Shoulder Joint. This is the joint between the head of the humerus and the glenoid cavity of the scapula. It is a ball-and-socket joint.

The Elbow Joint. This involves three specific articulations:

1. *The humeral-ulnar articulation:* This is a hinge type of joint in which the trochlea of the humerus fits into the trochlear (semilunar) notch of the ulna.

2. *The humeral-radial articulation:* The capitulum of the humerus articulates with a slight depression, the *fovea,* on the head of the radius. It is a gliding type of joint.

3. *The proximal radio-ulnar articulation:* In this joint, the circumference of the disclike head of the radius turns in the radial notch of the ulna. It is held in position by a ringlike *annular ligament.* It is a pivot joint, permitting the radius to turn on its long axis, thus allowing pronation and supination of the hand.

Intermediate and Inferior Radio-ulnar Articulations. The *intermediate articulation* is between the shafts of the radius and ulna, which are united by an *interosseous membrane,* a syndesmosis type of joint.

The *inferior articulation* is between the head of the ulna, which fits into the ulnar notch of the radius. It is a pivot joint, permitting pronation and supination. An *articular disc* of fibrocartilage lying below the ulnar head excludes the ulna from the wrist joint.

Wrist Joint. This joint involves the articulation between the lower end of the radius and two carpal bones, the lunate and navicular, a condyloid type of joint permitting flexion, extension, abduction, and adduction of the hand. To a slight extent, the articular disc of the inferior radio-ulnar joint and the triquetrum is involved in wrist movement.

Carpometacarpal Joints. The joint between the trapezium and the metacarpal bone of the thumb is a saddle joint, giving great freedom of movement and permitting grasping action. The remaining carpometacarpal joints are gliding joints with limited movement.

Metacarpophalangeal and Interphalangeal Joints. The articulations of the rounded heads of the metacarpal bones with the slightly concave proximal ends of the 1st phalanges are condyloid joints; that with the 1st phalanx of the thumb and the interphangeal joints are hinge joints.

The Hip Joint. In this joint the head of the femur fits into a deep socket, the *acetabulum* of the hipbone (os coxae). It is a ball-and-socket joint.

The Knee Joint. This articulation is between the two large, rounded condyles of the femur and the two superior surfaces of the condyles of the tibia, upon which rest the two *semilunar cartilages* or *menisci.* It is a double condyloid joint, but because of the structure of the joint, especially the presence of collateral ligaments, movement is hingelike, and the knee is regarded as a hinge joint.

The Tibiofibular Joints. An interosseous membrane joins the tibia and the fibula together throughout their entire length, a syndesmosis type of articulation. At the upper end, the head of the fibula forms a gliding joint with the lateral condyle of the tibia. At the lower end, the tibia and fibula are tightly bound together by tibiofibular ligaments, a syndesmosis type of joint.

The Ankle Joint. This joint is formed by the distal end of the tibia and its malleolus and the lateral malleolus of the fibula, which together form a boxlike structure that receives the convex surface of the talus and its two facets, the lateral and medial. It is a hinge joint.

The Metatarsophalangeal Articulations. These are condyloid joints formed by the concave articular surfaces of the proximal ends of the 1st phalanges articulating with the rounded distal ends of the metatarsal bones.

The Interphalangeal Joints. These are simple hinge joints between the phalanges of the digits.

DISORDERS AND DISEASES OF THE BONES AND JOINTS

Acromegaly. Hypertrophy of bones of the face, hands, and feet; due to hypersecretion of the growth hormone (somatotrophin) by the anterior lobe of the pituitary gland.

Ankylosis. Fusion or consolidation of bones of a joint; this results in reduced movement or complete immobility.

Arthritis. A group of ailments involving joints, muscles, and tendons; formerly referred to as "rheumatism". Some common types are (1) *infectious arthritis,* a form which accompanies diseases such as tuberculosis, syphilis, gonorrhea, and rheumatic fever; (2) *metabolic arthritis,* commonly called "gout" (due to the presence of excessive quantities of uric acid in the blood and to deposition of urates in or around joints, especially those of the hallux); (3) *osteoarthritis* or degenerative joint disease, a chronic disease of the joints characterized by degeneration of articular cartilage and hypertrophy of bone and cartilage; common in older persons; (4) *rheumatoid arthritis,* either gradual or sudden in onset; may last for years or disappear quickly; in advanced stages joints may swell to twice normal size, cartilage between bones may disappear and joints fuse and become immobile; affects young as well as old. Etiology is unknown.

Bunion. (Also called *hallux valgus.*) Displacement of the great toe laterally toward the other toes; often results from wearing of shoes or socks that are too short, the pressure causing enlargement of the metatarsophalangeal joint and thickening of the joint capsule.

Bursitis. Inflammation of a bursa. "Housemaid's knee" (inflammation of the patellar bursa) and "tennis elbow" (inflammation of the olecranon bursa) are examples.

Clubfoot. See *Talipes.*

Dislocation. Displacement of the ends of the bones of a joint; results in disarrangement of the parts.

Fracture. A break or crack in a bone. Types of fractures are *Colles's,* in which the radius is broken about ½ to 1 inch from the wrist; *compound,* in which a bone is broken and there is an external wound at the point of fracture; *greenstick,* in which a bone is partially broken, with a splintering effect; *Pott's,* a break of the tibia just above the ankle; and *simple,* in which a bone is broken but there is no external wound.

Gout. See *Arthritis.*

Hallux valgus. See *Bunion.*

Kyphosis. Abnormal increase in the thoracic curve of the vertebral column; gives rise to the condition called "hunchback."

Lordosis. The abnormal forward curvature of the spine in the lumbar region; results in protrusion of the buttocks.

Lumbago. Backache in the lumbar and lumbosacral regions.

Neoplasms. These comprise various tumors of bone. Common *benign tumors* include osteomas or extoses, giant-cell tumors, and various types of cysts. *Malignant tumors* include osteogenic sarcoma, Ewing's tumor of the bone marrow, chondrosarcomas, and metastatic tumors from sources outside of bone.

Osteitis. Inflammation of bone tissue. In *osteitis deformans* (Paget's disease), decalcification and softening occur, followed by deposition of calcium, with resultant thickening and development of abnormalities. Its cause is unknown. *Osteitis fibrosa cystica* is characterized by resorption of bone brought about by excessive secretion of the parathyroid hormone.

Osteoarthritis. See *Arthritis.*

Osteomalacia. Softening of bones due to a deficiency of vitamin D in the diet or inadequate exposure to sunlight. Mineral content of bone is reduced, resulting from inadequate absorption of calcium and phosphorus from the intestinal tract. Called "adult rickets."

Osteomyelitis. Inflammation of bone and bone marrow due to infection by pyogenic bacteria, especially staphylococci. It may be localized or it may spread to and involve other tissues.

Osteoporosis. Softening of bone due to defective bone formation. It commonly involves the spine, pelvis, and femur and occurs most frequently in elderly persons. It is a common cause of low back pain and spontaneous fractures.

Paget's disease. See *Osteitis.*

Pott's disease. Tuberculosis of the spine; bodies of the vertebrae are often eroded, resulting in abnormal curvature.

Rickets. Condition in children in which the bones become bent and distorted; may result in knock-knee (*genu valgum*) or "bowleg" (*genu varum*). It is caused by defective calcification, which results from failure (attributed to absence of vitamin D) of the digestive tract to absorb calcium and phosphorus.

Scoliosis. Abnormal increase in lateral curvature of the vertebral column.

Spina bifida. A development defect in which there is incomplete closure of the vertebral arches; it results in herniation (protrusion) of the spinal cord.

Spondylitis. Inflammation of the vertebrae, takes many forms, due to many different causes.

Spondylolisthesis. Forward displacement of one vertebra over another, usually the fifth lumber over the sacrum, or the fourth lumbar over the fifth.

Sprain. The wrenching of a joint, with stretching or tearing of its ligaments.

Strain. Excessive stretching of the ligaments of a joint capsule.

Talipes. Any of a number of deformities of the foot, such as *talipes valgus* (flatfoot) and *talipes varus* (clubfoot). They are usually congenital.

Tenosynovitis. Inflammation of a tendon sheath.

6: THE MUSCULAR SYSTEM

The muscular system comprises the organs that, by their contraction and relaxation, produce the movements of the body as a whole and of its parts. The muscles constitute the "red flesh" of the body. They account for 42 percent of the total body weight in males and 36 percent in females. Also related to the muscular system are tendons, fasciae, and aponeuroses, all of which serve to secure the ends of the muscles and to determine the direction of their pull.

FUNCTIONS OF MUSCLES

The functions performed by muscles may be regarded as of two kinds: voluntary and involuntary. There is no strict dichotomy between the two; certain functions classed as "voluntary" may also, and quite readily, take place involuntarily, whereas some of the "involuntary" functions are amenable to conscious (voluntary) control.

Principally Voluntary Functions of Muscles. The functions served by muscular action that are principally voluntary in nature are as follows:

1. *Maintenance of posture,* whereby the general "attitude" of the entire framework of the body is controlled.
2. *Accomplishment of various movements in which action is externally visible,* shown in the following table.

Limbs	locomotion, exertion of force
Fingers	grasping, handling, manipulation
Toes	balance, leverage, locomotion
Diaphragm	respiration
Pharynx	swallowing
Tongue and lips	food manipulation, vocalization
Abdominal wall	respiration, defecation, vomiting
Head	bringing sense organs into more favorable position
Face	expression of mood and emotion

Principally Involuntary Functions of Muscles. The functions that are principally involuntary are as follows:

1. *Propulsion of substances through body passages,* as food through the digestive tract, blood through the vessels, and germ cells through the reproductive ducts.
2. *Expulsion of stored substances,* such as bile from the gallbladder, urine from the kidneys and the urinary bladder, and feces from the rectum; the last two are also readily capable of voluntary control.
3. *Regulation of the size of openings,* such as the pupil of the eye, the

pylorus, the anus, and the neck of the urinary bladder.

4. *Regulation of the diameter of tubes,* such as blood vessels and bronchioles.

TYPES OF MUSCULAR TISSUE

Muscle tissue is classified into three types on the basis of structure and function: smooth, skeletal, and cardiac. All three types develop from mesoderm. Smooth muscle arises specifically from mesenchyme, skeletal muscle from mesenchyme and myotomes of embryonic somites, and cardiac muscle from splanchnic mesoderm.

Smooth Muscle Tissue. This type of muscle, which is nonstriated, is found principally in structures in which action is involuntary (not under voluntary control). It exists as single cells arranged in small groups, in bundles, or in sheets or layers. It is divided into two types, based on location and contractile properties: *visceral* and *multiunit. Visceral muscles* are found principally in the walls of hollow visceral organs; *multiunit muscles* are found in the iris and ciliary body of the eye, in the walls of blood vessels, in precapillary sphincters, and in the piloerector muscles attached to hair follicles.

The cells of fibers of smooth muscle (Fig. 6-1) are spindle-shaped, long and narrow. Each cell has a single nucleus located in its middle portion. Its cytoplasm consists of *sarcoplasm* in which lie *myofilaments* made up of *myofibrils,* the contractile elements. A sarcolemma is lacking. In certain regions, the plasma membranes of adjacent cells come into close contact with each other, forming a union or *nexus.* These connections probably play a role in the transmission of the impulses for contraction from one cell to another.

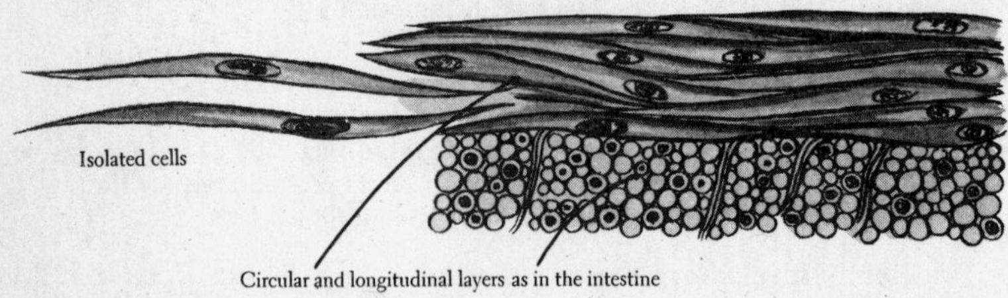

Fig. 6-1. Smooth muscle cells.

LOCATION. Smooth muscle tissue is found in visceral organs, as the wall of the digestive tract, trachea, bronchi, urinary bladder, gallbladder, and urinary and genital ducts. It is also found in the walls of blood vessels, the capsule of the spleen, the ciliary body and iris of the eye, and the nipple of the mammary gland. Smooth muscle fibers are abundant in the dermis of the skin, especially in the erector muscles attached to hair follicles and in the dartos muscle of the scrotum.

INNERVATION. Smooth muscles are innervated by efferent fibers of both divisions of the autonomic nervous system.

TYPE OF ACTION. Smooth muscles are not subject to voluntary control, their action being involuntary.* Their control depends upon (1) chemical substances (hormones) in the tissue surrounding the cells and (2) impulses received through autonomic nerves. Some cells, such as those of the intestine and uterus, are *autorhythmic;* that is, they contract rhythmically and spontaneously in the absence of an external stimulus. Most cells, however, respond only to nerve impulses over the autonomic nervous system, which may stimulate or inhibit activity. Since nerve endings are present in relatively few cells, the nervous excitation is apparently passed between points of close contact between two cells, each known as a *nexus* or *tight junction.*

Skeletal Muscle Tissue. This type of tissue, also called *striated* or *striped* muscle, comprises the skeletal or voluntary muscles. It consists of *fibers,* which are large, multinucleated cells. The fibers are grouped into bundles called *fasciculi.*

Skeletal muscle fibers (Fig. 6-2) vary in length from 1 to 40 mm and in diameter from 10 to 100 μm. Each fiber is invested by a *sarcolemma* or cell membrane, which encloses a semifluid matrix, the *sarcoplasm,* in which the contractile elements, *myofibrils,* are embedded. Each fiber contains numerous *nuclei,* which lie at the periphery of the cell, close to the sarcolemma. Numerous *mitochondria* or *sarcosomes* and small *Golgi complexes* are also present.

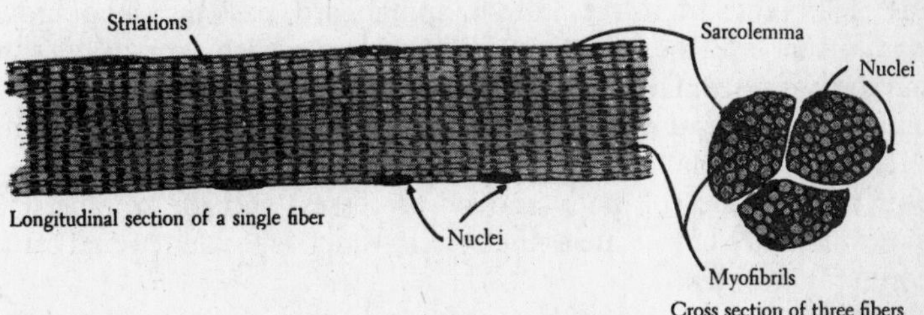

Fig. 6-2. Skeletal muscle fibers.

The electron microscope has also revealed the presence of a system of delicate, transverse tubules (*sarcotubules*) that extend inward from the sarcolemma, their walls being continuous with the cell membrane, and a *sarcoplasmic reticulum,* an anastomosing network of tubules and vesi-

*Recent research is modifying the view that organs innervated by the autonomic nervous system are not subject to voluntary control. Experiments on laboratory animals and humans indicate that, by training, individuals can be taught to control such autonomic responses as glandular secretion, rate of heartbeat, and constriction of blood vessels. See *Biofeedback* in Vol. 2.

cles that lie between and around the myofibrils. The myofibrils consist of thick and thin *myofilaments,* the former containing the protein *myosin,* the latter, *actin.* A system of transverse tubules (T tubules) permits the movement of fluids between the sarcoplasmic reticulum and the myofibrils.

Each myofibril (Fig. 6-3) is marked by alternating light and dark bands that give the fiber its striated appearance. The bands of one fibril are at the same level as those in adjoining fibers, so the striations seem to cross the entire fiber. The two main striations are the *A* and *I bands.* The A or *anisotropic* band is dark; the lighter band is the I or *isotropic band.* In the middle of the A band is a lighter region, the *H band,* and in the middle of the I band is a dark *Z band.* The region between two Z bands or discs comprises a *sarcomere,* the contractile unit.

Muscle fibers are grouped together to form a primary bundle or *fasciculus,* which is surrounded by a connective tissue membrane, the *perimysium.* Within a fasciculus, each muscle fiber is surrounded by a delicate sheath of reticular connective tissue fibers, called the *endomysium.* A muscle, such as the biceps, is composed of many groups of fasciculi. A connective tissue sheath, the *epimysium,* encloses an entire muscle.

LOCATION. Skeletal muscle tissue comprises all muscles attached to the skeleton, the muscles of the tongue and the soft palate, those that move the scalp, those in the pharynx and the upper part of the esophagus, and the extrinsic eye muscles.

INNERVATION. Skeletal muscle is innervated by afferent and efferent fibers of cerebrospinal nerves. The axons of *efferent motor neurons,* whose cell bodies lie in the brain or spinal cord, pass by way of cranial or spinal nerves to the muscles. Within a muscle, each axon divides into a number of terminal branches, each of which terminates in a single muscle fiber. The number of terminal branches of an axon varies from one or a few in small muscles to several hundred in larger muscles. A motor neuron (the cell body and its processes) and the muscle fibers that it innervates constitute a *motor unit,* which is the functional unit of the muscular system.

Each terminal branch of the axon ends in a specialized structure called a *motor end plate,* a slightly elevated structure marked by an aggregation of nuclei. Each nerve fiber ends in a slight invagination called a *synaptic trough* or *gutter.* At this point, nerve impulses bring about the release of acetylcholine, which initiates contraction. The entire structure constitutes the *myoneural* or *neuromuscular junction.*

Afferent or *sensory fibers* ending in skeletal muscle are myelinated fibers (axonlike dendrites) of sensory neurons whose cell bodies lie in the dorsal root ganglia of spinal nerves or sensory ganglia of cranial nerves. They terminate in specialized receptor structures called *neuromuscular spindles,* which are stimulated by the stretch of muscle fibers within the spindle. Afferent impulses arising from stimulation of the

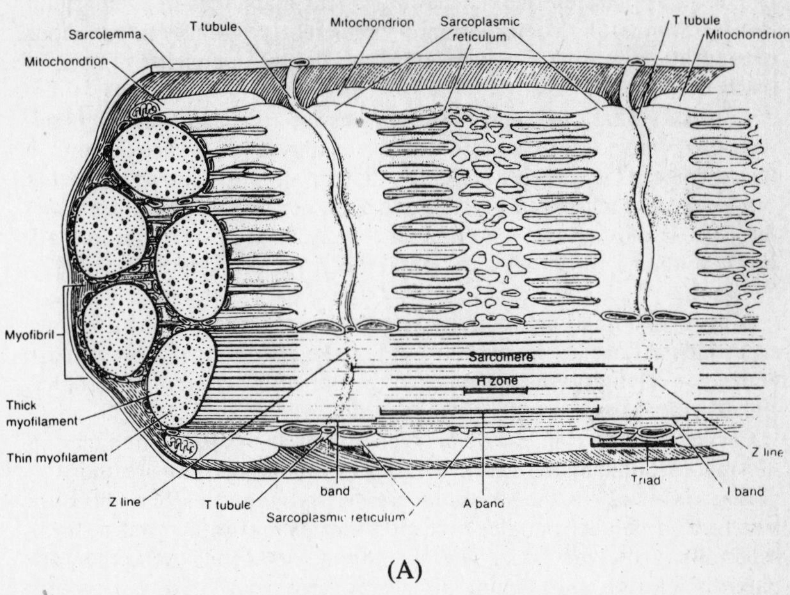

(A)

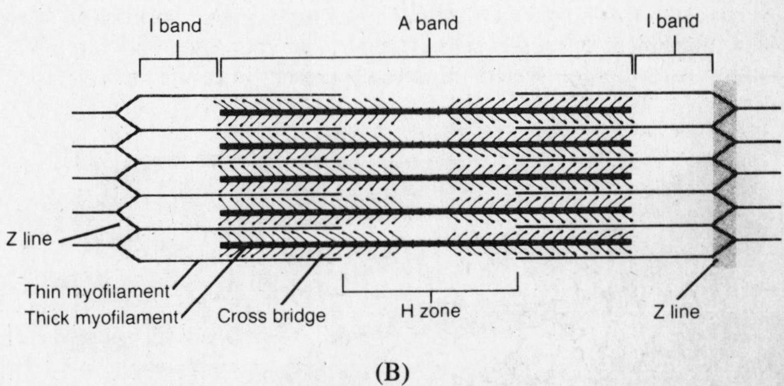

(B)

Fig. 6-3. Microscopic structure of skeletal muscle. (*A*) Enlarged aspect of a muscle fiber based on an electron micrograph. (*B*) Details of a sarcomere showing thin and thick myofilaments and various internal zones. (Reprinted with permission of Harper & Row, Publishers, Inc., from G. J. Tortora and N. P. Anagnostakos, *Principles of Anatomy and Physiology,* 3rd ed., © 1981.)

afferent nerve endings are of primary importance in the initiation of such activities as the stretch reflex, relaxation of antagonistic muscles, and the transmission of information to the brain that the cerebellum utilizes in muscular coordination or the cerebrum utilizes in conscious recognition of muscular movement or the change in position of parts of the body.

TYPE OF ACTION. Although skeletal muscles are under voluntary control, the brain is not conscious of the contraction of specific muscle fibers. Instead, it initiates activities that bring about movements that can be seen or felt such as raising the arm, bending the fingers, closing the jaw, or moving the lips. By controlling an activity by initiating a movement or inhibiting it, voluntary control of skeletal muscles is effected.

Most of our muscular activity is reflexly controlled, and in that respect such actions are involuntary. A few skeletal muscles, however, are entirely involuntary, such as those in the upper one-third of the esophagus and the stapedius muscle of the middle ear.

Cardiac Muscle (Fig. 6-4). This type of muscle tissue is found only in the myocardium of the heart. The fibers are striated, as are those of skeletal muscle, but the striations are less pronounced. The fibers are arranged in parallel bundles or bands that extend in a spiral fashion, especially in the ventricles. The fibers branch and join each other, apparently forming a syncytium, but electromyographic studies show that this is not the case; distinct cell membranes separate the individual cells. A *sarcolemma* is present, and the nuclei occupy a central position in the fiber. In addition to the regular striations, distinct, dark, transverse bands called *intercalated discs* are present. These mark the end-to-end attachment of muscle cells. *Myofibrils* having essentially the same structure as those in skeletal muscle are present in cardiac muscle.

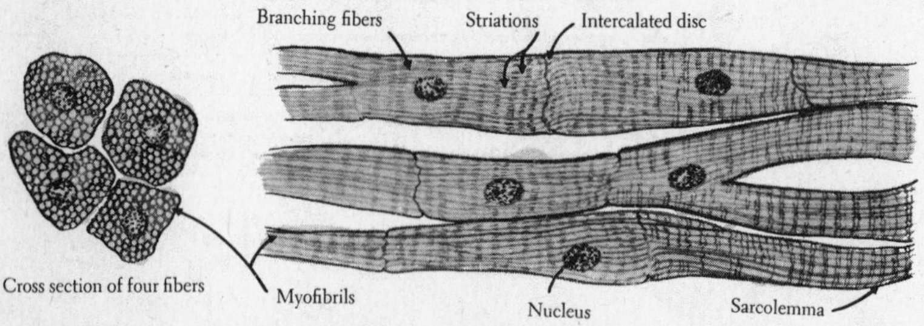

Fig. 6-4. Cardiac muscle.

Atypical fibers, called *Purkinje fibers,* serve to provide an impulse-conducting system; these fibers differ from the usual muscle fibers in having fewer fibrils, less distinct striations, considerable sarcoplasm about the nuclei, and a large amount of glycogen. Such fibers are concentrated in the SA node, the AV node, the atrioventricular bundle (bundle of His), and branches of these structures, all of which are described in Chapter 10.

LOCATION. Cardiac muscle is found only in the heart.

INNERVATION AND TYPE OF ACTION. Cardiac muscles are innervated by fibers from the autonomic nervous system. Their action is involuntary, automatic, and rhythmic.

CHARACTERISTICS OF MUSCLE TISSUES

Type	*Where Found*	*Structural Characteristics*	*Functional Characteristics*
Smooth muscle	In walls of tubes and hollow organs, as digestive tract, blood vessels, bladder	Long, spindle-shaped cells; no striations; usually arranged in layers	Contractions are slow, involuntary, often rhythmic
Skeletal (striated) muscle	Attached to skeleton, in wall of pharynx and esophagus	Long, cylindrical, multinucleated fibers; nuclei located at periphery of fiber; striations prominent	Contractions may occur rapidly and may be of a forceful nature; under voluntary control
Cardiac (heart) muscle	In myocardium	Short, branching, fibers; nucleus centrally located; striations not distinct	Contractions are involuntary, automatic, rhythmic

PHYSIOLOGY OF MUSCLES

Properties of Muscle Tissue. Four properties of muscle tissue facilitate the functioning of muscles: contractility, extensibility, elasticity, and irritability.

CONTRACTILITY. Muscle tissue has the ability to shorten or attempt to shorten. This is *contractility.* The ability to exert tension between two points of attachment is the basis of the work accomplished by muscles. When muscles shorten *(isotonic contraction)* they may produce movement, as in walking, running, or lifting weights. If a muscle is prevented from shortening *(isometric contraction)*, its contraction produces tension but no movement, as in the contraction of muscles that maintain posture. No external work is done, the energy being expended in the form of heat.

EXTENSIBILITY. Muscle tissue also has the ability to stretch upon the application of force. Smooth muscle cells of the visceral organs, such as the stomach, urinary bladder, and uterus, can stretch to a remarkable degree as these organs become filled. Cardiac muscles and muscles of the blood vessels stretch when these structures become filled with blood. Striated muscles stretch when the opposing muscles contract, as when the flexor muscle of the arm contracts, forcing the extensor muscle to lengthen. This permits movement at a joint.

Elasticity. This property of muscle tissue enables muscles to regain their original size and shape after having been stretched. Through elasticity, muscle tone is maintained in skeletal muscles, and organs such as the urinary bladder and the uterus resume their minimum size after they have been emptied. When arteries, especially the muscular arteries, have been stretched by blood forced into them by contraction (systole) of the heart, they tend to resume normal size during diastole. The resulting tension acts to maintain a steady flow of blood through the circulatory system.

Irritability. Muscle tissue has the property of being able to respond to a stimulus. This stimulus is ordinarily a nerve impulse transmitted through the cranial, spinal, and autonomic nerves, but other stimuli may induce muscle contraction. Examples of the latter are the application of an electric current, the action of irritating chemical substances, and a physical stimulus, such as pinching.

Contraction of an Individual Muscle (Experimental Observations). The contraction of a muscle can be studied experimentally by removing a single muscle from the body of an animal, subjecting it to various kinds of stimuli, and recording the results obtained. For this purpose, the gastrocnemius of a frog is the most generally used muscle.

Types of Stimuli. Four types of stimuli are commonly employed:

1. *Mechanical:* tapping, pinching, stretching.
2. *Chemical:* such agents as an acid or a salt.
3. *Thermal:* application of heat by various means.
4. *Electrical:* application of various strengths of electric current.

For experimental purposes, the electrical type of stimulus is most frequently used because its strength is easily controlled and there is a minimum of injury to the tissues. An *inductorium,* using a current from dry cells, has been commonly used as a stimulator. However, new *electronic stimulators* by which the frequency, amplitude, and duration of the stimuli can be readily controlled have largely replaced the inductorium in experimental work.

Intensity of a Stimulus. To provoke a response (in this case, muscle contraction) a stimulus must be of a certain intensity. A weak stimulus will elicit no contraction; such a stimulus is said to be *subminimal.* Should the intensity of the stimulus be increased gradually, a point will be reached at which the muscle will barely respond; this is called a *threshold* or *minimal stimulus.* Now, if the intensity of the stimulus is gradually increased further, the extent of contraction will also increase gradually until a point is reached beyond which further increase in intensity of the stimulus produces no increase in the extent of contraction; this is called a *maximal stimulus.*

The All-or-None Principle. If cardiac muscle is stimulated, the whole muscle contracts to its fullest extent. This illustrates the "all-or-none principle," namely, that *each contraction is maximal for the condi-*

tions under which the muscle is stimulated. The principle holds true also for a *single* striated muscle fiber. When a single fiber is stimulated, it contracts to its fullest extent. A muscle consisting of a great many individual fibers does not follow this principle. A minimal stimulus applied to an entire muscle may cause one or a few of the fibers to contract, and the result is a slight response. But as the stimulus is increased, more and more fibers are stimulated, and a greater contraction of the muscle is brought about. Finally, a point may be reached at which all the fibers are being stimulated, and, since each is then contracting to its fullest extent, a *maximal response* has been attained, with no further contraction possible.

It is not to be inferred from the foregoing remarks that under the all-or-none principle the response of a muscle to a stimulus is always the same. Not only do muscle cells vary, but the physiological conditions under which they respond may also change. In brief, then, the all-or-none principle means that *for a given muscle or muscle fiber, the degree of response is independent of the strength of the stimulus.*

RECORDING OF MUSCLE RESPONSE. For obtaining records of many types of physiological activity, a *kymograph,* consisting of a revolving drum bearing paper upon which activities are recorded, was formerly widely used. It now has been largely replaced by multichannel graphic recorders, which produce much more reliable records.

MUSCLE-NERVE PREPARATION. To prepare a muscle for experimental purposes, a frog in good physical condition is pithed (its brain is destroyed). The skin is removed from the hind legs, exposing the muscles. The gastrocnemius (calf) muscle and its tendon of Achilles (which passes under the foot) are carefully separated from the shank. The sciatic nerve is then separated from the surrounding thigh muscles. All muscles and other tissues are cut away until only the gastrocnemius and the lower end of the femur (from which it originates) and the nerve remain. The femur is then placed in a muscle clamp, and the tendon of Achilles is attached by a hook to a lever that activates a transducer. A weight is attached to the lever to provide the tension that normally exists in the body. Stimuli may be applied to the nerve or directly to the muscle through stimulating electrodes. A typical setup is shown in Fig. 6-5.

PHENOMENA OF MUSCLE CONTRACTION. Some of the phenomena of muscle contraction that have been observed experimentally are (1) the effect of gradually increasing the strength of the stimulus; (2) a single muscle twitch recorded on a rapidly moving drum; (3) the effects of rapidly repeated stimuli; and (4) fatigue, or the effects of long-continued stimulation.

Effect of Increasing the Strength of a Stimulus. Subminimal stimuli (those with insufficient strength) fail to elicit any muscular response. A threshold stimulus is one that produces a minimal response. In experiments stimuli of gradually increasing strength are applied to a muscle.

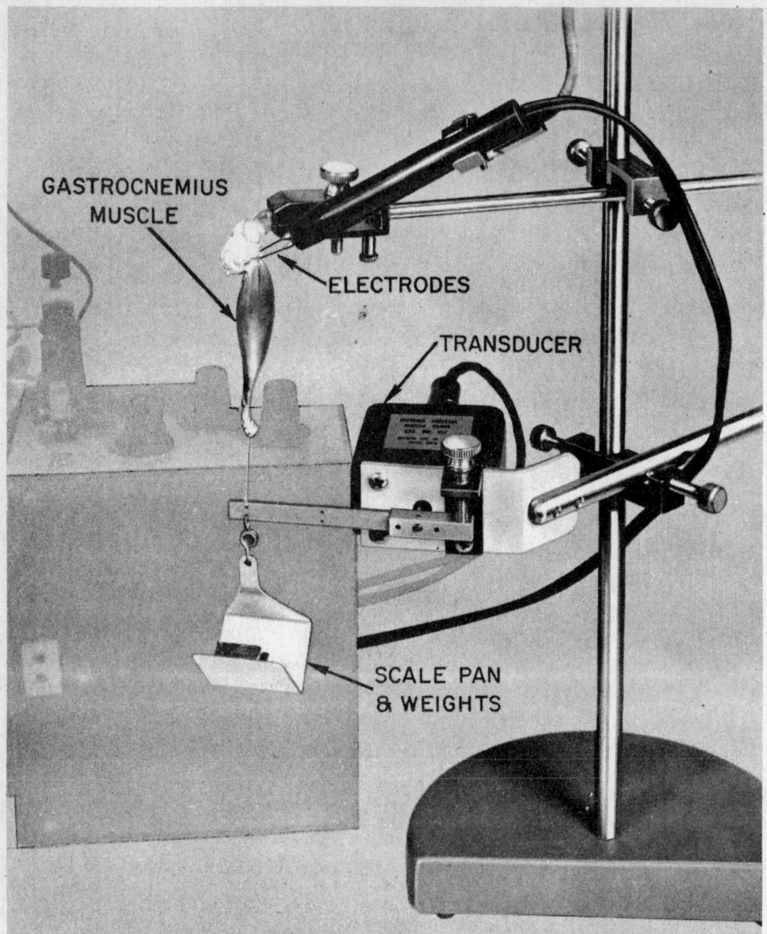

Fig. 6-5. Gastrocnemius preparation for electronic recording. Contraction of the muscle displaces the lever of the transducer. Note the scale pan and weights for muscle loading and the electrode set used for stimulation. (Reprinted with permission from *Bioinstrumentation: Experiments in Physiology,* Harvard Apparatus Foundation, Inc., Millis, Mass., 1971.)

As the stimuli increase in strength, muscular reactions increase up to a point of maximal response. A further increase in the strength of the stimulus does not increase the response beyond this point; it merely continues to elicit the same maximal reaction.

A Single Muscle Twitch. The response of a muscle when stimulated is a single quick contraction known as a *muscle twitch.* If the muscle shortens, it is an *isotonic contraction;* if the muscle does not shorten, the contraction is *isometric.*

When a single adequate stimulus is applied, there is a pause before a response is noted. This is the *latent period.* There then follows a period in which tension increases and the muscle shortens, the *contraction period.* Following this is a *period of relaxation,* during which tension returns to zero. The length of a muscle twitch for a frog gastrocnemius is about one-tenth of a second (100 msec), the length of each phase being as follows: latent period, 10 msec; contraction period, 40 msec; relaxation period, 50 msec (Fig. 6-6).

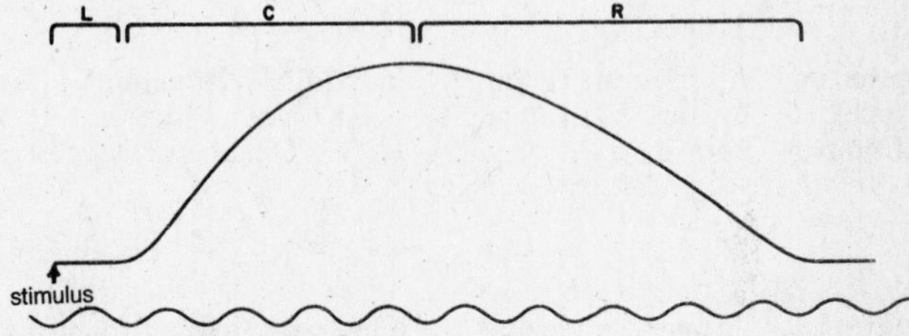

Fig. 6-6. Record of a single muscle contraction. (*L*) Latent period. (*C*) Contraction phase. (*R*) Relaxation phase. Time is indicated by the lower wavy line made by a tuning fork vibrating at 100 double vibrations per second. (Reprinted with permission of Wm. C. Brown Company from E. B. Steen, *Laboratory Manual and Study Guide for Anatomy and Physiology,* 3rd ed., 1976.)

The duration of a muscle twitch varies, depending on the type of muscle, the physiological condition of the muscle, and environmental conditions. Some muscles contain fast-acting fibers; others react at a slower rate.

Effects of Rapidly Repeated Stimuli. Rapidly repeated stimuli produce summation, tetanus, a refractory period, and a "staircase" effect (Fig. 6-7).

When a single subminimal stimulus does not elicit a contraction, rapidly repeated subminimal stimuli may do so; this effect is called *summation.*

When stimuli above threshold strength are repeated fairly rapidly and only partial relaxation occurs, *incomplete tetanus* exists. When, however, the stimuli are repeated very rapidly, so that no relaxation takes place and the muscle twitches are blended into a single sustained contraction, the condition is known as *complete tetanus.* Practically all skeletal muscle contractions are tetanic in nature. Even a rapid movement, such as the blinking of an eyelid, involves tetanus.

After a muscle has been stimulated, for a very short period it loses its irritability and will not respond to a second stimulus while in this condition. This is known as the *absolute refractory period.* It is brief for

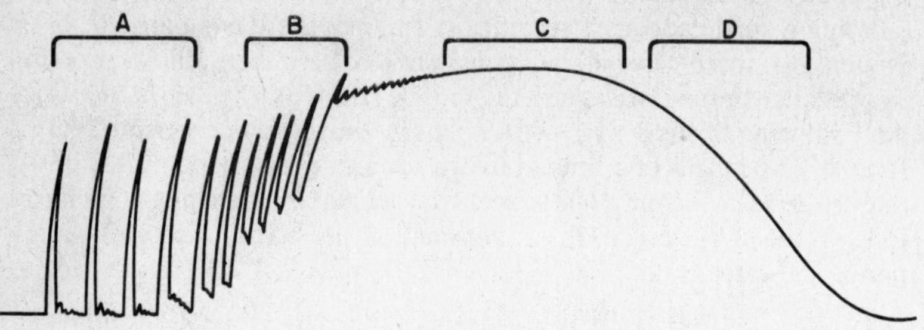

Fig. 6-7. Record of muscle contractions. (*A*) Individual contractions. (*B*) Summation (treppe). (*C*) Tetanus. (*D*) Fatigue. (Reprinted with permission of Wm. C. Brown Company from E. B. Steen, *Laboratory Manual and Study Guide for Anatomy and Physiology,* 3rd ed., 1976.)

striated muscles, longer for cardiac muscles, and longest for nonstriated muscles.

Following the application of repeated stimuli it is sometimes noted that the contractions are successively higher, even though there is no increase in the strength of the stimulus. The resultant recording resembles a staircase; hence it is called the "staircase" effect or *treppe.* It is believed that with each contraction the irritability of the muscle protoplasm increases, resulting in a slightly increased contraction at each successive stimulus. This is the basis of the "warming up" period required of athletes.

Fatigue. When a muscle has been maintained in a state of sustained contraction through repeated stimulation, it gradually loses its ability to contract, and relaxation occurs. This is the mechanism of *fatigue,* a condition characterized by decreased irritability and contractility. During fatigue the fibers tend to resume their original state and contractions are reduced in extent or fail to occur, even though the muscle is still being stimulated. In the latter condition the muscle is regarded as being completely fatigued. *Fatigue, then, is the reduced capacity or inability of a muscle to perform work.* This type of fatigue (muscle fatigue) is due to the inability of the muscle to synthesize adenosine triphosphate (ATP), its energy source, at a sufficiently rapid rate to maintain contractions.

Factors in Bodily Fatigue. The sensation of fatigue, experienced in the brain, is projected to (localized in) specific muscles. As already noted, it is followed by a diminished capacity or total inability of a muscle to perform work. *Acute* or *physiological fatigue* is that brought on by prolonged mental or physical activity. It is relieved by adequate rest and sleep. *Chronic* or *pathologic fatigue* is that resulting from a diseased or pathologic condition. It is not relieved by rest. The significant factors giving rise to fatigue are (1) excessive activity; (2) malnutrition; (3) circulatory disturbances; (4) respiratory disturbances; (5) metabolic disorders; (6) musculoskeletal disorders; (7) infections; (8) endocrine disturbances; and (9) psychogenic factors.

Excessive Activity. Energy-producing substances may be used up faster than the restorative processes can replenish them or faster than the circulatory system can supply essential raw materials or remove waste products. Recovery is achieved through rest.

Malnutrition. Lack of essential food materials, especially proteins, minerals, or vitamins, may deprive muscle cells of the chemical substances vital to contraction. Depletion of sodium chloride (table salt) is a common factor in fatigue.

Circulatory Disturbances. Circulatory disturbances prevent the adequate supply of oxygen and energy materials (e.g., glucose) to muscles and the effective removal of waste products. This is seen in anemia, in which fatigue is a consequence of reduced hemoglobin. Heart disorders such as those involving congenital defects or coronary occlusion, impair circulation and cause chronic fatigue.

Respiratory Disturbances. Nearly all respiratory disorders in which there is a disturbance in gaseous exchange are characterized by fatigue, as in pneumonia, tuberculosis, and emphysema.

Metabolic Disorders. Disturbances in carbohydrate metabolism, as in hyperglycemia in diabetes mellitus and hypoglycemia due to hyperinsulinism, and disturbances in acid-base and electrolyte balance are characterized by fatigue. Accumulation in the body of waste products, such as ketone bodies in diabetes or protein residues in kidney disorders, are characterized by chronic fatigue. Alcoholism and drug addiction also cause fatigue.

Musculoskeletal Disorders. Diseases that involve the muscles, such as muscular dystrophy, myasthenia gravis, and myositis, are characterized by fatigue. Skeletal and joint disorders, skeletal abnormalities, trauma, and infection are also causative factors. Poor posture, which puts an undue strain on ligaments and muscles, is a common cause of fatigue.

Infections. In nearly all infectious diseases, fatigue is one of the commonest symptoms. It is thought to be due to the presence of metabolic or toxic products manufactured by invading organisms. Fatigue may be considered as a protective mechanism whereby bodily energy is directed toward overcoming the infection; the enforced rest facilitates the restorative processes.

Endocrine Disturbances. Hormone imbalance that interferes with normal metabolism may result in symptoms of fatigue. Examples are the fatigue states that accompany the menopause, diabetes, and thyroid disorders.

Psychogenic Factors. Conditions that have a psychological or psychosomatic background, such as emotional conflicts, frustration, grief, worry, and boredom, are often characterized by fatigue.

Changes Occurring During Muscular Activity. When a skeletal muscle contracts, physical and chemical changes occur. The shape, consistency, and appearance of a muscle are altered, and phenomena associated with the release of energy, movement, electrical changes, and production of heat are manifested. A summary of these changes follows.

Physical Changes. When a muscle contracts isotonically, a shortening occurs and diameter increases. Overall volume, however, is not changed. The physical consistency of a muscle also changes. When relaxed, the fibers are soft and pliable; when contracted, the fibers may become hard and rigid. Certain structural changes occur in the light and dark bands that constitute the striations. These changes are the result of ongoing chemical reactions that are involved in the release of energy.

Chemical Changes. Research in the field of muscle physiology has led to the development of the *sliding-filament theory* of muscle contraction. According to this theory, the shortening of muscle fibers results from changes in the arrangement of molecules within the fiber. Two substances of primary importance are the proteins *actin* and *myosin.* The thin *actin filaments* lie parallel to and between the thicker *myosin filaments,* to which they are connected at each end by cross-bridges. Actin filaments extend from the midsection of one A band to the midsection of the next A band; that is, their ends extend from one H zone to the next H zone but do not pass through an H zone. Myosin filaments extend the entire length of the A band, in the center of which is the H zone. The cross striations of a muscle fiber arise from the arrangement of these filaments within the fiber.

In a state of relaxation, the myosin bridges are prevented from combining with actin by two proteins, *troponin* and *tropomyosin,* which inhibit the process of contraction. The initiation of an action potential by the arrival of a nerve impulse at a muscle fiber initiates the development and spread of a membrane potential through the muscle fiber. This triggers the release of *calcium* from sacs of the sarcoplasmic reticulum. Calcium reacts with troponin, producing a change that is transmitted through the tropomyosin molecule to the actin molecules, enabling cross-bridges of the myosin molecule to attach to the actin filaments. The actin filaments are then pulled toward the middle of the myosin filaments of each sarcomere. This sliding action shortens the sarcomeres and is the basis for the shortening or contraction of the myofibrils of a muscle fiber.

In addition to the changes in shape and structure just described, chemical changes also account for the release of energy that enables a muscle to perform work. The numerous chemical reactions involved in muscle contraction are extremely complex, but the basic processes involved will be summarized briefly.

Release of Energy for Muscle Contraction. A nucleotide, *adenosine triphosphate* (ATP), the source of energy for most cellular activities, is the primary source of energy for muscle contraction. A muscle contraction may be initiated by a nerve impulse or a chemical agent such as a hormone, or it may arise spontaneously within the muscle. In skeletal muscle, nerve impulses are the usual source of stimulation.

When a nerve impulse arrives at the myoneural junction, in the region of the motor end plate, *acetylcholine* (ACh) is released. Depolarization of the end plate occurs, and an end plate potential (EPP) is established. Almost immediately (within 5 msec), ACh is destroyed through the action of acetylcholinesterase, an enzyme present in the region of the end plate. Polarity is reestablished at the end plate, and the resting condition is resumed.

Upon the establishment of an EPP at the end plate, an action potential in the cell membrane of the muscle fiber develops; this spreads throughout the fiber, initiating the contractile process. Calcium ions are released, and actin combines with myosin ATP. An enzyme, adenosine triphosphatase (ATPase), is activated, and the following reaction occurs:

$$\text{ATP} \rightarrow \text{ADP} + \text{P (inorganic phosphate)} + \text{energy}$$

In this reaction, an energy-rich terminal phosphate group is separated from ATP with explosive rapidity. The energy activates the movement of the thin filaments of actin, and contraction occurs.

There is present in the muscle a second high-energy compound, creatine phosphate (CP). The ATP is resynthesized by the transfer of a phosphate group from creatine phosphate to ADP, as shown in the following reaction:

$$\text{ADP} + \text{creatine phosphate} \rightarrow \text{ATP} + \text{creatine}$$

Neither of these reactions requires oxygen; thus, no oxygen is utilized in muscle contraction. There is sufficient ATP in a normal muscle to enable it to contract a considerable number of times, but unless ATP is quickly restored, it is soon exhausted. For a muscle to contract repeatedly, a supply of ATP is needed to replace that used. This is accomplished by creatine phosphate synthesis, oxidative phosphorylation, and glycolysis.

During periods of rest, creatine phosphate accumulates in muscle cells and is available for immediate use. If contractions continue, creatine phosphate levels drop and muscle cells must obtain their ATP from other sources. Oxidative phosphorylation is then resorted to, but its effectiveness is limited by the rate of oxygen delivery to a muscle, the availability of glucose or glycogen, and the effectiveness of the enzymes involved. The most important factor is delivery of oxygen to the tissues. In prolonged activity, such as running, the circulatory system may be unable to deliver enough oxygen to sustain continuous activity. Even with an adequate supply of oxygen, ATP production may be inadequate. Anaerobic glycolysis is then resorted to.

Through glycolysis, ATP is produced anaerobically in the breakdown of glucose to lactic acid. Pyruvic acid is also produced in the process. Glycolysis proceeds in the absence of oxygen, but it requires large amounts of glucose from the blood or from the breakdown of glycogen stored in muscle cells. Of the lactic acid produced, a part of it, about one-fifth, is oxidized to carbon dioxide and water. The remaining four-fifths enters the circulation and passes to the liver or to muscles, where it is resynthesized to glucose or glycogen.

Muscle fibers differ in their speed of contraction, which depends upon their ability to produce and utilize ATP. Three types of fibers are as follows:

1. *High-oxidative, slow-twitch fibers,* which are usually red fibers of large diameter, with a large number of actin and myosin filaments. They are slow-acting and fatigue slowly.

2. *High-oxidative, fast-twitch fibers,* which can synthesize ATP rapidly through oxidative phosphorylation. These fibers are fast-acting and fatigue at a moderate rate. They contain a protein, *myoglobin,* that resembles hemoglobin and aids in the utilization of oxygen. These fibers predominate in red muscle.

3. *Low-oxidative, fast-twitch fibers,* which produce ATP largely through the glycolytic process. These fibers are rich in glycogen but contain little myoglobin. They constitute white muscle and fatigue rapidly.

Oxygen Debt. The release of energy for muscle contraction is an *anaerobic process.* This means that it does not directly require the presence of oxygen. Since oxygen is not stored in muscle tissue, it must be brought to muscle cells by the bloodstream as it is needed. When skeletal muscles are contacting continuously, as in strenuous exercise, the chemical changes involving energy release require little or no oxygen. The organic phosphates provide the energy needed for contraction. Their supply, however, is soon exhausted and must be replenished. This is accomplished by their resynthesis. The energy for the reactions in this process comes from the oxidation of lactic acid to carbon dioxide and water, an *aerobic process* requiring oxygen. The consequence is that, during active contraction, the muscle acquires an *energy debt* that is repaid after contraction and during the resting period through the utilization of oxygen. This energy debt is regarded, and is usually referred to, as *oxygen debt.* The commonest manifestation of it is the increased respiratory activity that occurs and persists for some time after energy has been released in vigorous muscular activity such as running.

Thermal Changes. Of the total energy expended in muscle contraction, only about 35 percent is utilized for the performance of work; the balance is liberated in the form of heat, which is employed to maintain body temperature. In cold weather, the production of body heat can be increased through voluntary muscular activity (rubbing hands together, walking) or involuntarily by shivering. Conversely, in warm weather, muscular activity is deliberately curtailed to reduce heat production.

Initial heat is the heat produced in the contraction of muscles; *delayed* or *recovery heat* is that produced in the restorative processes.

ELECTRICAL CHANGES. When a muscle contracts, a minute electric potential develops. The result is an electric discharge that lasts a few milliseconds and is of an amplitude measured in microvolts. These changes can be detected and recorded by the use of a suitable apparatus.

In effect, a muscle is like a tiny battery; when contracting, it generates a current called the *action current*. This current flows from the positive to the negative pole. The active region of a muscle is electrically negative to the inactive regions. When the muscle is resting, no current is generated.

The current generated by the human heart is strong enough to be detected by a galvanometer. When electrodes are attached to the arm and the leg, the current generated by cardiac activity can be conducted to an *electrocardiograph,* which records the nature of the impulses. Its printed record, the *electrocardiogram,* enables a physician to analyze the heartbeat and to detect and diagnose various types of cardiac dysfunction.

The study of the electric properties of muscles comprises *electromyography.* Electrodes are applied to the skin over a muscle or inserted into a muscle. When the muscle contracts, either voluntarily or following stimulation, a current is generated that is transmitted to an *electromyograph.* This instrument amplifies the current, and responses are recorded by a pen-writing recorder, on motion picture film, in a cathode ray oscilliscope or by a speaker system. A written record is called an *electromyogram.* Electromyography is utilized in the determination of the exact function of a muscle, the detection of abnormal conditions within a muscle, and the study of neuromuscular relationships and disorders.

Factors Involved in Muscle Dysfunction. There are several sites that may be the seat of the trouble in muscle failure or muscle malfunction: (1) the muscle itself, (2) the connection between nerve and muscle (myoneural junction), (3) the nerve supplying the muscle, (4) the spinal cord, and (5) the brain.

THE MUSCLE ITSELF. Factors that may involve the muscle directly are improper development, injury (as from a cut or blow), improper or inadequate nutrition, the presence of an abnormal structure (such as a tumor), the presence of foreign organisms (such as bacteria or trichinella), and the effects of toxic substances.

THE CONNECTION BETWEEN NERVE AND MUSCLE. A factor in muscular dysfunction is the failure of acetylcholine formed at the myoneural junction to induce muscle contraction, a condition that prevails in myasthenia gravis. Curare, a drug used by South American Indians for poison darts and arrows, produces its effects by acting on this junction.

THE NERVE SUPPLYING THE MUSCLE. Muscle function depends on

the maintenance of normal nerve functioning. Injury to, or severance of, motor nerve fibers results in paralysis and muscle atrophy.

THE SPINAL CORD. Impulses originating in the brain pass through nerve tracts in the spinal cord and thence through nerves to the muscle. Injury to, or disease in the conducting pathways of the cord, or destruction of motor cells in the gray matter of the cord (as in poliomyelitis), lead to muscle dysfunction.

THE BRAIN. Organic disorders in the motor centers of the brain, where impulses initiating movement arise, may bring about abnormal muscle functioning. *Cerebral palsy* in children is a congenital condition characterized by disturbances in voluntary movement. *Cerebral apoplexy,* or *stroke syndrome,* usually follows a cerebrovascular accident (CVA), which may result from an aneurysm, hemorrhage, embolism, thrombosis, or vascular insufficiency. Interference in blood supply to the motor areas of the brain results in destruction or malfunction. Hemiplegia (paralysis on one side of the body) is a common result.

Muscle Tone (Tonus). Both nonstriated and striated muscles are normally in a mild state of contraction, the condition called *tonus,* or *muscle tone.* In skeletal muscles, muscle tone depends on a steady stream of impulses to the muscles. Its basis is apparently a reflex action (stretch reflex) in which impulses initiated in receptors in muscles and tendons are being continually carried by afferent nerves to the central nervous system, where connections are made with efferent fibers. These efferent fibers then conduct impulses to muscles, stimulating them to contract slightly.

If a muscle exhibits little resistance to stretch, it is said to be *flaccid,* and a state of *hypotonia* exists. If resistance to stretch is high, the muscle is *spastic,* and a state of *hypertonicity* exists.

Muscle tone varies in accordance with existing conditions. It disappears when the nerves leading to a muscle are severed or destroyed. It is diminished or may be entirely lacking during sleep or in anesthesia. Emotional states have a profound effect on muscle tone, increasing it in periods of excitement, anticipation, and anxiety, decreasing it in grief, worry, or depression. The state of health also greatly affects tonus, which is lower in ill health. Muscle tone is also amenable to conscious regulation, as when one "braces" oneself for a leap or to receive a shock (increase in tonus), or in relaxation (decrease in tonus).

In smooth muscles, tone depends only to a limited extent upon nervous impulses from the central nervous system. Impulses over the two divisions of the autonomic nervous system normally result in a balanced action, excessive contractions induced by activity of one division being counteracted by impulses over the other. However, smooth muscles that have been deprived of their nerve supply are still capable of exhibiting various degrees of tone, probably due to substances in the fluids bathing the cells. Hormones, especially epinephrine and oxytocin, and other substances, such as prostaglandins, markedly influence the degree of contractility. The salt composition of body fluids is also a factor.

Effects of Training and Work on the Muscles. Activity of skeletal muscles brings about profound effects in nearly all other systems of the body. The rate and force of the heartbeat, the rate of respiration, the amount of perspiration produced, the appetite, and the development of the skeletal framework all respond to activity by the skeletal muscles. These effects are either temporary or permanent, depending on the intensity or the duration of the activity. But of equal importance is the fact that the amount of training and work performed by a muscle is reflected in the muscle itself: in its size and structure, its strength, and its efficiency.

EFFECT OF TRAINING ON SIZE AND STRUCTURE OF MUSCLES. Muscles that remain unused tend to atrophy; those that are used excessively tend to hypertrophy. This is due to changes in the *sarcoplasm* of the individual fibers and *not* to an increase in the *number* of fibers. Muscle fibers that have been destroyed can regenerate only to a limited degree. In injuries of extensive nature, muscle tissue is replaced by connective (scar) tissue. In muscles that are used excessively, the amount of connective tissue between the fibers increases, making the surrounding structure tougher.

EFFECT OF TRAINING ON STRENGTH OF MUSCLES. Strength (capacity to do work) is increased by training. This is accounted for in several ways: (1) increase in the size of the muscle; (2) improved coordination, in which antagonistic muscles are completely relaxed at the proper time and thus do not impede the functioning of the acting muscles; and (3) improved functioning in the cortical region of the brain, where the impulses that bring about contraction of the muscles are initiated.

EFFECT OF TRAINING ON THE EFFICIENCY OF MUSCLES. Muscular efficiency is attained by (1) performance of work at the proper rate and with average loads, (2) improved coordination of all muscles involved in an activity, (3) improved adjustments of the circulatory and respiratory systems to the requirements of the muscular activity, (4) improved movement at joints involved in the activity, and (5) elimination of excess fat.

THE SKELETAL MUSCLES

The skeletal (voluntary) muscles include all the muscles that are attached to and that move the skeletal parts (Fig. 6-8). Skeletal muscles also form the walls of the oral, abdominal, and pelvic cavities and are attached to movable structures such as the lips, the eyelids and eyeballs, the skin, and the tongue. As previously shown, there are two principal muscle groups:

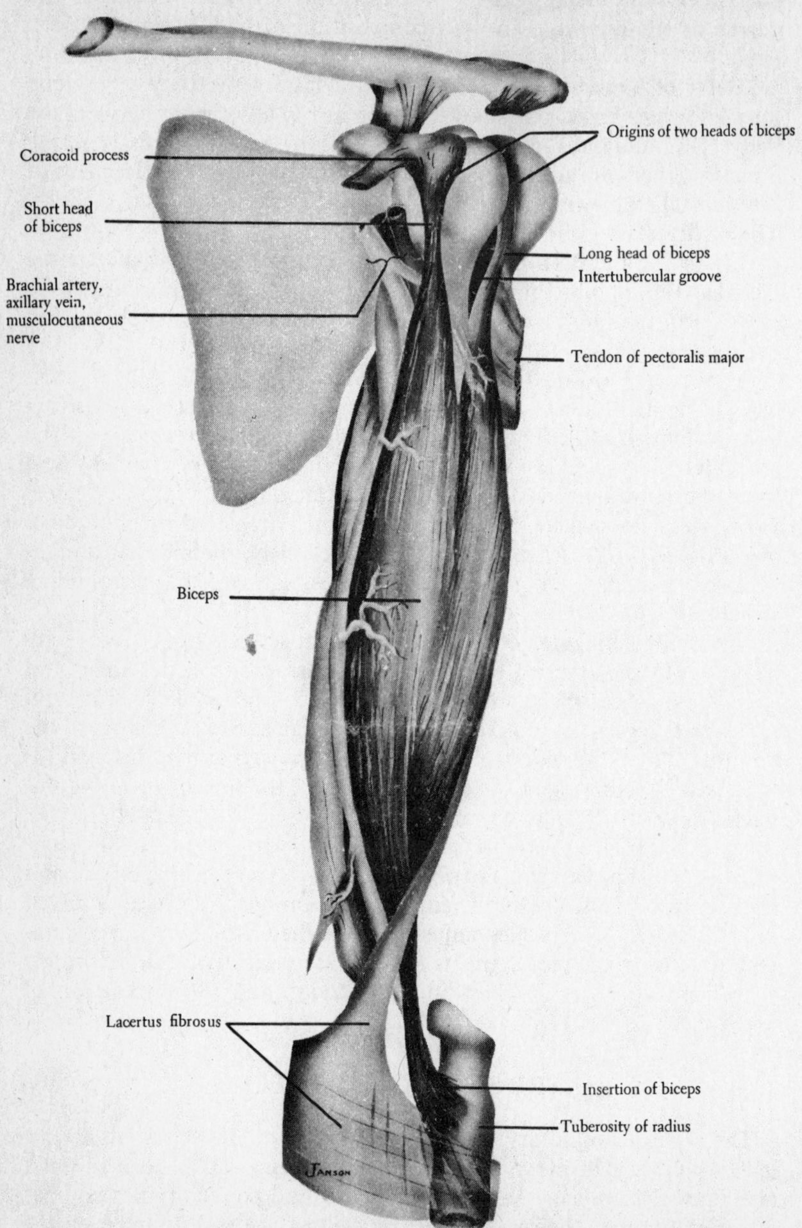

Fig. 6-8. An entire muscle.

1. *Axial muscles:* those of the head, face, neck, and trunk.
2. *Appendicular muscles:* those of the extremities.

Structure of a Skeletal Muscle (Fig. 6-9). A skeletal muscle contains many fibers bound together in primary bundles called *fasciculi.* Each fiber is invested by delicate reticular tissue, the *endomysium;* the sheath of connective tissue that surrounds the bundle is the *perimysium.* Primary bundles are grouped together to form secondary bundles, and these form tertiary bundles. The entire muscle is enclosed in a sheath of connective tissue, the *epimysium,* a part of the *deep fascia.*

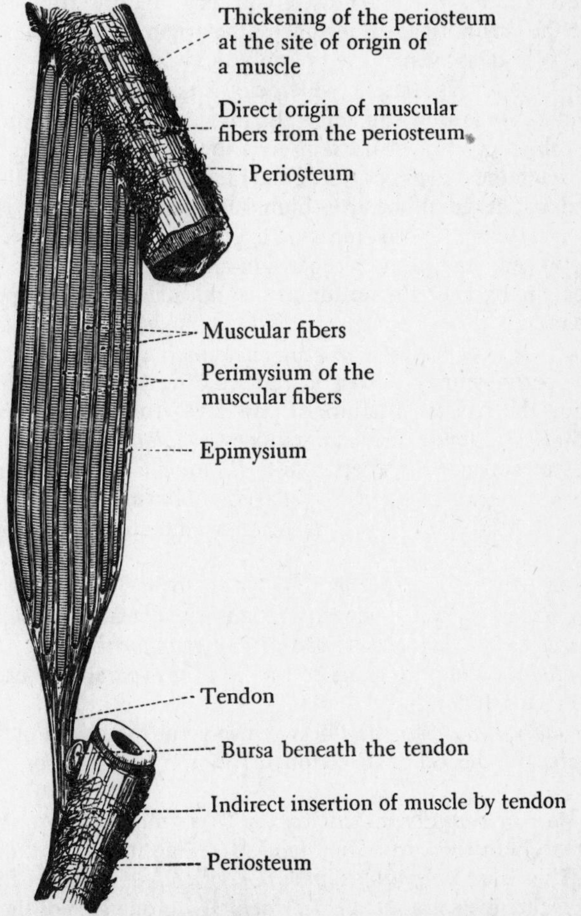

Fig. 6-9. Diagram of the structure of a muscle and modes of attachment to bone. (Reprinted with permission of The Macmillan Company from C. V. Toldt and A. D. Rosa, *Atlas of Human Anatomy for Students and Physicians,* 1948.)

Blood and Nerve Supply of Skeletal Muscles. Muscles are abundantly supplied with *blood vessels.* Capillaries entwine the individual fibers and, through anastomoses, form an interlacing network throughout the muscle. In general, the vessels run parallel to the muscle fibers. *Nerve fibers* are very numerous in striated muscles; they are myelinated afferent and efferent fibers from the cerebrospinal nerves.

Attachments of Skeletal Muscles. To perform its work, a muscle requires attachment at two points: an *origin,* the relatively fixed point of attachment (usually the one nearer to the midline of the body), and an *insertion,* the movable end of the muscle (usually the one farther from the midline). When a muscle contracts, the insertion is brought nearer to the origin. In some instances, the origin is fixed absolutely and only one type of movement is possible in contraction. An example of this is the temporalis muscle, which closes the jaw. In other instances, either end of the muscle can act as the insertion, depending on which is held in a fixed position. An example is the biceps muscle; in ordinary activity, when the biceps contracts, the radius, on which it is inserted, is supinated and flexed on the arm (humerus). But when one is "chinning" oneself on a bar, the radius remains fixed and the arm is flexed on the forearm. In general, however, the origin and the insertion are predetermined on the basis of the customary or normal movement performed by the muscle.

The *means* of attachment are *direct* and *indirect.* In direct attachment, the perimysium or fascia is attached directly to the periosteum of a bone or the perichondrium of cartilage. Indirect attachment is achieved through tendons or aponeuroses. *Tendons* are dense, round or flat cords of collagenous fibers, white in color and varying in length; they are extremely strong and inelastic. *Aponeuroses* are broad, flat sheets of connective tissue, white in color and similar in structure to tendons.

Shape of Muscles. The general shape or form of a muscle depends upon the arrangement of fibers in the muscle. In general there are three types: *parallel, fan-shaped,* and *pennate* or *penniform.*

In *parallel muscles,* the fibers run more or less parallel to each other, as in the rectus abdominis.

In *fan-shaped muscles,* the fibers converge at the origin or insertion and diverge in the other direction, as in the trapezius or pectoralis major.

In *pennate* or *penniform muscles,* the fibers may be short and attach obliquely to both sides of a long tendon, giving it the appearance of a feather. This is a *bipinnate (bipenniform) muscle,* such as the rectus femoris. If the fibers are attached principally to one side of the tendon, the muscle is *unipennate,* such as the extensor digitorum longus. If there are several tendons of origin that interdigitate with several tendons of insertion, with the fibers passing obliquely from one to the other, the muscle is *multipennate,* such as the deltoid.

FASCIAE. The fasciae, superficial and deep, differ in structure and in function.

Superficial Fascia. Underneath the skin and over most of the surface of the body there is a more or less continuous sheath of connective tissue, the *superficial fascia.* It contains a considerable amount of fat, the amount varying in different parts of the body. When fat cells predominate, it becomes *adipose tissue.* In some places the fascia is dense; in others it is loose. In some regions of the body it can be separated into layers as in the lower abdominal wall; in others it is closely bound to the surrounding tissue.

The superficial fascia serves the following *functions:* (1) It connects the skin with underlying parts; (2) it forms a protective and insulating covering for the body; (3) it plays an important role in inflammatory processes by which foreign or noxious agents are counteracted; and (4) it provides a principal reservoir for the storage of food and water.

Deep Fascia. The *deep fascia* is the usually dense, inelastic fibrous tissue that forms the investing sheaths of muscles. In certain parts of the body, the fascia is continued between the individual muscles as *intermuscular septa.* It is thick and well developed in some regions; in others it is thin and hardly distinguishable from the epimysium.

The splitting and fusing of deep fascia into compartments, planes, and clefts makes it possible to predict the course of the movement of pus, the localization of pathology, and surgical procedure.

The deep fascia serves the following *functions:* (1) It forms investing coverings for the muscles; (2) in some cases, it attaches the muscle to bone or cartilage; (3) it separates individual muscles, thus facilitating their movement by avoiding mutual interference; and (4) it encloses blood and lymphatic vessels and nerves.

The deep fascia is given a specific name appropriate to the region of the body in which it is found: *axillary fascia* of the armpit, *orbital fascia* of the eye, *cervical fascia* of the neck, *palmar fascia* of the hand.

Subserous Fascia. The *subserous fascia* or *visceral fascia* forms the fibrous layer of serous membranes (pleura, pericardium, peritoneum) covering and supporting the viscera and serving to attach the parietal layers of the serous membranes to the deep fascia of the internal surface of the body wall.

Bursa. A bursa ("pocket") is a space within the areolar connective tissue that forms a closed sac containing a small amount of fluid *(synovia).* Bursae are generally located at places where muscles or tendons pass over hard structures. They lessen friction and facilitate movement. Bursae are classified as follows:

1. *Subtendinous:* found where tendons rub against bone, cartilage, or other firm structures (e.g., between the tendon of the triceps and the olecranon).

2. *Articular:* found between bones in certain joints (e.g., between the dens and the transverse ligament of the atlas).

3. *Subcutaneous:* found between the skin and bony prominences (e.g., at the tip of the elbow, over the kneecap, and over the acromion).

4. *Submuscular:* found between muscles and bony prominences (e.g., between the gluteus maximus and the great trochanter).

Tendon Sheaths. When a tendon passes over a bony surface or traverses an area of possible friction or compression before attaching to its insertion, it may pass through a tubular bursa called a *tendon sheath.* Such a sheath is actually a double tube whose inner section (the *visceral tube*) lies next to and adheres to the tendon and is separated from the outer or *parietal tube* by a space filled with synovial fluid. Tendon sheaths are found in the hands and the feet and in the intertubercular groove.

The Mechanics of Skeletal Muscular Action. To recapitulate, muscles produce movement by contraction. When a muscle shortens (contracts), ordinarily the insertion is pulled nearer the origin, and the part on which the muscle is inserted moves. If the part is a bone of the skeleton, movement will occur at the joint located between the origin and the insertion.

LEVERAGE. Most skeletal muscles involve the principle of leverage, the bones serving as levers. A *lever* is a simple machine by which work is accomplished. It consists of a barlike structure that moves on a fixed point or *fulcrum.* When an *effort* or a force is applied to one part of the lever, another part moves. Any force or weight that opposes this movement constitutes the *resistance.*

Types of Levers. There are three classes of levers.

Class I: The fulcrum lies between the point where force is applied and the resistance.

Class II: The resistance lies between the fulcrum and the point where force is applied.

Class III: The force is applied at a point between the fulcrum and the point of resistance.

The *advantages* gained through leverage, depending of course on the position of the fulcrum in relation to the point where force is applied, are (1) speed, gained at the expense of power; (2) power, gained at the expense of speed; and (3) a change in the direction of movement.

Fig. 6-10 shows examples of the three types of levers. Nearly all levers involved in skeletal action in the body are of classes I and III.

THE EFFECT OF GRAVITY UPON MUSCLE ACTION. Many movements of the body result from the effects of gravity, and muscles often act to resist such forces. For example, when the forearm is flexed, the biceps and brachialis muscles contract to maintain its position. If these muscles relax, the forearm is extended as a result of the effects of gravity; the triceps is not functioning. However, the biceps and brachialis relax grad-

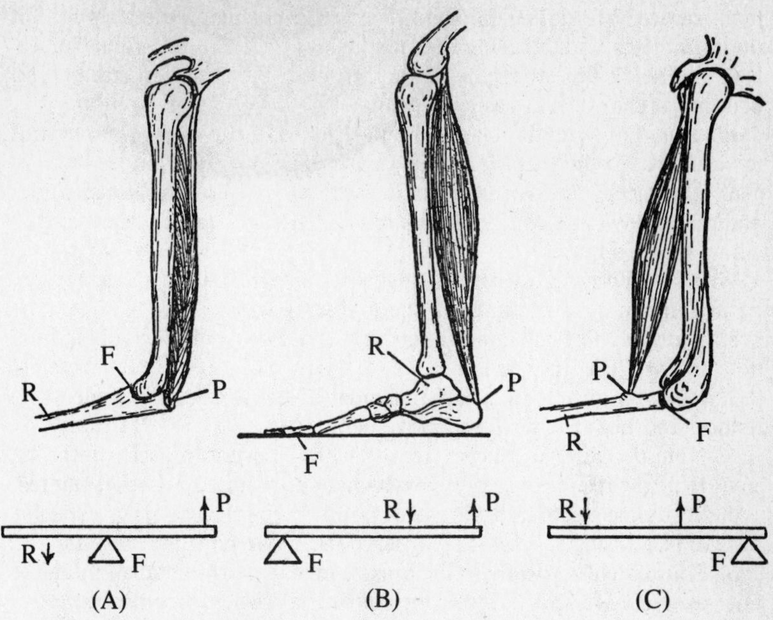

Fig. 6-10. Simple levers. *(A)* Class I lever (triceps muscle extending arm at elbow). *(B)* Class II lever (gastrocnemius muscle raising weight of body on toes). *(C)* Class III lever (biceps muscle flexing arm at elbow). Abbreviations: *F,* fulcrum. *P,* power. *R,* resistance or weight lifted. (Adapted from G. Alexander and D. G. Alexander, *Biology,* 9th ed., 1970, Barnes & Noble Books, a division of Harper & Row, Publishers.)

ually, thus preventing too precipitous a movement. In sitting down, the quadriceps femoris of the thigh and the gluteus maximus of the buttocks act in the same way to prevent the forceful, jolting movement that would otherwise occur. An understanding of the action of gravity is essential to an understanding of muscle function since the forces of gravity may aid, hinder, or sometimes replace the forces operating in muscle contraction.

Classification of Skeletal Muscles as to Type of Action. In accomplishing their primary objective of performing work, muscles fall into four classes: prime mover, antagonist, synergist, and fixator.

Prime Mover. This muscle bears the principle responsibility for a specific action, as the deltoid muscle in the abduction of the arm.

Antagonist. This muscle produces the opposite movement. The latissimus dorsi and pectoralis, for example, are antagonists of the deltoid; the triceps is the antagonist of the biceps.

Synergist. When a prime mover passes over two joints, the synergistic muscle helps to stabilize the movement of one joint so that the movement desired at another joint may be accomplished. Synergists inhibit undesirable or encumbering movements. As an example, when the fingers are being flexed by the flexor muscles, the muscles that inhibit bending at the wrist act as synergists.

Fixator. This muscle "fixes" the position of a limb when movement is occurring in the distal portion. In the example given under "Synergist," the biceps and triceps muscles are fixators, holding the shoulder and the elbow in position when movements are taking place at the elbow or the wrist.

WAYS OF DETERMINING THE SPECIFIC ACTION OF MUSCLES. Muscular action can be determined in the following ways:

1. Watch or feel the contraction of a muscle in a living subject and note the resulting movement.
2. Isolate a muscle in an animal cadaver, then pull the muscle or its tendon and note the resulting movement.
3. Note the clinical effects of paralysis of a certain muscle.
4. Note the results of surgery in which tendons have been transplanted.
5. Stimulate electrically the motor point of a nerve leading to a specific muscle in a living subject, and observe the movement that results.
6. Stimulate electrically living muscle in an experimental animal, and note the electrical changes in action potentials when it contracts.
7. In electromyography, note action potentials and the resulting activity correlated with the specific muscle stimulated.

Language for Description of Skeletal Muscles. For the sake of universal understanding and ease of reference, it is essential to adopt a special terminology and a uniform pattern of presentation in describing specific muscles.

DESCRIPTIVE TERMINOLOGY. The special terminology for the description of muscles includes the names of the muscles and terms referring to their action.

Names of Muscles. The etymological background of the name of a muscle often reveals one or more facts about it, as in the following examples:

Shape—rhomboideus, trapezius, serratus, deltoid
Position—pectoralis, intercostal, latissimus dorsi, supraspinatus
Shape and Position—external oblique, rectus femoris
Size and Position—vastus lateralis
Number of Heads of Origin—biceps, triceps, quadriceps
Origin and Insertion—stylohyoideus, sternothyroideus
Action—levator scapulae, flexor carpi radialis, tensor tympani

Terms for Muscular Action. The action of a muscle is identified by the nature of the movement that occurs when it contracts. For explanations of the principal terms, see Fig. 5-37.

PATTERN OF PRESENTATION. Ease of reference is attained when the elements of description are presented consistently in the same sequence. The correct sequence is seen in the following description of the biceps brachii.

1. *Name:* Biceps brachii.

2. *Location:* In the (upper) arm, anterior to the humerus. It forms the bulk of the arm.

3. *Origin:* By two heads arising from the scapula. The *short head* arises by a flat tendon from the tip of the coracoid process. The *long head* arises from the supraglenoid tuberosity and glenoid ligament by a long tendon that passes over the head of the humerus within the joint capsule. It continues distally within the intertubercular groove of the humerus.

4. *Insertion:* By a large tendon that passes over the anterior surface of the elbow joint and is inserted on the tuberosity of the radius.

5. *Action:* Flexion and supination of the forearm. The latter of these actions is especially pronounced when the forearm is flexed and pronated. The position of the muscle is such that it can bring about movements at three joints: the shoulder, elbow, and radio-ulnar articulations. In addition to the foregoing, which are its principal actions, the biceps brachii can act as a flexor and medial rotator of the humerus. It also acts to pull the humerus toward the scapula, thus stabilizing the head of the humerus in the glenoid cavity. The long head, acting alone, abducts the humerus; the short head adducts the humerus.

6. *Nerve Supply:* Via the musculocutaneous nerve, a branch of it supplying each head. The nerve fibers come from the 5th and 6th cervical nerves.

7. *Blood Supply:* Via arteries that branch from the brachial artery. Branches of the axillary vein carry blood away from the muscle.

Table of Skeletal Muscles (Figs. 6-11 and 6-12). There are 656 muscles in the body, 327 pairs and 2 unpaired muscles. (The two unpaired muscles are the orbicularis oris and the diaphragm.) The actual number varies with different authorities, some listing as separate muscles what others regard as parts of a single muscle. The tables on the next several pages are organized into main divisions and subdivisions in the following manner:

I. *Muscles of the Head*
   A. Muscles of expression
   B. Muscles of mastication
   C. Muscles of the tongue
   D. Muscles of the pharynx
   E. Muscles of the soft palate
II. *Muscles of the Neck*
   A. Muscles that move the head
   B. Muscles that move the hyoid bone and the larynx
   C. Muscles that act on the upper ribs

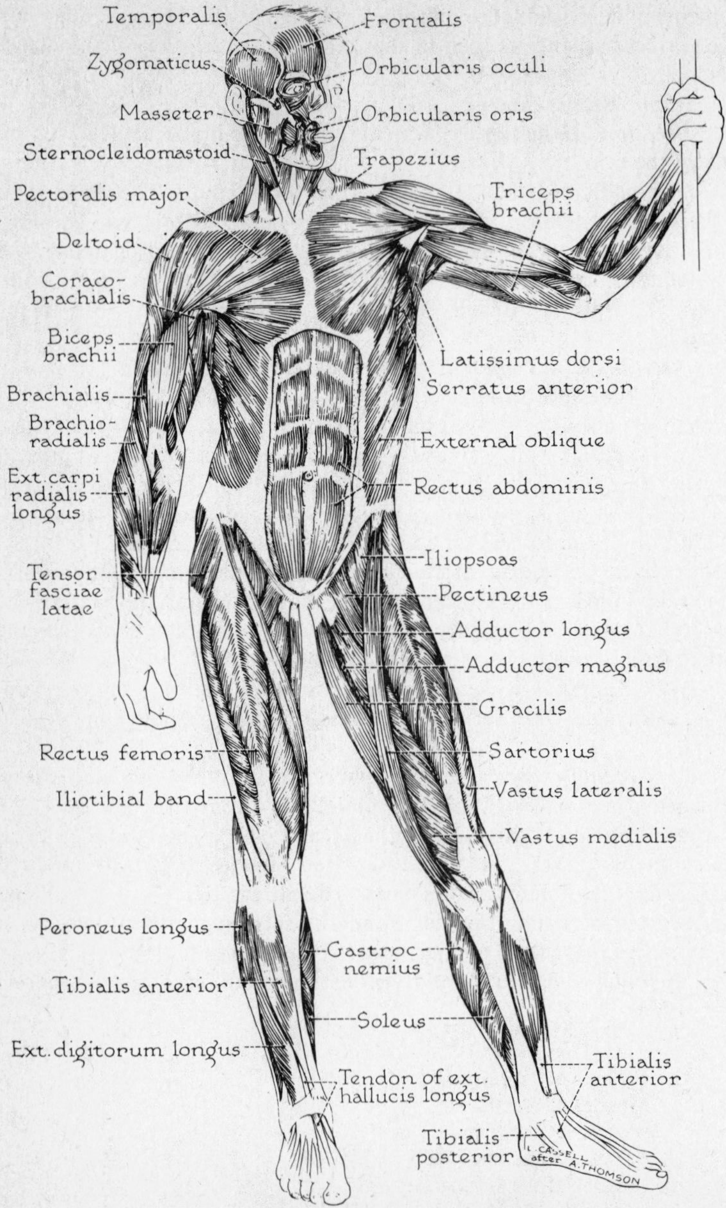

Fig. 6-11. Muscles of the body, anterior view. (Reprinted with permission of W. B. Saunders Co., Philadelphia, from B. G. King and M. J. Showers, *Human Anatomy and Physiology,* 6th ed., 1969.)

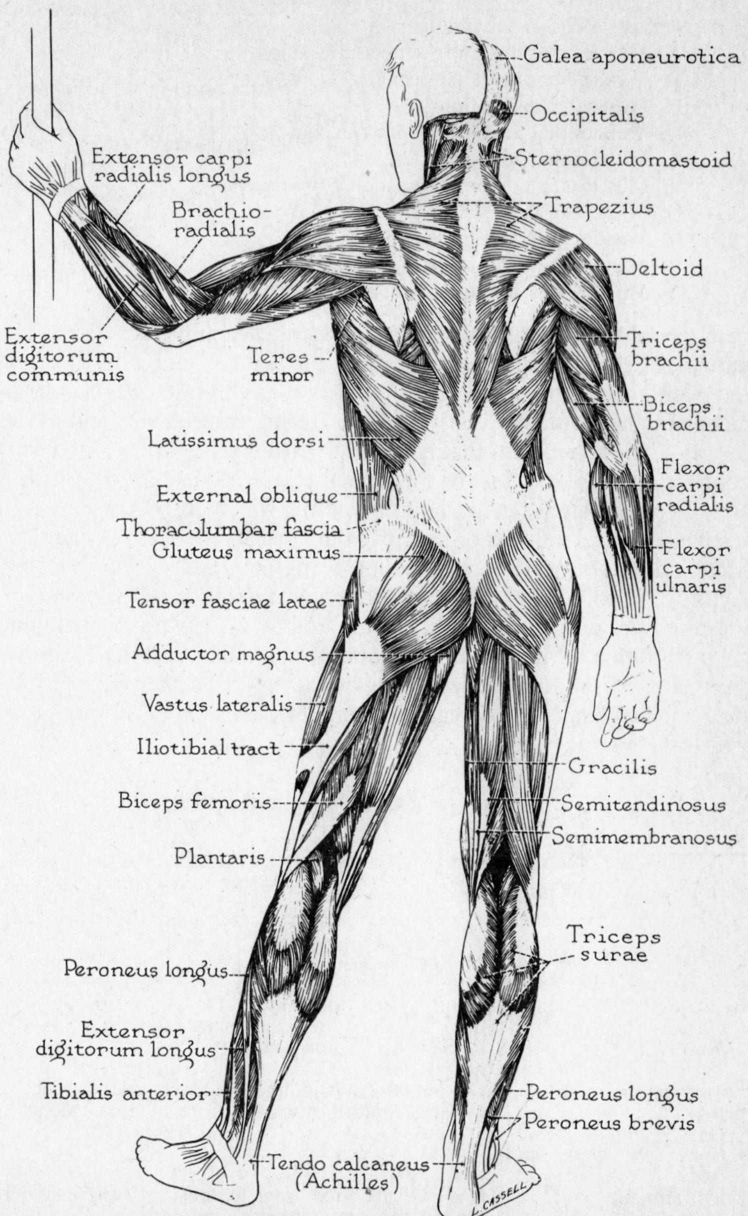

Fig. 6-12. Muscles of the body, posterior view. (Reprinted with permission of W. B. Saunders Co., Philadelphia, from B. G. King and M. J. Showers, *Human Anatomy and Physiology,* 6th ed., 1969.)

III. *Muscles of the Trunk and the Extremities*
   A. Muscles that move the vertebral column
   B. Muscles that move the scapula
   C. Muscles of respiration
   D. Muscles that act on the humerus
   E. Muscles that act on the forearm
   F. Muscles that act on the hand and the fingers
   G. Muscles of the pelvic outlet
   H. Muscles that act on the femur
   I. Muscles that act on the leg
   J. Muscles that act on the foot and the toes

In the tables only the principal muscles of each group are listed. The data included are location, origin, insertion, and action. Under *location,* the position or a brief description will be found; under *origin,* the attachment that remains relatively fixed during contraction; and under *insertion,* the attachment that is pulled toward the point of origin during contraction. The *action* is the movement performed as though the muscle were acting alone. In activities involving widely distributed areas of the body, several muscles or their parts are usually involved. As a rule, a muscle does not act alone, its action being modified by the following factors or conditions: (1) whether all or a part of the muscle is contracting (the trapezius may adduct, elevate, rotate, or depress the scapula); (2) the position of the part acted upon (the brachioradialis may be either a pronator or a supinator, depending on the position of the forearm); (3) the action of fixators and synergists; and (4) the number of joints over which the muscle passes.

I. MUSCLES OF THE HEAD

A. Muscles of Expression (Fig. 6-13). These muscles are used in the expression of fear, anger, pleasure, pain, joy, grief, and other emotions.

Name, *Location*	O—Origin I—Insertion	Action
Epicranius (frontal belly) *On each side of forehead*	O—By galea aponeurotica I—Skin of eyebrow and root of nose	Elevates eyebrow; wrinkles forehead
Epicranius (occipital belly) *On occipital bone above superior nuchal line*	O—Superior nuchal line I—Galea aponeurotica	Tenses galea aponeurotica and pulls scalp backward
Corrugator supercilii *Under frontalis, on medial side of eyebrow*	O—Frontal bone near junction with nasal bone I—Skin of eyebrow	Pulls eyebrow medially and downward; wrinkles forehead vertically
Orbicularis oculi *Encircles orbit underlying eyebrow; fibers occupy the eyelids*	O—Lacrimal bone; frontal and maxillary bones; inner canthus of eye I—Lateral palpebral raphe	Closes eyelids; tightens skin of forehead

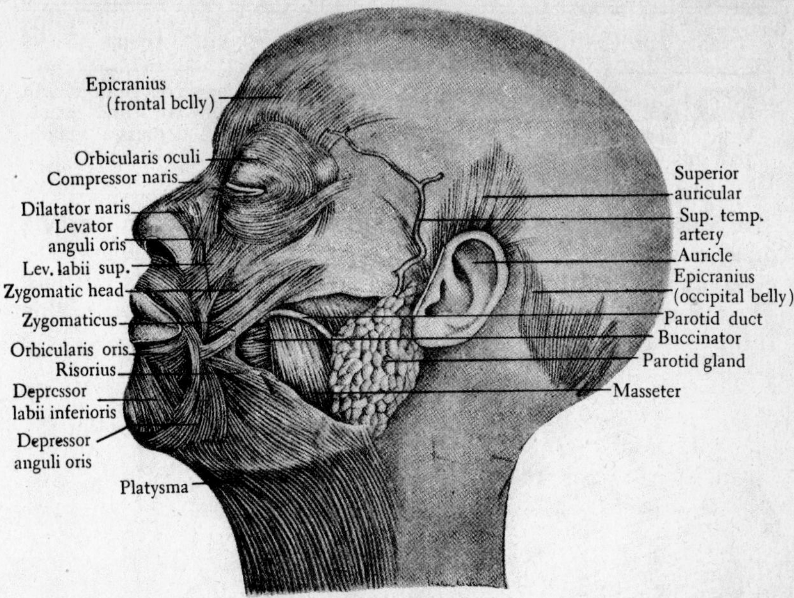

Fig. 6-13. Muscles of expression. (Reprinted with permission of C. V. Mosby Company and the author from C. C. Francis, *Introduction to Human Anatomy,* 2nd ed., 1954.)

Name, *Location*	**O**—Origin **I**—Insertion	Action
Orbicularis oris *Encircles mouth, forming fleshy portion of lips*	(A sphincterlike muscle with no definite origin or insertion)	Closes oral orifice and protrudes lip
Zygomaticus *Extends diagonally upward from corner of mouth*	**O**—Zygomatic arch **I**—Corner of mouth	Elevates corner of mouth
Triangularis (Depressor anguli oris) *Extends along side of chin*	**O**—Oblique line of mandible **I**—Orbicularis oris muscle	Depresses corner of mouth
Risorius *Extends across middle of cheek*	**O**—Fascia over masseter muscle **I**—Corner of mouth	Draws corner of mouth laterally
Quadratus labii inferioris (Depressor labii inferioris) *Extends along chin beneath lower lip*	**O**—Oblique line of mandible **I**—Lower lip	Depresses lower lip
Buccinator *Deep within skin of cheek*	**O**—Alveolar process of mandible and maxilla, and pterygomandibular raphe **I**—Lips near angle of mouth	Compresses cheek and retracts angle of mouth

Name, *Location*	O—Origin I—Insertion	Action
Platysma *Broad, thin, sheet of muscle covering side of neck and lower jaw*	O—Fascia over deltoid and pectoralis muscles I—Angle of mouth, inferior border of mandible, and skin of cheek	Draws corner of mouth downward and backward; depresses mandible

B. Muscles of Mastication (Fig. 6-14). These muscles act on the mandible, elevating it to close the jaw, depressing it to open the jaw. The elevators include the temporalis, masseter, and pterygoideus internus. The pterygoideus externus is the depressor, the digastricus (see section II.B.).

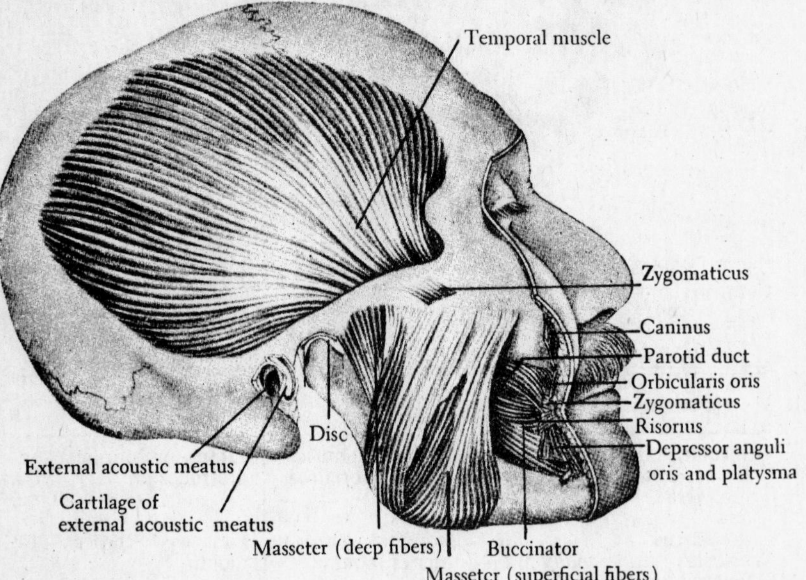

Fig. 6-14. Muscles of mastication. (Reprinted with permission of C. V. Mosby Company and the author from C. C. Francis, *Introduction to Human Anatomy,* 2nd ed., 1954.)

Name, *Location*	O—Origin I—Insertion	Action
Temporalis *Occupies temporal fossa of skull*	O—Bones forming the temporal fossa, except zygomatic I—Coronoid process of mandible	Elevates mandible (closes jaw); its posterior fibers retract mandible
Masseter *Covers lateral surface of ramus of mandible*	O—Zygomatic arch I—Ramus, angle, and posterior portion of body of mandible	Elevates mandible; fibers in deep portion assist in retraction; fibers in superficial portion assist in protraction

Name, *Location*	**O**—Origin **I**—Insertion	Action
Pterygoideus internus *Lies on medial side of ramus of mandible*	**O**—Pterygoid fossa of sphenoid; palatine, and maxilla **I**—Medial surface of ramus of mandible	Elevates and protracts mandible
Pterygoideus externus *Lies medially to temporalis*	**O**—Great wing and lateral pterygoid plate of sphenoid bone **I**—Neck of condyloid process of mandible	Protracts and depresses mandible; assists in opening mouth

C. Muscles of the Tongue. The movements of the tongue are complex. Protraction, retraction, elevation, depression, and lateral movements are possible. Changes in the shape of the tongue, whereby it becomes broad or narrow, flattened or pointed, or grooved, are accomplished to a considerable extent by *intrinsic musculature* of the tongue itself. The muscles in the following table, part of the *extrinsic musculature,* produce the larger movements of this organ.

Name, *Location*	**O**—Origin **I**—Insertion	Action
Genioglossus *Beneath anterior portion; forms bulk of tongue*	**O**—Superior genial tubercules of mandible (posterior surface of symphysis) **I**—Inferior surface of tongue from root to tip; some fibers extend to hyoid bone	Posterior fibers protract and depress tongue; anterior fibers retract tongue; elevates hyoid bone
Hyoglossus *Beneath posterior portion of tongue*	**O**—Hyoid bone, body, and greater cornu **I**—Along sides of tongue	Retracts and depresses tongue
Styloglossus *Along lateral ventral surface of tongue*	**O**—Styloid process of temporal bone **I**—Sides of tongue	Retracts tongue; unilateral contraction pulls tongue laterally

D. Muscles of the Pharynx. The muscles of the pharynx constitute the muscular portions of its wall and bring about swallowing by their constricting action.

Name, *Location*	**O**—Origin **I**—Insertion	Action
Constrictor pharyngis superior *Wall of pharynx*	**O**—Pterygoid processes of sphenoid bone, pterygomandibular ligament, and mylohyoid ridge of mandible **I**—Posterior median line of pharynx; occipital bone	Constricts pharynx during swallowing

Name, *Location*	**O**—Origin **I**—Insertion	Action
Constrictor pharyngis medius *Wall of pharynx*	**O**—Greater and lesser cornua of hyoid and stylohyoid ligaments **I**—Posterior median line of pharynx	Constricts pharynx in swallowing
Constrictor pharyngis inferior *Wall of pharynx*	**O**—Thyroid and cricoid cartilages of larynx **I**—Posterior median line of pharynx	Constricts pharynx in swallowing
Stylopharyngeus *Wall of pharynx*	**O**—Styloid process of temporal bone **I**—Lateral wall of pharynx; posterior border of thyroid cartilage	Elevates larynx and pharynx during swallowing

E. Muscles of the Soft Palate. These muscles play an important role in swallowing (deglutition) in that they (1) constrict the isthmus of the fauces, thus preventing the bolus of food from reentering the buccal cavity, and (2) close off the nasal portion of the pharynx, thus preventing food from entering the nasal cavity.

Name, *Location*	**O**—Origin **I**—Insertion	Action
Palatoglossus (Glossopalatinus) *In anterior palatine pillar (glossopharyngeal arch), which forms anterior border of tonsillar recess, and in lateral border of isthmus of fauces*	**O**—Lateral side of tongue **I**—Soft palate	Constricts faucial isthmus
Palatopharyngeus (Pharyngopalatinus) *In posterior palatine pillar (pharyngopalatine arch), which forms posterior border of tonsillar recess, and in lateral border of pharyngeal isthmus*	**O**—Soft palate and cartilage of auditory tube **I**—Lateral wall of larynx and posterior border of thyroid cartilage of larynx	Constricts fauces and elevates larynx and pharynx in swallowing
Levator veli palatini *In dorsolateral portion of soft palate*	**O**—Petrous portion of temporal bone and cartilage of auditory tube **I**—Soft palate	Elevates palate, closing off nasal pharynx
Tensor veli palatini *Lateral to levator veli palatini*	**O**—Scaphoid fossa and angular spine of sphenoid bone and cartilage of auditory tube **I**—Soft palate	Tenses velum of soft palate and opens auditory tube
Musculus uvulae *Within the uvula*	**O**—Posterior nasal spine of palatine bone **I**—Uvula	Draws uvula upward; assists in closing nasal pharynx

II. MUSCLES OF THE NECK

A. Muscles that Move the Head (Figs. 6-15 and 6-16). Flexion, extension, and rotation are the principal movements of the head. Flexion and extension occur at the atlanto-occipital articulation (between the occipital bone and the atlas); lateral flexion (bending the head and neck sideways) is also possible. Rotation occurs at the atlanto-epistrophic articulation.

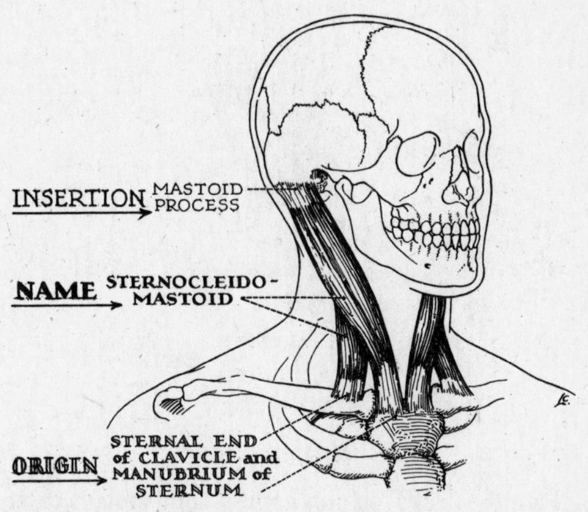

Fig. 6-15. Muscles of the neck: sternocleidomastoid. (Reprinted with permission of W. B. Saunders Co., Philadelphia, from B. G. King and M. J. Showers, *Human Anatomy and Physiology*, 6th ed., 1969.)

Name, *Location*	O—Origin I—Insertion	Action
Sternocleidomastoideus *Large muscles extending diagonally across sides of neck; when head is rotated, one stands out prominently*	O—Clavicle (median third) and sternum (manubrium) I—Mastoid process of temporal bone	When both muscles contract, head is extended (tipped backward); if head is fixed, vertebral column is flexed in cervical region (head and neck pulled forward); unilateral contraction turns face toward opposite side
Splenius capitis *Extends diagonally across posterolateral side of neck*	O—Spinous processes of lower cervical and upper thoracic vertebrae I—Mastoid process of temporal bone and lateral portion of superior nuchal line	When both contract, head is extended; when one contracts, head is rotated and face tipped upward

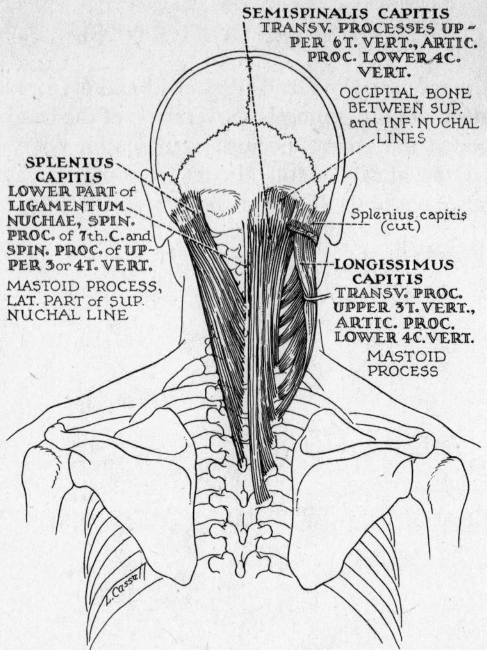

Fig. 6-16. Muscles of the neck: splenius capitis, semispinalis capitis, and longissimus capitis. (Reprinted with permission of W. B. Saunders Co., Philadelphia from B. G. King and M. J. Showers, *Human Anatomy and Physiology,* 6th ed., 1969.)

Name, *Location*	O—Origin I—Insertion	Action
Semispinalis capitis *Band of muscles consisting of several slips lying along spines of cervical and thoracic vertebrae under splenius capitis*	O—Transverse and spinous processes of lower cervical and upper thoracic vertebrae I—Lower surface of occipital bone between superior and inferior nuchal lines	When both contract, head is extended; when one contracts, head is rotated slightly and face tipped upward
Longissimus capitis *Long, bandlike muscle lying under splenius capitis and lateral to semispinalis capitis*	O—Transverse processes of upper four thoracic vertebrae and articular processes of last four cervical vertebrae I—To mastoid process by a short tendon	When both contract, head is extended; when one contracts, head is rotated and inclined
Rectus capitus anterior *A short, flat, muscle superior to lateral mass of atlas*	O—Lateral mass of atlas I—Inferior surface of occipital bone anterior to foramen magnum	Flexes head; unilateral contraction rotates head

B. Muscles that Move the Hyoid Bone and the Larynx (Fig. 6-17). These muscles, which play an important role in swallowing movements, are classified as either suprahyoid or infrahyoid. The *suprahyoid group,* which lies superior to the hyoid bone, comprises the stylohyoideus digastricus, mylohyoideus, and geniohyoideus. When the *hyoid bone* is fixed, they depress the mandible (open the mouth); when the *mandible* is fixed, they elevate the hyoid and with it the larynx, thus raising the base of the tongue and forcing the food into the pharynx, the first step in deglutition; when both the hyoid bone and the mandible are fixed, these muscles assist in flexing the head. The *infrahyoid group,* which lies below the hyoid bone, comprises the sternohyoideus, omohyoideus, sternothyroidns, and thyrohyoideus. The action of the muscles is to move the hyoid and the larynx downward.

Name, *Location*	O—Origin I—Insertion	Action
Suprahyoid group		
Stylohyoideus *Long, slender muscle located within the angle of the mandible*	O—Styloid process of temporal bone I—Body of hyoid bone	Pulls hyoid bone upward and posteriorly
Digastricus *Possess two bellies with a common rounded tendon which perforates the stylohyoideus and lies below the mandible on the mylohyoid and hyoglossus muscles in the anterior triangle*	O—Anterior belly on inner surface of mandibular symphysis; posterior belly on medial surface of mastoid process of temporal bone I—By an intermediate tendon to body and cornua of the hyoid	Anterior belly pulls hyoid upward and forward; depresses mandible if hyoid is fixed. Posterior belly pulls hyoid upward and posteriorly; assists in extending head if hyoid is fixed
Mylohyoideus *Thin, flat muscular sheet forming floor of mouth*	O—Medial surface of body of mandible I—Hyoid bone and median raphe; extends from mandibular symphysis to hyoid	Elevates hyoid bone and raises floor of mouth; when hyoid is fixed, depresses mandible
Geniohyoideus *Parallels anterior belly of digastricus; covered by mylohyoideus*	O—Inferior mental spine of mandible I—Body of hyoid bone	Elevates hyoid bone and pulls it anteriorly

Name, *Location*	O—Origin I—Insertion	Action
Infrahyoid group		
Sternohyoideus *Medially located in anterior portion of neck*	O—Medial end of clavicle and posterior surface of manubrium of sternum I—Inferior border of body of hyoid bone	Depresses hyoid bone; in forced inspiration, assists in raising sternum

Name, *Location*	O—Origin I—Insertion	Action
Omohyoideus *Slender muscle lying lateral to sternohyoideus and obliquely crossing sternothyroideus*	O—Superior border of scapula I—Body of hyoid bone	Depresses hyoid bone
Sternothyroideus *Lateral and internal to sternohyoideus; lower end partially covered by sternocleidomastoideus*	O—Posterior surface of manubrium and 1st costal cartilage I—Thyroid cartilage of larynx	Depresses larynx and hyoid bone
Thyrohyoideus *Short, broad muscle between hyoid bone and larynx*	O—Thyroid cartilage of larynx I—Body and greater cornua of hyoid bone	Draws hyoid bone toward larynx or larynx toward hyoid

C. Muscles that Act on the Upper Ribs (Fig. 6-18). These include the scalenus anterior, scalenus medius, and scalenus posterior. All three are located in the neck beneath the sternocleidomastoideus. When the neck is fixed, these muscles elevate the first two ribs, assisting in forced inspiration. When the thorax is fixed, bilateral action flexes the neck, and unilateral action bends the neck and head to the side.

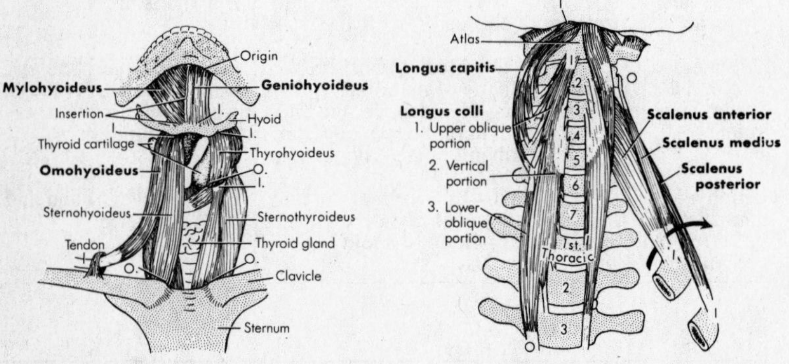

Fig. 6-17. Muscles that act on the hyoid bone and the larynx.

Fig. 6-18. Muscles that act on the upper ribs. The arrow indicates the course of the subclavian artery.

(Reprinted with permission of Lea & Febinger, Philadelphia, from J. E. Crouch, *Functional Human Anatomy,* 1965.)

Name, *Location*	O—Origin I—Insertion	Action
Scalenus anterior	O—Transverse processes of 3rd to 6th cervical vertebrae I—1st rib	(See text)
Scalenus medius	O—Transverse processes of 2nd to 6th cervical vertebrae I—1st and 2nd ribs	(See text)
Scalenus posterior *The three muscles together constitute a triangular mass extending from the first two ribs to the transverse processes of the 2nd to the 6th cervical vertebrae*	O—Transverse processes of 5th and 6th cervical vertebrae I—2nd rib; sometimes also 3rd rib	(See text)

III. MUSCLES OF THE TRUNK AND EXTREMITIES

A. Muscles that Move the Vertebral Column (Figs. 6-19 and 6-20). The principal movements of the vertebral column are flexion (forward movement or "bending") and extension (backward movement or "straightening"). These movements occur mainly in the cervical and lumbar regions. Unilateral contraction results in a sidewise movement or lateral flexion.

Name, *Location*	O—Origin I—Insertion	Action
Iliopsoas *(See* Iliacus *and* Psoas major *in section III.H)*		When femurs are fixed, acts to flex vertebral column on hips
Quadratus lumborum *Quadrilateral-shaped muscle forming portion of posterior wall of abdomen between pelvis and thorax*	O—Crest of ilium and transverse processes of lower lumbar vertebrae I—Transverse processes of upper lumbar vertebrae and inferior surface of 12th rib	Bilateral action extends vertebral column; unilateral action bends vertebral column sideways
Erector spinae (sacrospinalis) *Long muscular mass extending along vertebral column from sacrum to skull; consists of two main portions:* iliocostalis *(lateral) and* longissimus *(medial)*	O—Ilium, sacrum, and lumbar and thoracic vertebrae I—Lumbar, thoracic, and cervical vertebrae; the whole has a common origin, but its insertion is through numerous parts, some of which are given special names	Extends vertebral column; unilateral action bends vertebral column sideways or rotates it; assists in expiratory movements by depressing ribs

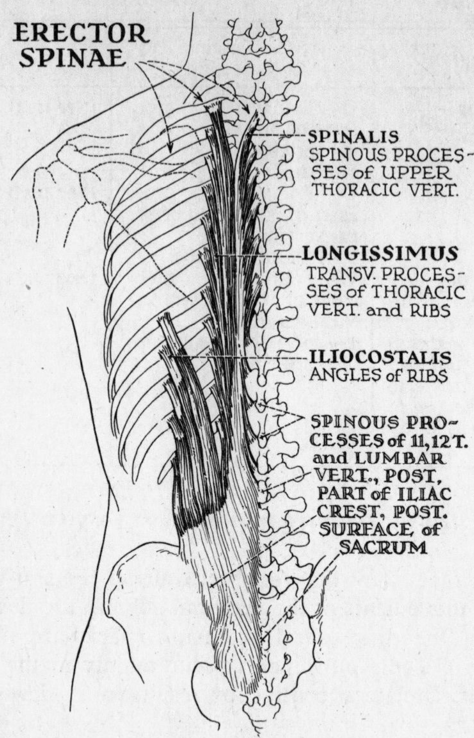

Fig. 6-19. Muscles that act on the vertebral column: erector spinae. (Reprinted with permission of W. B. Saunders Co., Philadelphia, from B. G. King and M. J. Showers, *Human Anatomy and Physiology,* 6th ed., 1969.)

B. Muscles that Move the Scapula (Figs. 6-21 and 6-22). Movements of the scapula are principally abduction (away from the midline) and adduction (toward the midline). The scapula may also be elevated and depressed, as well as rotated. The abductors include the serratus anterior and pectoralis minor. The adductors include the trapezius and rhomboideus. The elevator is the levator scapulae.

Name, *Location*	O—Origin I—Insertion	Action
Serratus anterior *Lateral surface of thoracic wall, covering posterior portions of ribs; anterior border has serrated or notched appearance*	O—Lateral surface of first nine ribs I—Vertebral border of scapula	Pulls scapula laterally and forward (abduction); elevates ribs when scapula is fixed

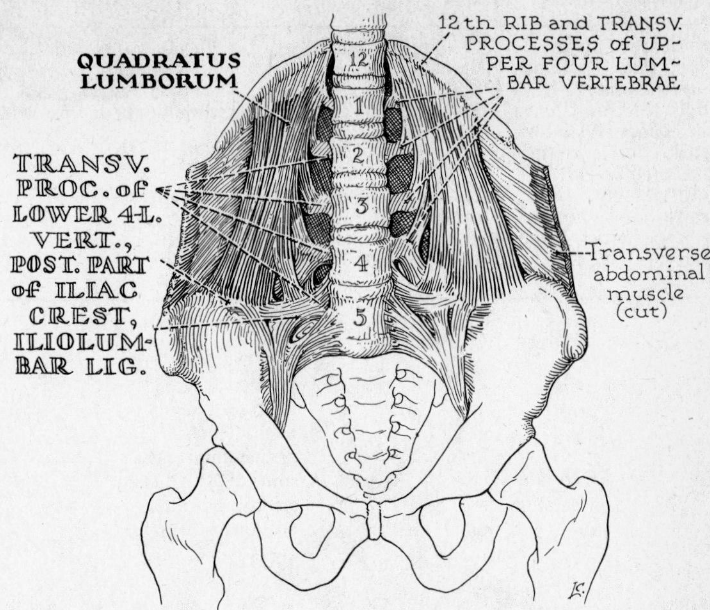

Fig. 6-20. Muscles that act on the vertebral column: quadratus lumborum. (Reprinted with permission of W. B. Saunders Co., Philadelphia, from B. G. King and M. J. Showers, *Human Anatomy and Physiology,* 6th ed., 1969.)

Name, *Location*	O—Origin I—Insertion	Action
Pectoralis minor *Under pectoralis major*	**O**—Sternal end of 2nd to 5th ribs **I**—Corocoid process of scapula	Abducts scapula and pulls it downward
Trapezius *Most superficial muscle on upper surface of back; large, triangular muscle consisting of three portions: superior, middle, and inferior*	**O**—External protuberance and superior nuchal line of occipital bone, ligamentum nuchae, spinous processes of cervical and all thoracic vertebrae **I**—Lateral third of clavicle; spine and acromion of scapula	The three portions acting as a whole adduct scapula; superior portion alone elevates and rotates scapula, raising shoulder; inferior alone depresses and rotates scapula, lowering shoulder; when scapula is fixed, contraction of superior portion extends head and pulls it laterally.

Name, *Location*	O—Origin I—Insertion	Action
Rhomboideus major *Thin, flat rhombus-shaped muscle lying under trapezius; fibers run obliquely downward and laterally from their origin*	O—Spinous processes of first four thoracic vertebrae I—Vertebral border of scapula	Elevates and pulls scapula medially; rotates scapula so that point of shoulder is depressed
Levator scapulae *Long, narrow muscle in neck under trapezius*	O—Transverse processes of first four cervical vertebrae I—Vertebral border of scapula	Elevates scapula

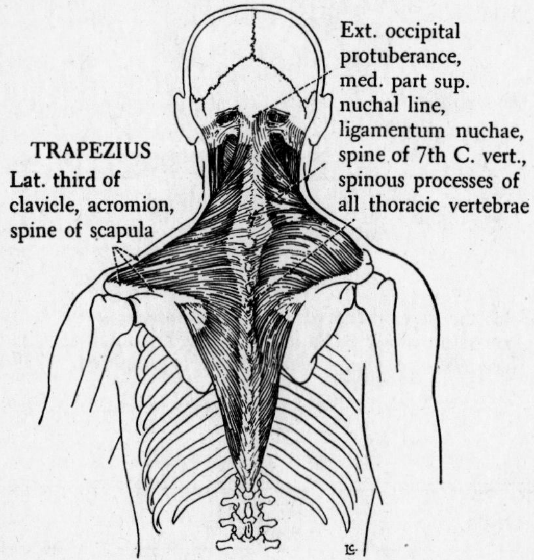

Fig. 6-21. Muscles that move the scapula: trapezius. (Reprinted with permission of W. B. Saunders Co., Philadelphia, from B. G. King and M. J. Showers, *Human Anatomy and Physiology,* 6th ed., 1969.)

C. Muscles of Respiration. The muscles involved in respiration are those that either directly or indirectly bring about changes in the volume of the thoracic cavity. Any muscle attached to the thoracic cage may be involved, such as the scalenes and the sternocleidomastoid. Those that increase the volume bring about *inspiration;* those that decrease it bring about *expiration.* The *inspiratory muscles* include the diaphragm and the external intercostals. The *expiratory muscles* include four abdominal muscles (rectus abdominis, obliquus abdominis exter-

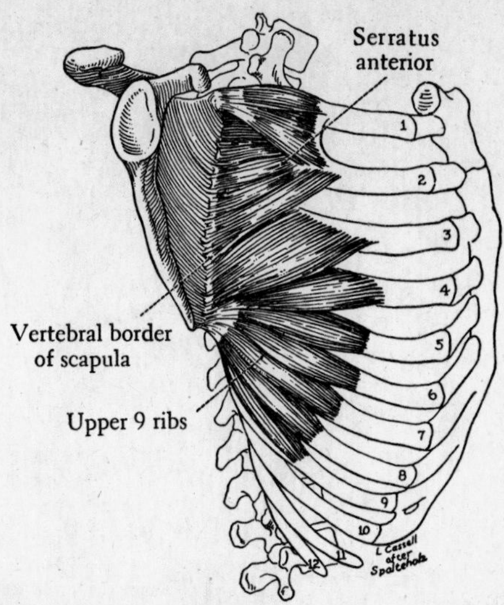

Fig. 6-22. Muscles that move the scapula: serratus anterior. (Reprinted with permission of W. B. Saunders Co., Philadelphia, from B. G. King and M. J. Showers, *Human Anatomy and Physiology,* 6th ed., 1969.)

nus, obliquus abdominis internus, and transversus abdominis) and the internal intercostals.

DIAPHRAGM (Fig. 6-23). The location and description, origin, and action of the diaphragm are as follows.

Location and Description. The diaphragm is a musculofibrous structure separating the thoracic and abdominal cavities. The convex portion forms the floor of the thorax; the concave portion forms the roof of the abdominal cavity. The central portion is fibrous, consisting of the *central tendon.* The diaphragm has three large openings: (1) the *aortic hiatus* , through which pass the aorta, thoracic duct, and azygous vein; (2) the *esophageal hiatus,* through which pass the esophagus and vagus nerve; and (3) the *vena caval foramen,* through which pass the inferior vena cava and branches of the phrenic nerve. Smaller openings transmit the splanchnic nerves and hemiazygous vein.

Origin. Inner surfaces of six lower ribs and their costal cartilages; bodies of lumbar vertebrae; posterior surface of xiphoid process of sternum.

Insertion. Central tendon of diaphragm.

Action. Depresses central tendon so that the diaphragm as a whole descends, thus increasing the vertical diameter of the thoracic cavity. This reduces intrathoracic pressure, and *inspiration* occurs; it also increases intra-abdominal pressure, which assists in expulsion of sub-

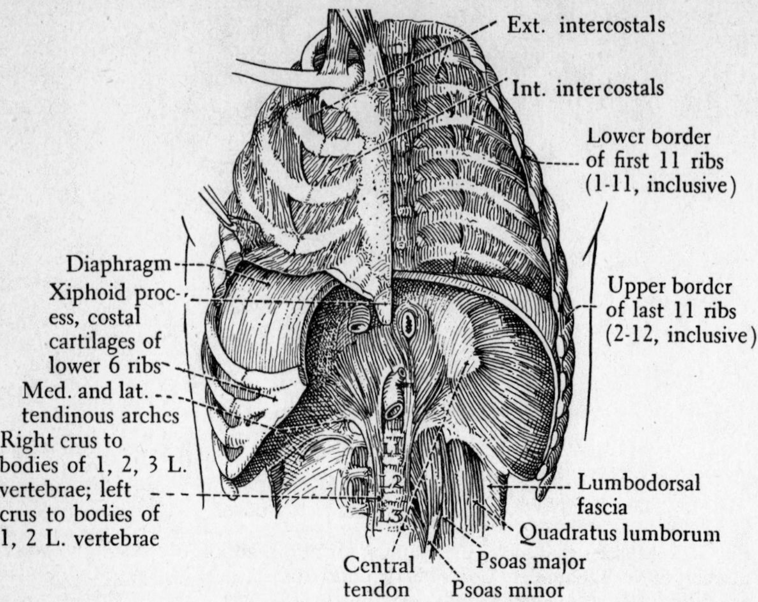

Fig. 6-23. Diaphragm and intercostal muscles. (Reprinted with permission of W. B. Saunders Co., Philadelphia, from B. G. King and M. J. Showers, *Human Anatomy and Physiology,* 6th ed., 1969.)

stances from the organs within the abdominal cavity, as in defecation, urination, vomiting, and childbirth.

Name, *Location*	O—Origin I—Insertion	Action
Intercostal Muscles		
Intercostales externi (external intercostals) *In intercostal spaces from tubercles of ribs to intercostal cartilages; fibers pass obliquely forward and downward*	O—Lower margin of each rib external to costal groove I—Upper margin of next lower rib	Elevate ribs, assisting in inspiration
Intercostales interni (internal intercostals) *Same as preceding, except that fibers pass obliquely downward and posteriorly*	O—Lower margin of each rib; near angles of ribs from internal lip of costal groove; near distal ends of ribs from both internal and external lips of groove I—Upper margin of next lower rib	Uncertain; probably depress ribs, assisting in expiration

MUSCLES OF THE VENTRAL ABDOMINAL WALL. With the exception of the rectus abdominis, these are thin sheets of muscle tissue whose fibers run at right angles to each other. They act to support the abdominal viscera and to bring about changes in intra-abdominal pressure. Compression of the abdominal viscera causes a pressure to be exerted on the underside of the diaphragm, resulting in expiration.

The rectus abdominis is a larger paired band of muscles that depresses the sternum and ribs and pulls the thorax downward, flexing the vertebral column.

Name, *Location*	O—Origin I—Insertion	Action
Obliquus abdominis externus (external oblique) *Most superficial layer; fibers pass diagonally, ventrally, and medially from their origin*	O—Surfaces of lower eight ribs I—By an extensive aponeurosis to linea alba and iliac crest; lower border of aponeurosis forms *inguinal ligament,* band extending from anterior superior iliac spine to pubic tubercle	Compresses abdomen and depresses ribs; unilateral contraction assists in flexion and rotation of vertebral column
Obliquus abdominis internus (internal oblique) *Directly beneath external oblique, its fibers running in the opposite direction (i.e., from their origin cranially and medially)*	O—Lumbodorsal fascia, iliac crest, lateral portion of inguinal ligament I—By an extensive aponeurosis to lower three ribs, linea alba, pubic tubercle, and symphysis	(Same as preceding)
Transversus abdominis *Directly beneath internal oblique, fibers running transversely*	O—Costal cartilages of lower six ribs, lumbodorsal fascia, iliac crest, and inguinal ligament I—By an aponeurosis to linea alba	Compresses abdomen and depresses sternum
Rectus abdominis *Paired band of muscles extending from sternum to pubis on either side of linea alba*	O—Crest of pubis I—Xiphoid process of sternum and costal cartilages of 5th to 7th ribs	Depresses sternum and ribs, thereby increasing intra-abdominal pressure; flexes vertebral column by pulling thorax downward; when sternum is fixed, pelvis is flexed on trunk

INGUINAL CANAL. Extending from the anterior iliac crest to the pubic tubercle is a thick band, the *inguinal ligament,* which forms the lower border of aponeurosis of the external oblique muscle. Just above the medial portion of this ligament is a weak spot in the abdominal wall, the *inguinal canal.* This canal forms a slitlike passageway that, in the male, transmits the spermatic cord and contains fibers of the *cremaster muscle,* which functions in elevation of the testis. In the female, it transmits the round ligament of the uterus. Its external opening is the *superficial* or *subcutaneous inguinal ring;* its internal opening is the *deep inguinal ring.* The canal is about 4 cm long. Protrusion of the contents of the abdominal cavity into or through this canal constitutes an *inguinal hernia.*

D. Muscles that Act on the Humerus (Figs. 6-24 to 6-29). Owing to the laxness of the ligaments and the shallowness of the glenoid fossa, the degree of mobility at this joint is great, and a wide range of movements is possible. The principal movements and the primary muscles by which they are accomplished are listed here.

Abduction: deltoideus, supraspinatus
Adduction: pectoralis major
Flexion: coracobrachialis
Extension: teres major
External rotation: infraspinatus, teres minor
Internal rotation: latissimus dorsi, subscapularis

Note the multiple action of some of these muscles, as shown in the table that follows. For example, the deltoideus and supraspinatus, which

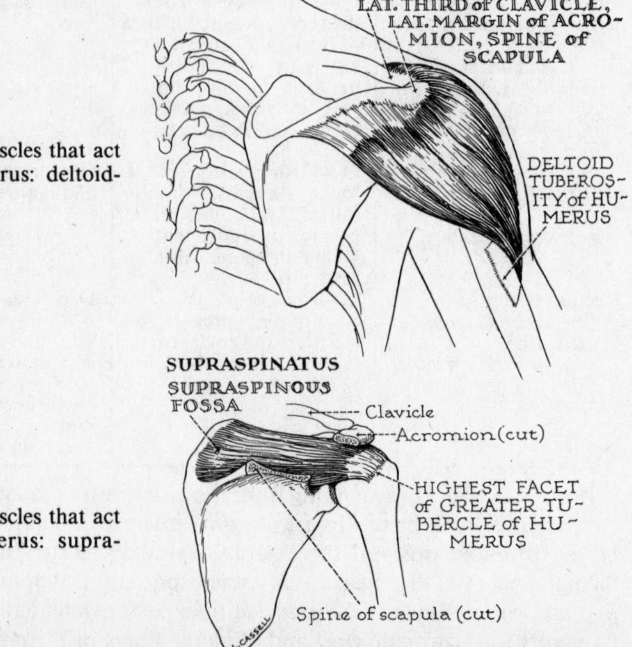

Fig. 6-24. Muscles that act on the humerus: deltoideus.

Fig. 6-25. Muscles that act on the humerus: supraspinatus.

(Reprinted with permission of W. B. Saunders Co., Philadelphia, from B. G. King and M. J. Showers, *Human Anatomy and Physiology,* 6th ed., 1969.)

abduct the arm, assist in lateral rotation. The pectoralis major, which adducts the arm, flexes and rotates it medially.

Name, *Location*	O—Origin I—Insertion	Action
Deltoideus *Thick, triangular muscle forming the shoulder prominence*	O—Acromium, spine of scapula, and lateral third of clavicle I—Deltoid tuberosity of humerus	Abducts humerus; anterior portion, acting singly, flexes and rotates humerus medially; posterior portion extends humerus and brings about lateral rotation
Supraspinatus *Occupies supraspinous fossa of scapula*	O—Surface of supraspinous fossa I—Uppermost facet of greater tubercle of humerus	Abducts humerus and rotates humerus laterally
Pectoralis major *Large, fan-shaped muscle covering upper portion of chest*	O—Medial half of clavicle; sternum, costal cartilages of 2nd to 6th ribs; aponeurosis of external oblique muscle I—Lateral lip of bicipital groove	Adducts humerus; flexes and rotates arm medially
Infraspinatus *Lies in infraspinous fossa of scapula*	O—Surface of infraspinous fossa I—Greater tubercle of humerus	Rotates humerus laterally or externally
Teres minor *Small muscle lying along axillary border of scapula*	O—Axillary border of scapula I—Greater tubercle of humerus distal to insertion of infraspinatus	Rotates humerous laterally; adducts humerus
Latissimus dorsi *Large, flat, triangular muscle superficially located on lateral posterior surface of back*	O—By lumbodorsal aponeurosis (fascia) to spinous processes of lower six thoracic and all lumbar vertebrae, posterior surface of sacrum, crest of ilium, and lower four ribs I—Floor of bicipital groove	Rotates humerus medially or internally; extends arm when arm is flexed; adducts arm when arm is abducted
Subscapularis *Occupies subscapular fossa on costal surface of scapula*	O—Surface of subscapular fossa I—Lesser tubercle of humerus	Rotates humerus medially
Coracobrachialis *Flat bandlike muscle lying in proximal portion of upper arm along short head of biceps*	O—Coracoid process of scapula I—Medial surface of humerus near middle of shaft	Flexes humerus; adducts humerus
Teres major *Lies along axillary border of scapula*	O—Axillary border of scapula at inferior angle I—Along medial lip of intertubercular groove just below lesser tubercle of humerus	Extends arm, assisting latissimus dorsi; adducts arm and rotates arm medially

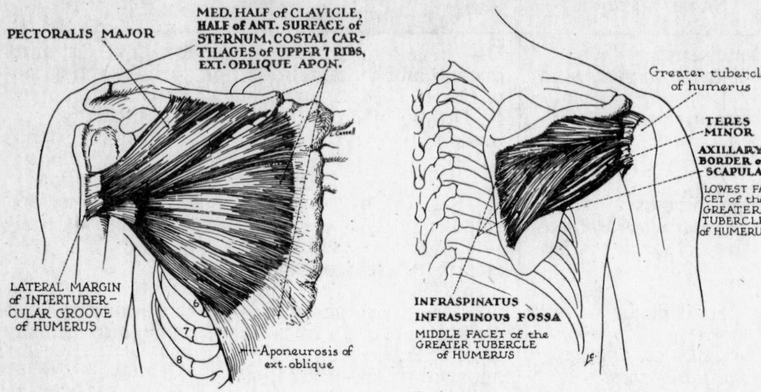

Fig. 6-26. Muscles that act on the humerus: pectoralis major.

Fig. 6-27. Muscles that act on the humerus: infraspinatus and teres minor.

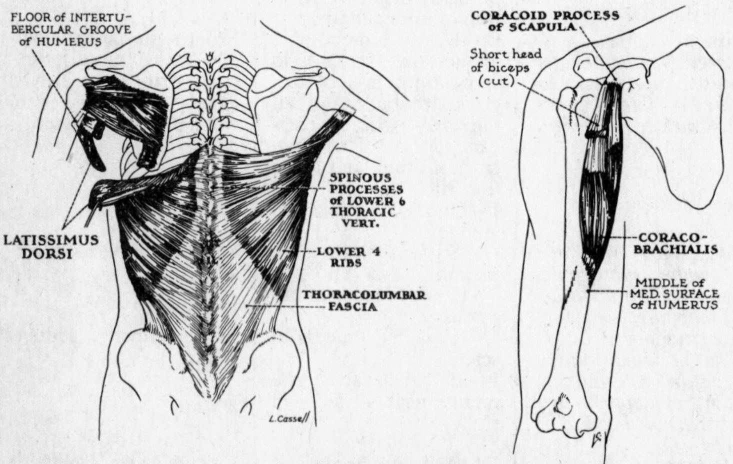

Fig. 6-28. Muscles that act on the humerus: latissimus dorsi.

Fig. 6-29. Muscles that act on the humerus: coracobrachialis.

(Reprinted with permission of W. B. Saunders Co., Philadelphia, from B. G. King and M. J. Showers, *Human Anatomy and Physiology,* 6th ed., 1969.)

E. Muscles that Act on the Forearm (Figs. 6-30 to 6-33). These muscles bring about flexion and extension at the humero-ulnar joint and pronation and supination at the radio-ulnar joint.

Flexors: biceps brachii, brachialis, brachioradialis
Extensors: triceps brachii
Pronators: Pronator teres, pronator quadratus
Supinators: Biceps brachii, supinator

Name, *Location*	O—Origin I—Insertion	Action
Biceps brachii *Large, spindle-shaped muscle forming major portion of bulge on anterior surface of upper arm; formed by union of two distinct heads*	O—*Long head:* by a long slender tendon into supraglenoid tuberosity of scapula (tendon passes through intertubercular groove); *short head:* by a tendon to coracoid process of scapula I—Into tuberosity of radius at proximal end and by an aponeurosis (semilunar fascia) to fascia of forearm	Flexes and supinates forearm; supination is especially pronounced when arm is flexed; acting singly, long head abducts humerus, short head adducts it; biceps acts to stabilize head of humerus in glenoid cavity
Brachialis *Anterior surface of humerus; partially covered by biceps*	O—Anterior surface of distal half of humerus I—Inferior portion of coronoid process of ulna	Flexes forearm
Triceps brachii *Large muscle occupying entire posterior surface of arm*	O—By three heads: middle or *long head* into infraglenoid tuberosity of scapula; outer or *lateral head* onto posterior surface of humerus above intertubercular groove; inner or *medial head* onto posterior surface of distal half of humerus I—Olecranon process of ulna	Extends forearm; long head extends and adducts arm
Brachioradialis *Long, spindle-shaped muscle lying along lateral surface of radius on volar aspect of forearm*	O—Upper two-thirds of lateral supracondylar ridge of humerus I—Lateral side of base of styloid process of radius	Flexes forearm; when forearm is extended and pronated, acts as a supinator; when forearm is flexed and supinated, acts as a pronator
Supinator *Short, flat, rhombus-shaped muscle on proximal, lateral surface of forearm*	O—Lateral epicondyle of humerus and proximal portion of ulna I—Volar and lateral surface of radius proximal to oblique line	Supinates forearm
Pronator teres *Spindle-shaped muscle extending diagonally across proximal surface of forearm on volar aspect*	O—Median epicondyle of humerus I—Lateral surface of radius	Pronates and flexes forearm

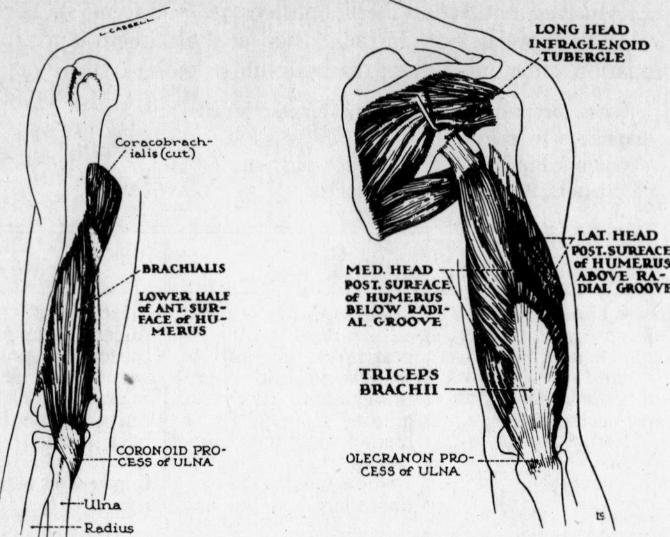

Fig. 6-30. Muscles that act on the forearm: brachialis.

Fig. 6-31. Muscles that act on the forearm: triceps brachii (posterior view of right arm).

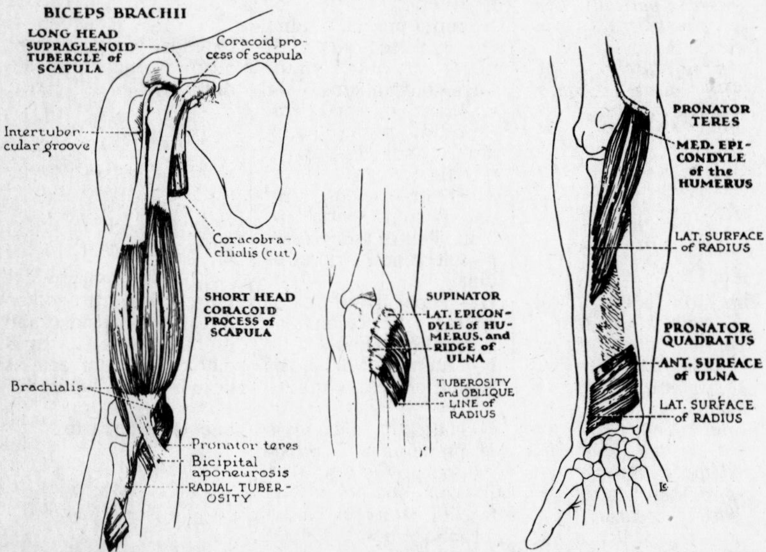

Fig. 6-32. Muscles that act on the forearm: biceps brachii and supinator muscles.

Fig. 6-33. Muscles that act on the forearm: pronator muscles (front of right forearm).

(Reprinted with permission of W. B. Saunders Co., Philadelphia, from B. G. King and M. J. Showers, *Human Anatomy and Physiology,* 6th ed., 1969.)

Name, *Location*	O—Origin I—Insertion	Action
Pronator quadratus *Short, square muscle lying at distal end of forearm; fibers are crossed by tendons of all flexors of hand*	**O**—Distal end of volar surface of ulna **I**—Distal end of volar surface of radius	Pronates forearm

F. Muscles that Act on the Hand and Its Digits (Figs. 6-34 and 6-35). There are many muscles in this group. Only a few of the superficial ones will be listed here. Flexion and extension may occur at the wrist and at the metacarpophalangeal and interphalangeal joints. Abduction and adduction of the hand may take place at the wrist; abduction and adduction of the digits may occur at the metacarpophalangeal joints. These muscles may be grouped as to their general location, as follows: those on the anterior aspect of the forearm (brachioradialis, flexor carpi radialis, palmaris longus, flexor carpi ulnaris); those on the dorsal aspect of the forearm (extensor carpi ulnaris, extensor digitorum communis, extensor pollicis brevis); and the intrinsic muscles of the hand (abductor pollicis brevis, flexor pollicis brevis, opponens pollicis, adductor pollicis brevis, abductor digiti minimi).

Name, *Location*	O—Origin I—Insertion	Action
Flexor carpi radialis *Medially located on anterior surface of forearm*	**O**—Median epicondyle of humerus **I**—Base of 2nd metacarpal	Flexes hand at wrist; flexes forearm
Palmaris longus *Medial to flexor carpi radialis*	**O**—Median epicondyle of humerus **I**—By palmar aponeurosis to volar surface of hand	Flexes hand and tightens palmar aponeurosis; flexes forearm weakly
Flexor carpi ulnaris *Most medial muscle of volar aspect of arm*	**O**—Median epicondyle of humerus and proximal two-thirds of ulna **I**—Chiefly on pisiform bone but with tendinous extensions to palmar aponeurosis and bases of 3rd, 4th, and 5th metacarpals	Flexes and adducts hand at wrist; flexes forearm
Extensor carpi ulnaris *Along dorsal surface of ulna*	**O**—Lateral epicondyle of humerus and proximal portion of ulna **I**—Base of 5th metacarpal	Extends and adducts hand; assists in extending and supinating forearm
Extensor digitorum communis *Broad, flat muscle on posterior surface of arm; gives rise to four tendons*	**O**—Lateral epicondyle of humerus **I**—By four tendons into dorsal surface of phalanges of fingers	Extends wrist and fingers; produces movements at all joints; extends forearm

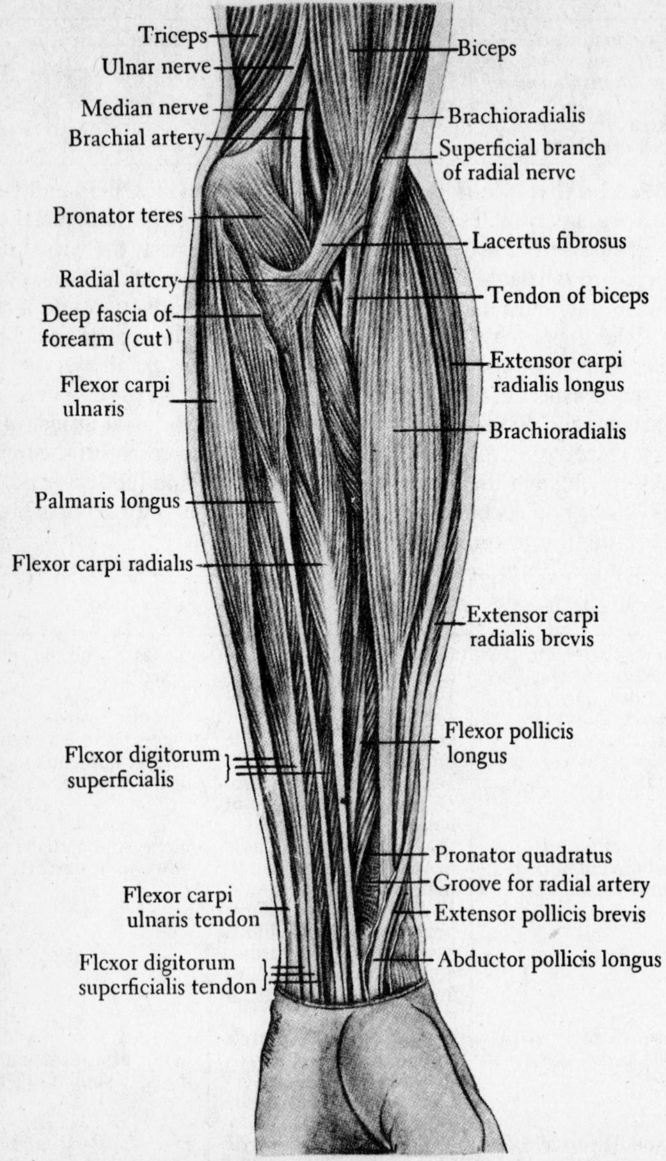

Fig. 6–34. Muscles that act on the hand and digits: superficial muscles of left forearm. (Reprinted with permission of C. V. Mosby Company and the author from C. C. Francis, *Introduction to Human Anatomy,* 2nd ed., 1954.)

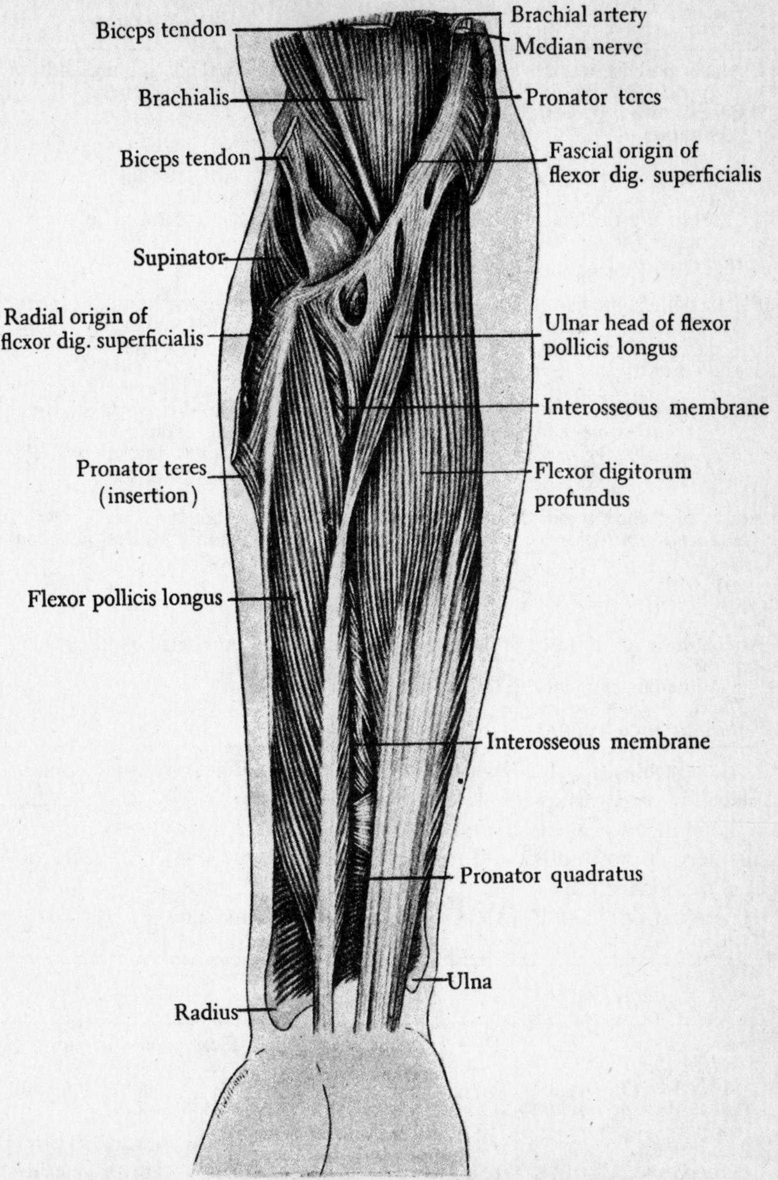

Fig. 6–35. Muscles that act on the hand and digits: deep muscles of right forearm. (Reprinted with permission of C. V. Mosby Company and the author from C. C. Francis, *Introduction to Human Anatomy,* 2nd ed., 1954.)

Name, *Location*	O—Origin I—Insertion	Action
Extensor pollicis brevis *Deep muscle lying under extensor digitorum communis*	O—Midportion of radius on dorsomedian surface I—Base of 1st phalanx of thumb	Extends thumb; adducts 1st metacarpal
Abductor pollicis brevis *Thin, flat muscle lying superficially in* thenar eminence *(fleshy portion of thumb)*	O—Trapezium and volar surface of transverse carpal ligament I—Radial side of base of proximal phalanx of thumb	Abducts thumb; flexes proximal phalanx; extends terminal phalanx
Flexor pollicis brevis *Medial to and partially covered by adductor pollicis brevis*	O—Distal row of carpal bones and transverse carpal ligament I—Base of 1st phalanx of thumb	Flexes thumb at metacarpophalangeal joint; adducts thumb
Opponens pollicis *In thenar eminence, deep to adductor pollicis brevis*	O—Trapezium and transverse carpal ligament I—Palmar surface of 1st metacarpal	Flexes 1st metacarpal and adducts it medially; rotates thumb medially in opposition
Adductor pollicis brevis *In thenar eminence*	O—Trapezium and transverse carpal ligament I—Radial side of base of proximal phalanx of thumb	Abducts and rotates thumb; flexes proximal phalanx
Abductor digiti minimi *Principal muscle of* hypothenar eminence *(fleshy portion on radial surface of palm)*	O—Pisiform bone I—Base of 1st phalanx of little finger	Abducts little finger

G. Muscles of the Pelvic Outlet (Fig. 6-36). These muscles can be placed in three groups: those of the *pelvic diaphragm* (coccygeus, levator ani, sphincter ani externus); that of the *urogenital diaphragm* (transversus perinei profundus); and the *external genital muscles* (bulbocavernosus, ischiocavernosus).

MUSCLES OF THE PELVIC DIAPHRAGM. These muscles are as follows.

Name, *Location*	O—Origin I—Insertion	Action
Coccygeus *Triangular muscle lying between spine of ischium and sacrum and coccyx*	O—Spine of ischium I—Sacrum and coccyx	Flexes sacrum and coccyx; assists levator ani in supporting pelvic and abdominal organs
Levator ani *Broad, muscular sheet forming major portion of pelvic floor; has three parts:* iliococcygeal, pubococcygeal, *and* puborectal	O—Inner side of pubis along a *tendinous arch* extending from pubic symphysis to spine of ilium I—Coccyx and raphe between coccyx and anus	Supports pelvic organs; flexes coccyx; elevates rectum during defecation; in female, constricts vagina

Name, *Location*	O—Origin I—Insertion	Action
Sphincter ani externis *External sphincter of anus*	O—Anococcygeal raphe I—Central tendinous point of perineum	Closes anal canal and anus and fixes central tendinous point of perineum

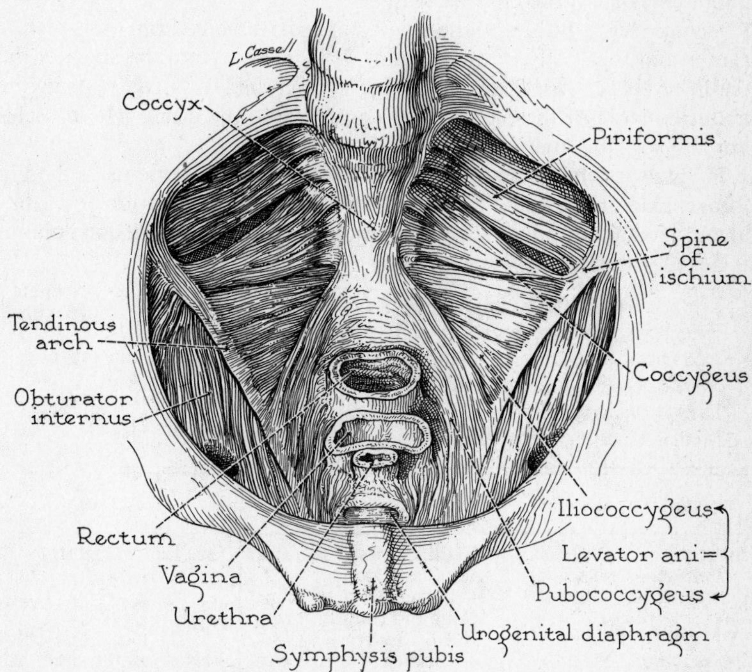

Fig. 6–36. Muscles of the pelvic floor, viewed from above. (After T. Jones and F. Netter.) (Reprinted with permission of W. B. Saunders Co., Philadelphia, from B. G. King and M. J. Showers, *Human Anatomy and Physiology,* 6th ed., 1969.)

MUSCLES OF THE UROGENITAL DIAPHRAGM. The *urogenital diaphragm* consists of the *transversus perinei profundus,* covered above and below by fascial membranes. The transversus perinei profundus originates from the junction of the pubic and ischial rami and unites with its fellow in a median raphe below and behind the membranous part of the urethra. It acts as a *sphincter* that compresses the membranous portion of the urethra in the male and the supravaginal portion of the urethra in the female.

EXTERNAL GENITAL MUSCLES. The principal external genital muscles are as follows.

Bulbocavernosus, consisting of two symmetrical halves. In the *male* it surrounds the corpus cavernosum penis and is inserted by an aponeurosis onto its superior surface. It compresses the urethra, acting to empty it; it also compresses the erectile tissue, thus assisting in bringing about erection. In the *female* its two halves lie on each side of the vestibule. By contraction they act to close the vaginal orifice. They assist in bringing about erection of the clitoris.

Ischiocavernosus, a spindle-shaped muscle arising from the inferior ramus and tuberosity of the ischium. It is inserted into the crura of the penis or the clitoris near the pubic symphysis. It assists in bringing about erection of these organs by compressing the crura. The muscles are larger in the male than in the female.

H. Muscles that Act on the Femur (Fig. 6-37). Movements at the hip joint include flexion and extension, abduction and adduction and lateral and medial rotation. When the femurs are fixed, the vertebral column is flexed, extended, or rotated.

Flexors: iliacus, psoas major (together known as *iliopsoas*)
Extensor: gluteus maximus
Adductors: pectineus, adductor longus, adductor brevis, adductor magnus
Abductors: tensor fascia latae, gluteus medius
Lateral rotator: piriformis
Medial rotator: gluteus minimus

Name, *Location*	O—Origin I—Insertion	Action
Iliacus *Triangular muscle occupying iliac fossa*	O—Iliac fossa, sacrum, sacroiliac ligament I—Lesser trochanter of femur and shaft immediately below	(See following entry)
Psoas major *Fuses with iliacus, with which it has common tendon for insertion; they are commonly grouped together as the* iliopsoas muscle	O—Bodies, transverse processes and intervertebral discs of last thoracic and lumbar vertebrae I—Lesser trochanter of femur	Flexes femur and adducts and rotates it medially; when thighs are fixed, this and iliacus act together to flex vertebral column and pelvis on femurs; unilateral action tends to rotate vertebral column and pelvis to opposite side
Gluteus maximus *Large, fleshy muscle forming prominence of buttock*	O—Posterior portion of iliac crest and lateral surface of ilium posterior to posterior gluteal line; posterior surface of sacrum and coccyx and sacroiliac ligaments I—Iliotibial band and gluteal tuberosity of femur	Extends femur and rotates it laterally; when thighs are fixed, pelvis is drawn dorsally, thus aiding in extension of trunk

Name, *Location*	O—Origin I—Insertion	Action
Pectineus *Inner side of thigh; most superior of adductor muscles*	O—Crest of pubis I—Pectineal line of femur	Adducts and flexes femur
Adductor longus *Flat, triangular muscle lying medial to pectineus*	O—Superior ramus of pubis I—On medial portion of linea aspera	Adducts and flexes femur; rotates femur laterally
Adductor brevis *Short muscle lying beneath pectineus and adductor longus*	O—Inferior ramus of pubis I—Upper third of linea aspera of femur	Adducts femur; rotates femur laterally
Adductor magnus *Large, triangular muscle lying beneath the other adductor muscles*	O—Inferior ramus and tuberosity of ischium I—Along entire length of linea aspera, medial supracondylar line, and adductor tubercle of medial epicondyle	Adducts femur; rotates femur laterally; lower portion extends thigh
Tensor fasciae latae *Flat muscle lying along lateral upper surface of thigh; fibers pass about one-third the length of the thigh*	O—By a flat tendinous band to anterior portion of crest of ilium I—Iliotibial band, which passes down lateral side of thigh, where it is closely fused with fascia lata to its termination on anterolateral surface of upper end of tibia	Tenses fascia lata; when leg is free, flexes, abducts, and medially rotates thigh
Gluteus medius *Lateral surface of ilium*	O—Anterior three-fourths of iliac crest and iliac surface between anterior and posterior gluteal lines I—Greater trochanter of femur	Abducts femur; superior portion may flex and rotate femur medially; posterior portion may extend and rotate femur laterally
Piriformis *Underneath posterior portion of gluteus medius*	O—Anterior surface of sacrum, sciatic notch of ilium, and sacrotuberous ligament I—Tip of greater trochanter of femur	Rotates laterally and abducts thigh; extends thigh weakly
Gluteus minimus *Deep-lying hip muscle covered by gluteus medius and piriformis*	O—Lateral surface of ilium between anterior and inferior gluteal lines; capsule of hip joint I—Anterior surface of greater trochanter of femur	Rotates femur medially; abducts femur

HAMSTRING MUSCLES. Three flexor muscles *biceps femoris, semimembranosus, semitendinosus* located on the posterior side of the thigh are known as the *hamstring muscles;* they are antagonists of the quadriceps femoris. The tendons of these muscles, called *hamstrings,* bound the popliteal space behind the knee.

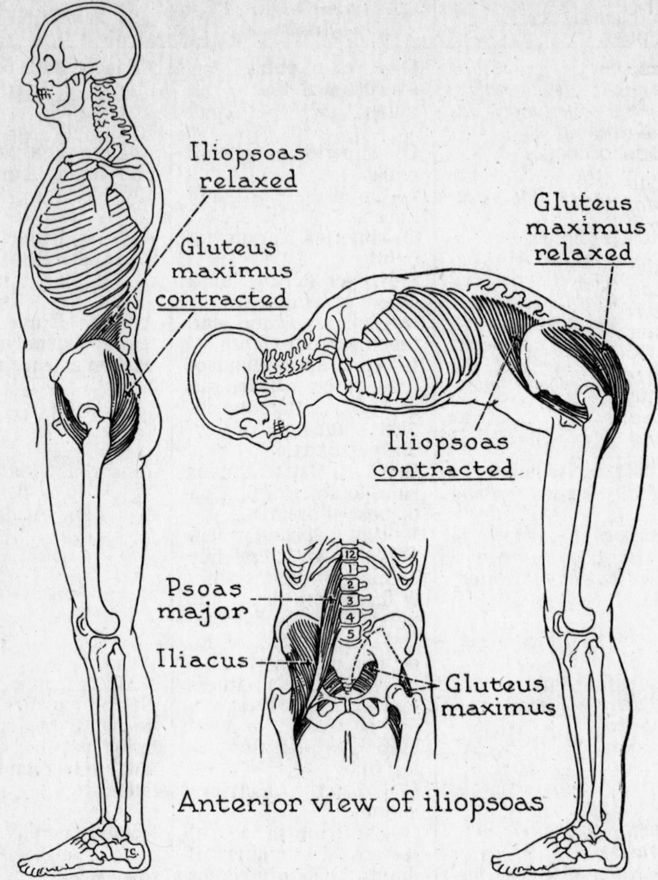

Fig. 6–37. Muscles that act on the femur and vertebral column: diagram showing action of iliopsoas and gluteus maximus with femurs fixed. (Reprinted with permission of W. B. Saunders Co., Philadelphia, from B. G. King and M. J. Showers, *Human Anatomy and Physiology,* 6th ed., 1969.)

I. Muscles that Act on the Leg. The principal actions produced at the knee joint are extension and flexion. However, some of the muscles have their insertions on the pelvic bone and consequently may bring about rotation, abduction, and adduction movements of the thigh.

Extensors: quadriceps femoris (four muscles)
Flexors: sartorius, the "hamstring" muscles, gracilis, popliteus

Quadriceps Femoris. This is a large muscular mass that forms the anterior portion of the thigh (Fig. 6-11). It is composed of four muscles that converge into a common tendon with a single insertion. The quadriceps is one of the largest and most powerful muscles of the body.

Name, *Location*	**O**—Origin **I**—Insertion	Action
Quadriceps femoris		
Rectus femoris *Most anterior of the group*	**O**—By two tendons, one to the anterior inferior iliac spine, the other to the rim of acetabulum	All four muscles extend leg; rectus femoris also flexes femur
Vastus lateralis *Lateral and posterior to rectus femoris*	**O**—Along lateral lip of upper half of linea aspera	
Vastus medialis *Medial and posterior to rectus femoris*	**O**—Medial lip of linea aspera throughout its entire length	
Vastus intermedius *Deep to rectus femoris and vastus medialis, adjacent to femur*	**O**—From anterolateral surface of femur and distal half of lateral margin of linea aspera **I**—All four muscles insert by a common tendon which passes over the knee, enclosing the patella and inserting on the tuberosity of the tibia; tendon from apex of the patella to the tuberosity is called *patellar ligament*	
Flexors		
Sartorius *Long, straplike muscle that passes diagonally across anterior and medial surfaces of thigh; called "tailor's muscle"*	**O**—Anterior superior spine of ilium **I**—Medial surface of tibia at proximal end, near tuberosity	Flexes leg at knee; flexes thigh at hip and rotates it laterally; functions in crossing of legs
Biceps femoris *On posterior lateral aspect of thigh*	**O**—*Long head:* ischial tuberosity; *short head:* lateral lip of linea aspera **I**—Lateral side head of fibula; lateral condyle of tibia	Flexes leg and rotates it externally; long head extends and adducts thigh
Semitendinosus *Medial to biceps femoris, its upper end being partially covered by it*	**O**—Ischial tuberosity **I**—Medial surface of proximal end of tibia	Flexes leg and rotates it medially; extends thigh; adducts thigh and rotates it medially
Semimembranosus *Internal to and partially covered by biceps femoris and semitendinosus*	**O**—Ischial tuberosity **I**—Posterior surface of medial condyle of tibia	Flexes leg and rotates it medially; extends thigh; adducts thigh and rotates it medially
Gracilis *Long, thin muscle on medial aspect of thigh*	**O**—Inferior rami of pubis and ischium **I**—Medial surface of tibia, just below medial condyle	Adducts and flexes thigh; rotates thigh medially; flexes leg at knee

Name, *Location*	O—Origin I—Insertion	Action
Popliteus *Short, flat, triangular muscle, deeply situated and hidden by heads of gastrocnemius*	**O**—Lateral condyle of femur **I**—Posterior surface of proximal end of tibia	Flexes leg and rotates it medially

J. Muscles that Act on the Foot and Its Digits. *Movements of the foot* occur at the ankle and the tarsal joints. *Extension* or *plantar flexion* is the downward motion in which the foot is "straightened"; *flexion* or *dorsiflexion* is the movement in which the foot is bent upward. These movements take place at the ankle joint between the malleoli of the tibia and fibula and the talus. Other movements of the foot are those in which the foot is turned inward and the sole partially upward *(inversion)* and the opposite movement, in which the foot is turned outward *(eversion).* These movements take place mainly between the tarsal bones. *Movements of the digits of the foot* occur at the metatarsophalangeal joints or the interphalangeal joints. The movements are flexion and extension. Abduction moves other digits away from the second digit; adduction is the opposite movement.

MUSCLES THAT ACT ON THE FOOT (Fig. 6-38). These may be grouped into extensors (plantar flexors) and flexors (dorsiflexors):

Extensors: gastrocnemius, soleus, plantaris, tibialis posterior, peroneus longus, peroneus brevis

Flexors: tibialis anterior, peroneous tertius

Name, *Location*	O—Origin I—Insertion	Action
Gastrocnemius *Large, superficial muscle forming major portion of calf*	**O**—By two heads from lateral and medial condyles of femur **I**—With soleus into calcaneus by tendo calcaneus (tendon of Achilles)	Extends foot; flexes leg at knee joint
Soleus *Broad, flat muscle under gastrocnemius*	**O**—From posterior surface of head and upper portion of shaft of fibula; medial surface of upper end of tibia below popliteal line **I**—Into calcaneus by tendon of Achilles with gastrocnemius	Extends foot
Plantaris *Long, slender muscle under gastrocnemius*	**O**—Just above lateral condyle of femur **I**—By a long, slender tendon lying along tendon of Achilles, to fibrous tissue above calcaneus	Extends foot; flexes leg

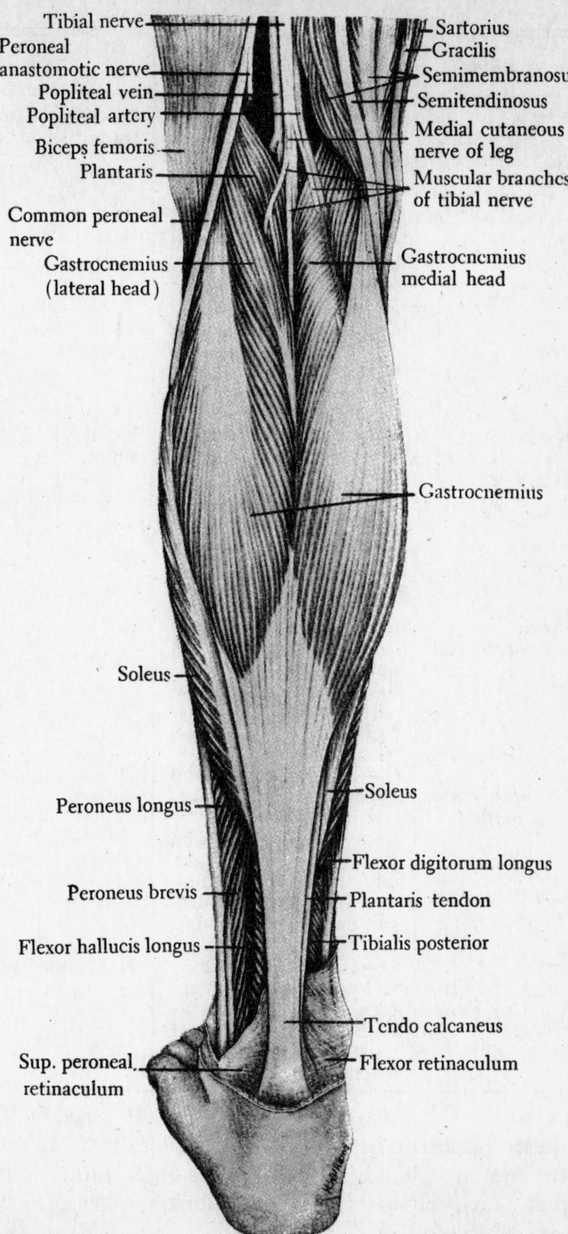

Fig. 6–38. Muscles that act on the foot. (Reprinted with permission of C. V. Mosby Company and the author from C. C. Francis, *Introduction to Human Anatomy,* 2nd ed., 1954.)

Name, *Location*	**O**—Origin **I**—Insertion	Action
Tibialis posterior *Deeply placed muscle of posterior side of leg*	**O**—Over an extensive area of posterior surfaces of interosseous membrane and proximal portions of tibia and fibula **I**—By a long tendon that passes through a groove behind medial malleolus to insert on tuberosity of navicular bone; fibrous extension connects with cuboid and cuneiform bones and with the bases of 2nd and 4th metatarsals	Extends and inverts foot; supports arches
Peroneus longus *Superficial muscle on lateral side of leg*	**O**—Lateral surfaces of proximal two-thirds of fibula and head of tibia **I**—By a long tendon passing behind lateral malleolus; inserts on plantar surface of 1st cuneiform and base of 1st metatarsal	Extends and everts foot; supports arches
Peroneus brevis *Distal and internal to peroneus longus*	**O**—Lower two-thirds of lateral surface of body of fibula **I**—By a long tendon that passes behind lateral malleolus to insert on base of 5th metatarsal	Extends and everts foot
Tibialis anterior *Superficial muscle lateral to shaft of tibia*	**O**—Lateral condyle and upper two-thirds of lateral surface of tibia **I**—By a long tendon that passes over anterior medial surface of foot to insert on 1st cuneiform and base of 1st metatarsal	Flexes and inverts foot
Peroneus tertius *Lateral to extensor digitorum longus, with which it functions*	**O**—Distal third of fibula and interosseous membrane **I**—By a long, slender tendon to base of 5th metatarsal	Flexes and everts foot

MUSCLES THAT ACT PRIMARILY ON THE DIGITS (Figs. 6-39 to 6-41). Some of these muscles lie in the lower leg and may assist in movements of the entire foot as well as acting on specific digits. Others are intrinsic muscles that lie entirely within the foot. Flexion, extension, abduction, and adduction are possible.

Name, *Location*	O—Origin I—Insertion	Action
Flexor digitorum longus *Medial side of leg, beneath soleus*	**O**—Posterior surface of shaft of tibia **I**—By long tendon that passes behind medial malleolus and continues obliquely forward and laterally into sole of foot; divides into four tendons which insert on terminal phalanges of four lateral toes	Flexes toes; extends and inverts foot
Flexor hallucis longus *Parallel to flexor digitorum on lateral side of leg*	**O**—Distal two-thirds of posterior surface of fibula **I**—By long tendon that passes posterior to medial malleolus and along lateral surface of sole of foot; inserted on hallux	Flexes great toe; extends and inverts foot
Extensor digitorum longus *Anterolateral surface of leg, lateral to tibialis anterior; in region of ankle, passes under transverse crural and cruciate ligaments and divides into four tendons which pass on dorsum of foot to digits*	**O**—Lateral condyle of tibia; proximal two-thirds of fibula and interosseous membrane **I**—Dorsal surface of phalanges of lateral four toes	Extends toes; flexes and foot
Extensor hallucis longus *Internal to and between tibialis anterior and extensor digitorum longus*	**O**—Middle portion of fibula; interosseous membrane **I**—By long tendon to base of distal phalanx of great toe	Extends great toe; flexes and inverts foot
Abductor hallucis *Large, superficial muscle along medial border of foot*	**O**—Calcaneus and adjacent ligaments **I**—On medial side of base of proximal phalanx of great toe	Abducts and flexes great toe; assists in maintaining longitudinal arch
Adductor hallucis *Muscle consisting of two heads on plantar surface of foot*	**O**—*Oblique head:* from bases of 2nd, 3rd, and 4th metatarsals, cuboid, and 3rd cuneiform; *transverse head:* from capsules of 2nd to 5th metatarsophalangeal joints **I**—To base of proximal phalanx and sheath of tendon of flexor hallucis longus	Adducts great toe and flexes proximal phalanx; aids in supporting arches

Name, *Location*	O—Origin I—Insertion	Action
Flexor digitorum brevis *Superficial muscle on sole of foot beneath plantar aponeurosis*	O—Tuberosity of calcaneus and plantar aponeurosis I—Into bases of phalanges of the four lateral toes	Flexes second row of phalanges on first row; supports longitudinal arch
Extensor digitorum brevis *Broad, thin muscle on dorsal surface of foot*	O—Calcaneus and cruciate ligaments I—Base of proximal phalanx of great toe; lateral surface of 2nd, 3rd, and 4th toes	Extends first four toes; abducts 2nd, 3rd, and 4th toes
Flexor hallucis brevis *Short muscle on medial side of plantar surface of foot; consists of two bellies that lie under 1st metatarsal bone*	O—1st, 2nd, and 3rd cuneiform bones and a fibrous extension of tendon of tibialis posterior I—Proximal phalanx of great toe	Flexes proximal phalanx of great toe
Extensor hallucis brevis *Most medial portion of extensor digitorum brevis*	O—Dorsal and lateral surface of calcaneus I—Base of proximal phalanx of great toe	Extends proximal phalanx of great toe
Quadratus plantae (Flexor accessorius) *Two-headed muscle, second layer of plantar surface of foot*	O—By a muscular head from medial surface of calcaneus and by a flat tendinous head from lateral border of inferior surface of calcaneus and from the long plantar ligament I—Lateral half of flexor digitorum longus	Flexes terminal phalanges of four small toes; as accessory to flexor digitorum longus, alters oblique pull of that muscle to pull in line with long axis of foot
Lumbricales *Four small muscles in second layer of plantar surface of foot on medial sides of four small toes*	O—From angles of division of flexor digitorum longus, each arising from two tendons except the first I—By tendons into dorsal surfaces of 1st phalanges of four small toes	Flex proximal phalanges and extend the two distal phalanges of the four small toes
Interossei dorsales *Four muscles situated between the metatarsals*	O—From adjacent sides of metatarsals I—The first into medial side of 2nd toe; the other three into lateral sides of 2nd, 3rd, and 4th toes	Flex proximal phalanges, extend middle and distal phalanges slightly; abduct toes from longitudinal axis of 2nd toe
Interossei plantares *Three muscles lying beneath the lateral three metatarsals*	O—From bases and medial sides of bodies of lateral three metatarsals I—Into medial sides of bases of 1st phalanges of lateral three metatarsals	Adduct three lateral toes toward axis of 2nd toe; otherwise, same as preceding

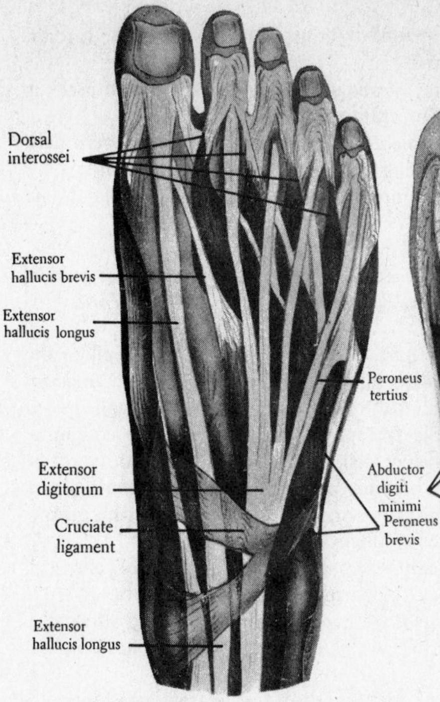

Fig. 6–39. Dorsal aspect of right foot.

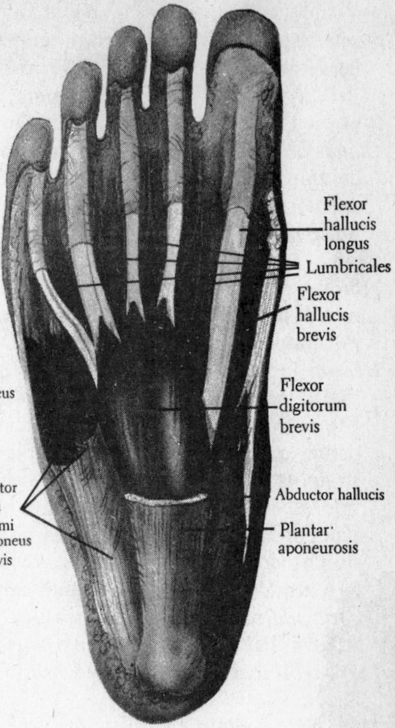

Fig. 6–40. Plantar view of right foot.

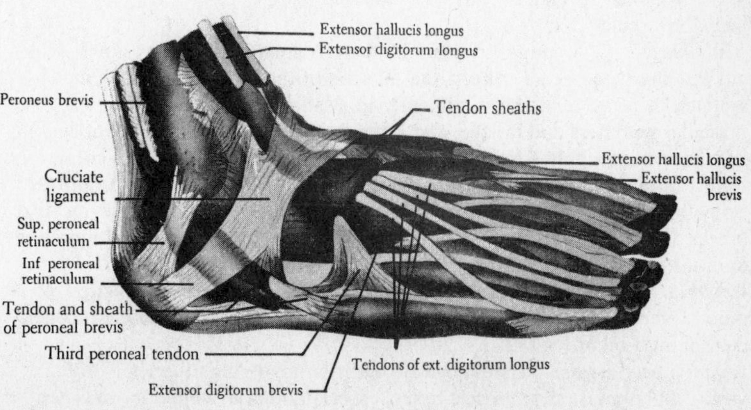

Fig. 6–41. Lateral view of right foot.

DISORDERS AND DISEASES OF THE MUSCULAR SYSTEM

Carpal tunnel syndrome. A condition resulting from compression of the median nerve within the carpal tunnel at the wrist.

"*Charley horse.*" An injury, common among athletes, in which a muscle is bruised or torn. It is accompanied by cramps and severe pain.

Claudication. Lameness or limping. In *intermittent claudication,* a severe pain develops in calf muscles during walking but disappears at rest. Symptoms are due to ischemia resulting from an interference in the blood supply to the muscles.

Cramp. A sustained spasm, usually accompanied by severe pain.

Fibromyositis. A group of nonspecific disorders characterized by pain or tenderness in muscles and related structures. See *Lumbago, Rheumatism, Torticollis, Pleurodynia, "Charley horse."*

Hernia. Protrusion of an organ or a part of an organ through the wall of the cavity that normally contains it. Also referred to as *rupture.* In an *inguinal hernia* (most common type of hernia) the intestine protrudes through the inguinal canal. Among the causative factors are extremely severe coughing, straining at defecation, lifting of heavy weights, and other physical overexertion, all of which may put excessive pressure on the abdominal contents. Other types of hernia take their names from the opening through which the protrusion occurs: *umbilical, diaphragmatic, femoral,* or *scrotal hernia.*

A *reducible hernia* is one that can be corrected by manipulation alone. A *strangulated hernia* is one in which the herniated structure is so tightly constricted that necrosis of tissue results if the condition is not relieved surgically.

Lumbago. General term applied to aches and pains in the lumbar or lumbosacral region.

Muscular atrophy. One of a number of muscular disorders characterized by muscular weakness and wasting, secondary to neural degeneration.

Muscular dystrophy. One of a number of inherited muscular disorders characterized by muscular weakness and wasting, often of a progressive nature. Occurs most frequently in males. Cause is unknown.

Myalgia. Muscular pain.

Myasthenia gravis. A condition in which there is great muscular weakness without atrophy. It especially affects the muscles of the face and neck. Symptoms are lack of facial expression, difficulty in swallowing and breathing, extreme muscular weakness and fatigue, and general prostration. It is due to the inability of nervous impulses to initiate contraction of muscle fibers. This condition results from the failure of the myoneural junction to produce acetylcholine or from the presence of excessive quantities of cholinesterase, an enzyme that breaks down acetylcholine. Recent evidence indicates that it is an autoimmune disease.

Myokymia. A muscular disorder characterized by persistent contraction of a muscle when at rest, or the widespread twitching of strips of muscle independent of each other.

Myopathy. Any disease or pathological condition involving muscles.

Myositis. Inflammation of muscle tissue, especially that of voluntary muscles.

Myotonia. One of a number of conditions in which there is an abnormally slow relaxation of muscle fibers following voluntary contraction.

Pleurodynia. Pain in the thorax, especially in intercostal spaces.

Rheumatism. General term applied to any of a number of disorders involving muscles, joints, tendons, ligaments, and fascia, characterized by soreness and stiffness of muscles and pain in joints and associated structures.

Spasm. A sudden, involuntary contraction of a muscle. When persistent, it is called a *tonic spasm;* if intermittent, consisting of alternate contractions and relaxations, it is called a *clonic spasm* or *clonus.* The term is also applied to constriction of a tube or of an opening, as *pyloric spasm* or *spasm of the bronchioles.*

Tetanus. A sustained contraction of a muscle. Also the name of a disease, commonly called *lockjaw,* which is characterized by sustained contraction of certain voluntary muscles. It is caused by the toxin of an infectious bacterium, *Clostridium tetani.*

Tetany. A condition characterized by painful, intermittent, tonic spasms of muscles. It is due to changes in pH and extracellular calcium, which increase nervous and muscular excitability. May be caused by parathyroid deficiency, vitamin D deficiency, or alkalosis.

Torticollis. A condition, also called *wryneck,* resulting from a spasm of the neck muscles, especially the sternocleidomastoideus. Characterized by twisting of the neck, resulting in an unnatural position of the head.

7: THE DIGESTIVE SYSTEM

The digestive system (Fig. 7-1) comprises the organs that act on ingested food, both mechanically and chemically, so that it may be absorbed and provide nutrition for the body. Digestion begins in the mouth with mastication and the mixing of food with saliva containing enzymes secreted by salivary glands. The passages and spaces from this point to the anus make up the alimentary canal, or gastrointestinal tract,

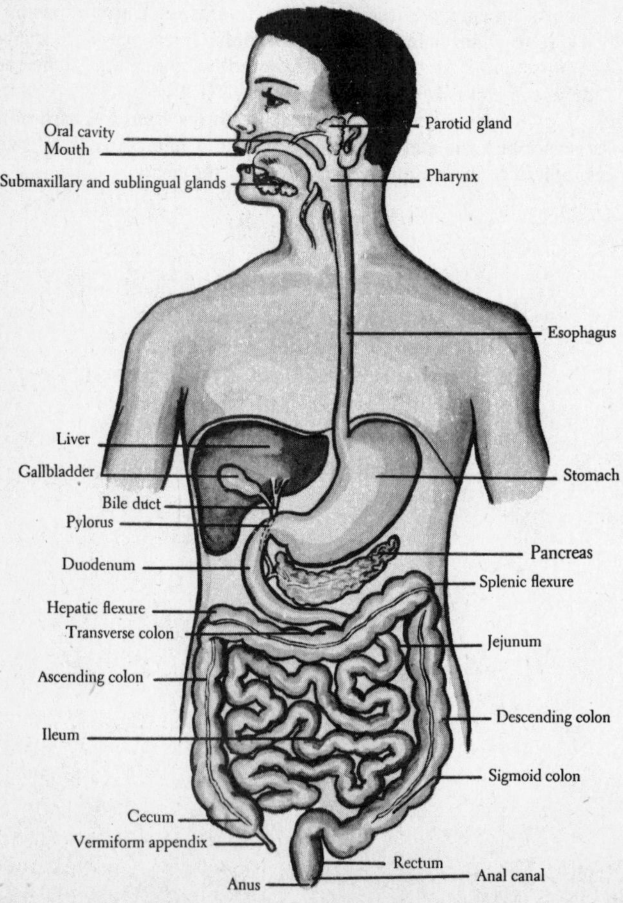

Fig. 7-1. The digestive system.

in which the complex compounds are reduced to soluble, absorbable substances, the usable food substances being absorbed and the indigestible and waste materials eliminated. The digestive glands secret enzymes and other chemical components essential to the breakdown of food substances and their absorption into the bloodstream.

ORGANIZATION OF THE DIGESTIVE SYSTEM

The two general divisions of the digestive system are the *alimentary canal* and the *accessory glands.* The following table shows the subdivisions and organic units within each of them:

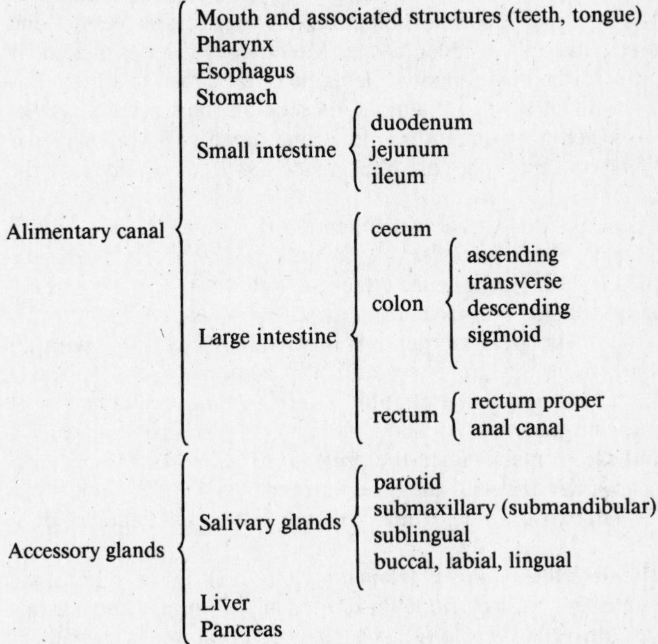

- Alimentary canal
  - Mouth and associated structures (teeth, tongue)
  - Pharynx
  - Esophagus
  - Stomach
  - Small intestine
    - duodenum
    - jejunum
    - ileum
  - Large intestine
    - cecum
    - colon
      - ascending
      - transverse
      - descending
      - sigmoid
    - rectum
      - rectum proper
      - anal canal
- Accessory glands
  - Salivary glands
    - parotid
    - submaxillary (submandibular)
    - sublingual
    - buccal, labial, lingual
  - Liver
  - Pancreas

THE BODY CAVITY AND ITS LININGS

Inasmuch as the digestive organs bear an intimate relationship to the body cavity, this cavity and its linings will be described here. In the embryo the body cavity, or *coelom,* develops by a splitting of the mesoderm. It is at first a single cavity, but with the development of the diaphragm it becomes separated into two parts, which form the *thoracic* and *abdominal cavities.* Subsequently, the thoracic cavity is further divided into two *pleural cavities* (containing the lungs) and the *pericardial cavity* (enclosing the heart). Each of the foregoing cavities is lined with a thin layer of epithelium called *mesothelium.* In the pleural cavities this layer is further identified as *pleura,* in the pericardial cavity as *pericardium,* and in the abdominal cavity as *peritoneum.*

Mesenteries and Ligaments. The peritoneal lining of each cavity is continuous with and reflected over the organs occupying the cavity (Fig. 1-3). The layer that lines the body wall is the *parietal* or *somatic* layer; that investing the organs is the *visceral* or *splanchnic* layer. The visceral layer is supplied by the autonomic nerves of underlying viscera. The parietal layer is supplied by neighboring wall nerves, so that many parts of it are sensitive to pain. A double layer of peritoneum with some connective tissue connects the parietal and visceral layers; this double layer is referred to as *mesentery.* A mesentery acts to support organs and hold them in position. It contains blood and lymph vessels, nerves, and sometimes considerable adipose tissue. Mesenteries are designated by special names, formed by prefixing the name of the organ to which they are attached with *meso-* (for example, mesocolon, mesorectum, mesoduodenum). When a mesentery connects one organ with another, it is called a *ligament;* examples are the *gastrosplenic ligament* and the *round ligament* of the uterus.

Omenta. The portion of the peritoneum that connects the stomach with the liver is called the *lesser omentum.* The *greater omentum* is a double-walled layer of peritoneum attached to the stomach and extending inferiorly a considerable distance to cover the intestines in the manner of an apron. The greater omentum is abundantly supplied with fat and forms an important protective and heat-conserving organ. It also plays a role in localizing inflammation within the abdominal cavity. It apparently has no pain nerves. The lesser and greater omenta enclose a space behind the stomach called the *omental bursa* or *lesser peritoneal sac.* This space connects with the true peritoneal cavity by means of an opening, the *epiploic foramen,* which lies between the liver and the duodenum.

Organs in Relation to the Peritoneum (Fig. 7-2). Some organs are *completely* invested with peritoneum (stomach, jejunum, ileum, transverse colon); others are *partially* invested (liver, cecum, ascending colon, descending colon, rectum, uterus); still others lie *behind* the peritoneum (urinary bladder, kidneys, pancreas, duodenum, and the great blood vessels, the aorta and the inferior vena cava) and are said to be *retroperitoneal.*

Peritoneal surfaces are moistened by a *serous fluid* that minimizes friction between the various organs and between the organs and the lining of the body wall.

LAYERS OF THE ALIMENTARY CANAL

The alimentary canal from the esophagus on possesses four layers (Fig. 7-3): mucous layer (mucosa), submucous layer (submucosa), mus-

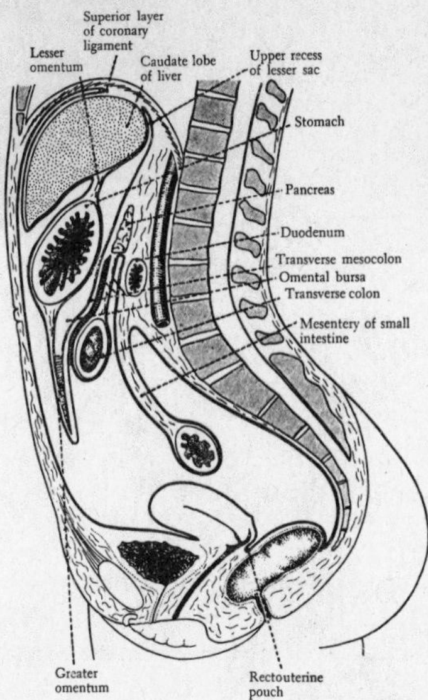

Fig. 7-2. Longitudinal section of body showing peritoneal relationships.

cular coat (muscularis externa), and a fibrous or serous layer (fibrosa or serosa).

Mucous Layer. The innermost layer of the alimentary canal, the *mucosa,* consists of a layer of epithelium (stratified or columnar), which forms a continuous layer from mouth to anus. Over its great length it is much invaginated to form tubular glands. Directly beneath this epithelium is the *lamina propria,* a thin layer of loose connective tissue on which the epithelium rests. Beneath the lamina propria is a thin layer of smooth muscle, the *muscularis mucosa,* consisting principally of longitudinal fibers.

Submucous Layer. This layer of loose connective tissue, the *submucosa,* contains the blood and lymph vessels and the nerves *(submucous plexus of Meissner).* Mast cells, lymphoid "wandering" cells, and eosinophils are present among the collagenous fibers. Sometimes glands from the mucosa extend into the submucosa. The submucosa provides an adjustable basis for movements and changes in the size of the tube.

Muscular Coat. The *muscularis externa* consists of cells arranged to form an inner *circular* layer and an outer *longitudinal* layer. A third *(oblique)* layer may be present. Between the circular and longitudinal layers lie nerve cells and fibers constituting the *myenteric plexus of Auerbach.*

Serous Layer. In the abdominal and pelvic cavities, a serous membrane, the *serosa,* forms the outermost layer of the alimentary canal. It

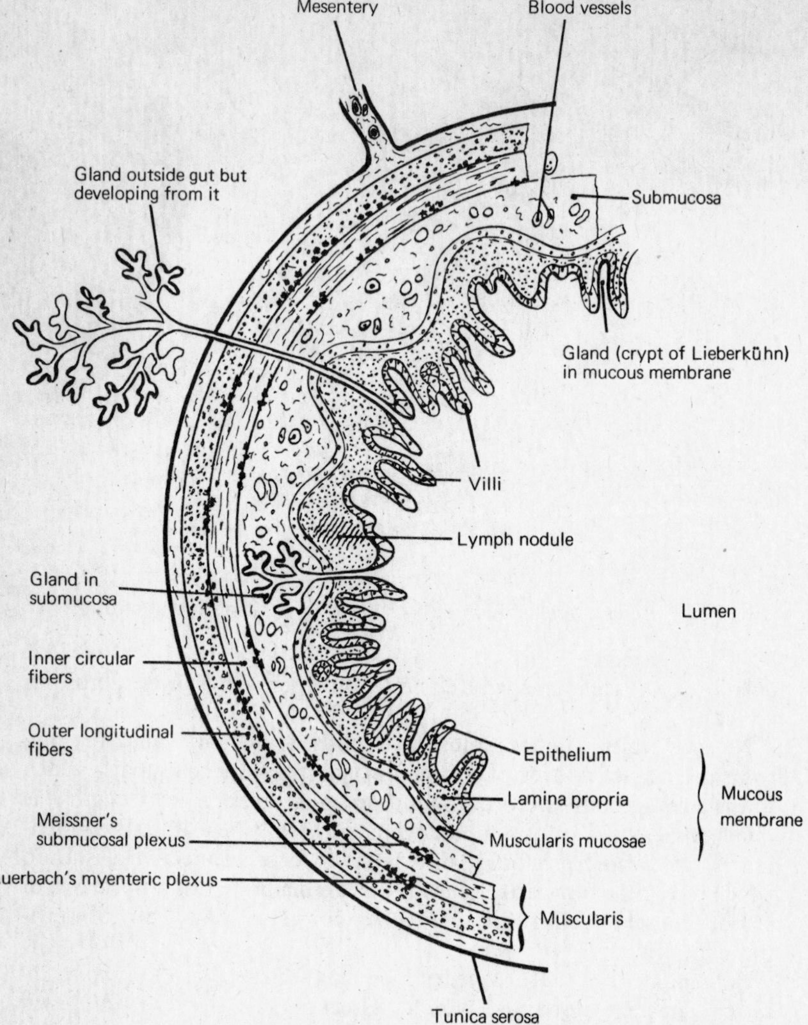

Fig. 7-3. Diagrammatic cross section of the alimentary canal. Modified from A. W. Ham, *Histology,* 6th ed., Lippincott, 1969. (Reprinted by permission of J. B. Lippincott Co. from E. M. Greisheimer and M. P. Wiedeman, *Physiology and Anatomy,* 9th ed., 1972.)

consists of a layer of mesothelium lying on a thin layer of loose connective tissue. It is continuous with the mesentery, which supports the intestines, and it contains blood and lymph vessels and sometimes adipose tissue. The serosa is lacking in the esophagus, the outer coat there being a fibrous coat, the *fibrosa.*

THE ALIMENTARY CANAL

The *alimentary canal* includes the mouth and associated structures, pharynx, esophagus, stomach, small intestine, and large intestine. The portion from the stomach to the anus comprises the *gastrointestinal* (GI) *tract.*

Mouth. The cavity of the mouth (Fig. 7-4) comprises two regions, the anterior *vestibule* or *labial cavity,* bounded externally by the lips and cheeks and internally by the teeth and gums, and the *mouth cavity proper* or *buccal cavity,* the region lying posterior to the teeth and gums. The latter is bounded superiorly by the hard and soft palates and inferiorly by the tongue, lower jaw, and intervening mucous membrane. Both cavities are lined with stratified squamous epithelium kept moist by the secretions of the large salivary and the smaller labial and buccal glands.

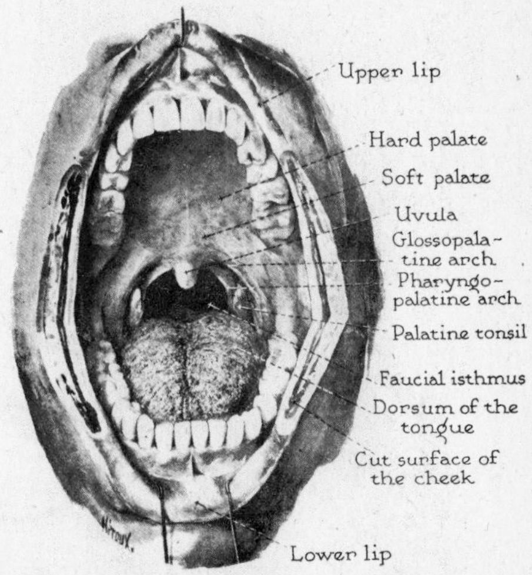

Fig. 7-4. The mouth cavity. (Gerrish.) (Reprinted with permission of The Macmillan Company from D. C. Kimber et al., *A Textbook of Anatomy and Physiology,* 13th ed., 1955.)

The *lips,* which border the mouth opening, contain a voluntary sphincter muscle, the *orbicularis oris.* Their external surface is covered with skin, their free margin and inner surface with a mucosa of stratified epithelium. The upper lip bears a median, vertical groove, the *philtrum.* The redness of the lips is due to the thinness and translucency of the epithelium through which the red color of the blood in the underlying capillary bed shows.

Each *cheek* contains a *buccinator muscle* and a subcutaneous pad of fat, the *buccal pad,* which is well developed in a baby. The lips and cheeks are of importance in sucking and aid in mastication by forcing food from the vestibule into the mouth cavity proper.

Within the cavity of the mouth are the teeth and the tongue, accessory digestive organs.

TEETH. These hard structures, which function in mastication, are borne on the alveolar processes of the maxillae and mandible. These processes, with the teeth and surrounding gums, form the *inferior* and *superior dental arches* (Fig. 7-5).

Teeth appear in two sets: first, the *deciduous* (temporary or "milk") teeth, of which there are 10 in each jaw and which are shed at the age levels shown in the accompanying table; subsequently the *permanent* teeth, of which there are 32, replace the deciduous teeth.

Eruption and Shedding of Teeth. The approximate age level at which deciduous teeth erupt and are shed and the period over which the permanent teeth erupt are shown on the chart on the facing page.

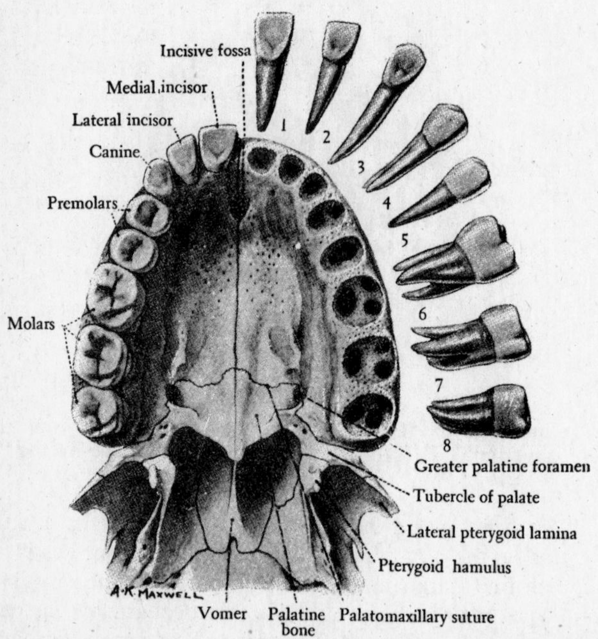

Fig. 7-5. Superior dental arch and hard palate as seen from below. Teeth on left side removed to show sockets for roots. (Reprinted with permission of Macmillan, London and Basingstoke, from W. J. Hamilton, *Textbook of Human Anatomy,* 1976.)

TEETH	DECIDUOUS		PERMANENT
	Erupt	*Shed*	*Erupt*
Central incisors	6–7½ mo	7 yr	6–8 yr
Lateral incisors	7–9 mo	8 yr	7–9 yr
Canines	16–18 mo	10 yr	9–11 yr
First premolars			9–10 yr
Second premolars			10–12 yr
First molars	12–14 mo	10 yr	6–7 yr
Second molars	20–24 mo	12 yr	11–13 yr
Third molars			17–25 yr

Permanent Teeth. In each jaw the following permanent teeth are found: four incisors (chisel-like in form), two canines (cuspids), four premolars (bicuspids), and six molars ("wisdom teeth"). A *cusp* is a raised portion on the chewing surface of a tooth.

Structure of a Typical Tooth. The *crown* of a tooth is the portion above the gum; the *root* is the portion embedded in a socket (alveolus) of a jawbone (mandible or maxilla). The *neck* is the slightly constricted region between the root and the crown; it is covered by the *gum* or *gingiva.* When a tooth has been cut in longitudinal section, the following parts can be seen (Fig. 7-6):

Hard parts, including enamel, dentine or "ivory," and cementum. *Enamel* is the material covering the crown. It consists of thin prisms or rods, twisted in form, that stand upright on the surface of the dentine. Enamel is the hardest substance in the body. *Dentine* forms the bulk of the tooth. It contains *canaliculi* and *dentinal canals,* but lacunae and bone cells are absent. Although it resembles bone in structure, it is harder than any bone. *Cementum,* another bonelike structure, covers the dentine of the root. It is continuous with the enamel at the neck of the tooth, forming a *cementoenamel junction.*

Soft parts, including the pulp and the periodontal membrane. The *pulp* consists of connective tissue containing a dense network of capillaries, lymph vessels, and nerve fibers; a layer of *odontoblasts* lies adjacent to the dentine. The *periodontal membrane* covers the root and lines the alveolus, where it serves as the periosteum. The periodontal membrane holds the tooth firmly in its socket.

TONGUE. The tongue is a freely movable muscular organ lying on the floor of the mouth. It consists of a *body* and a posterior *root.*

Functions of the Tongue. The tongue performs the following functions: (1) manipulation of food, (2) initiation of swallowing movements, (3) cleansing of the teeth and gums, (4) production of certain sounds in speech, and (5) provision of the sensation of taste through the taste buds on its surface.

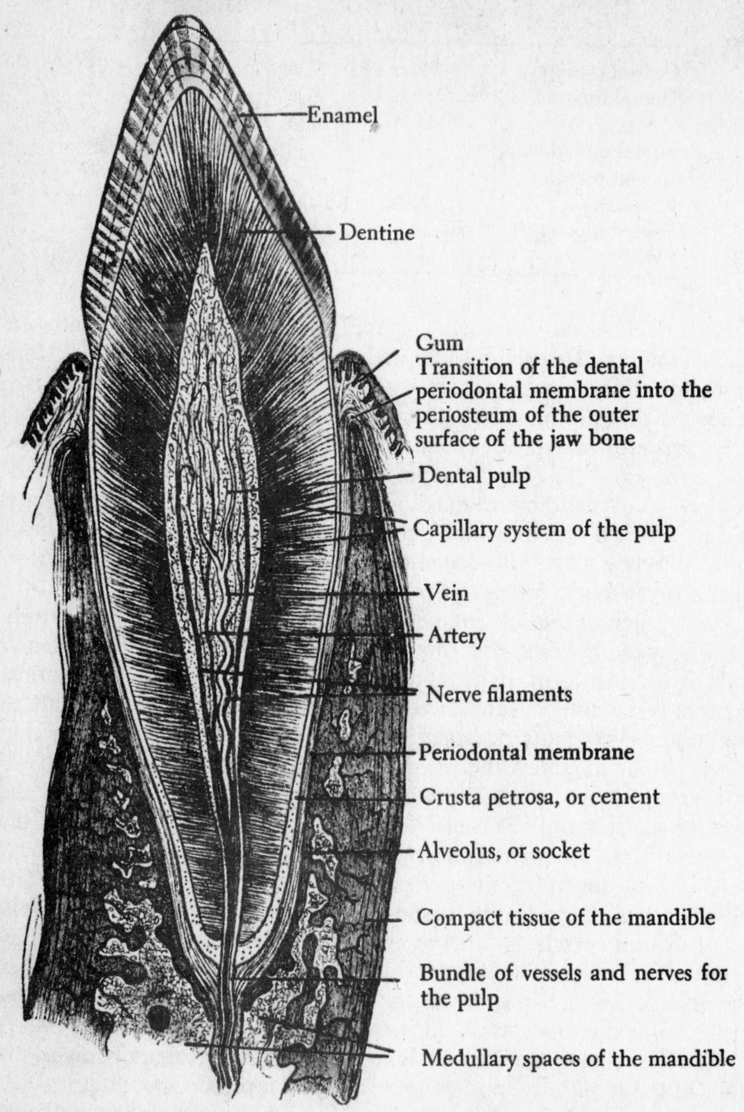

Fig. 7-6. Longitudinal section of a tooth (lower canine). (Reprinted with permission of The Macmillan Company from C. V. Toldt and A. D. Rosa, *Atlas of Human Anatomy for Students and Physicians,* 1948.)

Attachments of the Tongue. Posteriorly, the tongue is attached to the hyoid bone by its root, to the epiglottis by the glosso-epiglottic fold, and to the soft palate by the glossopalatine arches. *Anteriorly* and *inferiorly,* it is attached to the floor of the mouth by the frenulum *(frenulum linguae).* When the frenulum is unusually short, the person is said to be "tongue-tied" because of the speech defect to which this anomaly gives rise.

Surfaces of the Tongue. The upper surface or *dorsum* of the tongue is rough and covered with many elevations called *papillae.* The lower surface is covered with a smooth, thin mucous membrane.

Papillae of the Tongue (Fig. 7-7). The papillae are of three types: filiform, fungiform, and circumvallate.

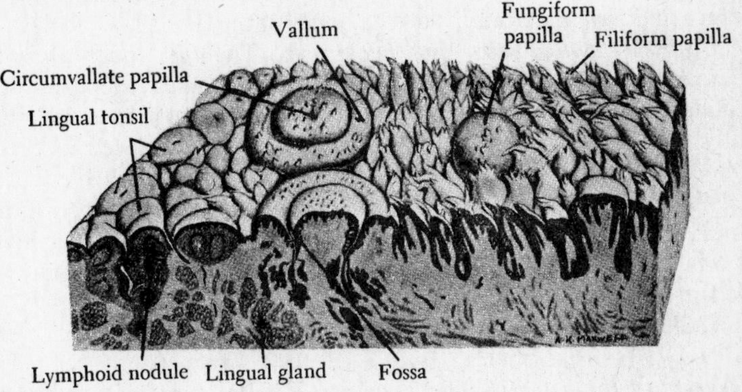

Fig. 7-7. Surface view of tongue to show papillae. (Based on Braus.) (Reprinted with permission of Macmillan, London and Basingstoke, from W. J. Hamilton, *Textbook of Human Anatomy,* 1976.)

The *filiform papillae* are 2 to 3 mm long and conical in shape, some of them having a double tip. Each consists of a slender core of connective tissue covered by stratified squamous epithelium, the superficial cells of which become transformed into hardened, scalelike structures that are constantly being shed. These papillae are quite numerous. They are arranged in more or less diagonal rows, extending laterally from the middle of the tongue.

The *fungiform papillae,* relatively few in number, measure 0.7 to 1.8 mm in length and 1.0 mm in thickness. Their distal portions are rounded and broader than their bases. They have a central core of connective tissue that is richly supplied with blood vessels which give the papillae a reddish color. Taste buds are present on some of these papillae.

The *circumvallate papillae* are 1 to 3 mm in diameter; they are 6 to 16 in number, the average number being 9 or 10. Each is surrounded by a deep circular cleft and a wall or *vallum*. The circumvallate papillae are arranged in a V-shaped row on the posterior dorsal surface of the tongue. Each contains a core of connective tissue and may bear secondary papillae. The stratified epithelium covering these papillae is smooth and, on their lateral walls, contains many taste buds; the number of buds on a single circumvallate papilla is estimated at about 250.

The *foramen cecum,* a small pit at the apex of the V formed by the circumvallate papillae, indicates the embryonic point of origin of the thyroid gland. The *terminal sulcus,* a groove roughly paralleling the circumvallate papillae, marks the junction of embryonic portions that give rise to the body and root of the tongue. Lateral to the terminal sulcus lie two masses of lymphoid tissue, the *lingual tonsils.*

Musculature of the Tongue. The muscles of the tongue are of two kinds, extrinsic and intrinsic. The *extrinsic muscles* are inserted into the tongue and have their origins on the bones of the skull. They include the *genioglossus, styloglossus,* and *hyoglossus.* The *intrinsic muscles* are contained entirely within the tongue. Muscle bundles extend in all directions, but fibers run in three more or less distinct planes: longitudinally, laterally, and vertically. The bundles are embedded in areolar connective tissue. Adipose tissue is fairly abundant.

Glands of the Tongue. These glands, called *lingual glands,* are of three types: *mucous glands,* found principally in the posterior region; *serous glands,* found deeply embedded in muscles and whose ducts empty into the furrows surrounding the circumvallate papillae; and *mixed glands,* which lie near the tip of the tongue, their ducts emptying on the inferior surface.

Nerves of the Tongue. The *hypoglossal nerve* supplies the intrinsic muscles and most of the extrinsic muscles except the glossopalatinus, which is supplied by the *vagus nerve.* Sensory fibers for touch, temperature, and taste are supplied by the *lingual nerve,* a branch of the mandibular division of the *trigeminal nerve,* which is distributed to the anterior two-thirds of the tongue. It also contains glandular fibers from the *chorda tympani,* a branch of the *facial nerve.* The posterior one-third receives sensory fibers from the *glossopharyngeal* and *vagus nerves,* fibers of the latter being restricted to the region near the epiglottis.

Blood Vessels of the Tongue. The *lingual artery,* a branch of the external caroid artery, is the principal source of blood; *lingual veins* convey blood to the internal jugular vein. *Lymphatic vessels* are numerous. They comprise four groups, apical, lateral, basal, and median, which drain into the superior cervical lymph nodes. Numerous masses of lymphatic tissue comprising the *lingual tonsils* are located in the posterior one-third of the tongue.

HARD PALATE. The hard palate is supported by portions of the maxillae and the palatine bones. The *incisive papilla,* a small fold of mucous membrane, is situated just posterior to the incisor teeth; openings of incisor canals are often found on either side. Other parts are the *median raphe,* the median line, usually somewhat lighter in color than the rest of the hard palate; *palatine rugae* (*plicae palatinae transversae*), transverse ridges in the anterior region; and *palatine glands,* in the submucosa under the mucous membrane.

SOFT PALATE. The soft palate is a muscular fold covered with mucous membrane extending posteriorly and laterally from the hard palate. Its free edge (*velum*) projects posteriorly and inferiorly and bears a median fingerlike projection, the *uvula.* The lateral portion of the soft palate forms the palatine arches. When swallowing occurs, the soft palate moves upward and closes the opening between the nasopharynx and the oropharynx.

ISTHMUS OF THE FAUCES. This is the opening between the oral cavity and the oral portion of the pharynx. Its *boundaries* are, superiorly, the soft palate; inferiorly, the dorsum of the root of the tongue; and laterally, the *glossopalatine arches.* These arches (also called *pillars of the fauces*) are downward continuations of the soft palate. The glossopalatine arch (anterior pillar) contains the glossopalatine (palatoglossus) muscle; the pharyngopalatine arch (posterior pillar) contains the pharyngopalatine (palatopharyngeus) muscle. A depression between them (the *tonsillar fossa*) lodges the palatine tonsil.

PALATINE (FAUCIAL) TONSILS. Two masses of lymphatic tissue, the *faucial tonsils,* lie between the two aforementioned pillars or arches. Each tonsil is enclosed in a capsule of connective tissue. The free surface is covered with stratified squamous epithelium, often infiltrated with leucocytes and containing several deep indentations, the *tonsillar crypts.* Adjacent to the crypts are nodules of lymphatic tissue similar to those in lymph nodes. Mucous glands are present. The only known function of the faucial tonsils is the formation of lymphocytes. Their removal, a practice usually inspired by anxiety over focal infection, is often unnecessary.

Pharynx. The pharynx is the second main component of the digestive tract. It serves as a common passageway for food and air. The pathways for these substances cross, but automatically controlled mechanisms operate to prevent the passage of food into the windpipe. The pharynx is a vertical, tubular structure extending downward from the base of the skull above to the openings of the larynx and esophagus (at about the level of the 5th vertebra) below. It lies immediately anterior to the spinal column. The pharynx is lined with mucous membrane; the epithelium in the upper portion is pseudostratified ciliated, and that in the middle and lower portions is stratified squamous.

REGIONS OF THE PHARYNX. The pharynx is divided into three regions: nasal, oral, and laryngeal.

The nasal portion, or *nasopharynx,* lies behind the nose, above the level of the soft palate. It communicates anteriorly with the nasal cavity through the two posterior nares or *choanae.* In its lateral walls are the openings to the *auditory tubes,* and on its posterior wall is a mass of lymphatic tissue, the *pharyngeal tonsil.* In childhood this frequently hypertrophies, forming an enlarged mass referred to as *adenoids.*

The oral portion, or *oropharynx,* lies below the soft palate and above the epiglottis. It communicates anteriorly with the oral cavity through the *isthmus of the fauces.* In its lateral walls between the two palatine arches are the palatine tonsils, lying in the *tonsillar fossae.*

The laryngeal portion, or *laryngopharynx,* is the lowermost portion, communicating directly with the larynx and the esophagus. The opening to the larynx, roughly triangular in shape, lies in its anterior wall, immediately behind an upward-projecting cartilaginous flap, the *epiglottis.* The *aryepiglottic folds* form its lateral boundaries. Inferiorly, at the level of the cricoid cartilage, the pharynx narrows and becomes continuous with the esophagus.

MUSCULATURE OF THE PHARYNX. The principal muscles of the pharynx are the constrictor muscles and the stylopharyngeus.

The constrictor muscles (superior, middle, and inferior) form the external or circular layer. They decrease the size of the pharynx, causing food particles to be grasped and forced downward into the esophagus (the act of swallowing).

The stylopharyngeus, with its origin on the styloid process and its insertion on the side of the pharynx and the thyroid cartilage, raises the pharynx and increases its diameter.

Esophagus. The esophagus is a muscular tube about 10 in. in length extending from the pharynx to the stomach. It lies between the levels of the 6th cervical and 7th thoracic vertebrae. The esophagus is situated in the neck, anterior to the bodies of the cervical vertebrae and posterior to the trachea. It passes inferiorly into the thoracic cavity, continuing through the mediastinal space to the diaphragm, where it passes through an opening, the *esophageal hiatus.* Turning to the left, it enters the stomach.

HISTOLOGIC STRUCTURE OF THE ESOPHAGUS. The *mucosa* consists of thick stratified epithelium thrown into folds and having a stellate lumen. A lamina propria is present. In the upper and lower portions, cardiac glands are usually found. The *submucosa* is a relatively thick layer containing collagenous and elastic fibers. In it lie the esophageal glands proper, blood vessels, and nerves. The *muscular coat* consists of two layers, an outer longitudinal and an inner circular, with Auerbach's plexus between them. Fibers are striated in the upper third, smooth in the lower third, and mixed in the middle third. A *fibrous coat* of loose connective tissue connects the esophagus to surrounding structures. Inasmuch as the peritoneal covering is lacking, there is no serosa.

The abdominal portion of the esophagus is short (about 2 cm long), and where it joins the stomach, fibers of its circular muscle layer, with some fibers from the stomach, form the *cardiac sphincter* muscle.

Stomach (Fig. 7-8). The stomach is an expanded portion of the alimentary canal lying in the upper left portion of the abdominal cavity. Its form is variable, but roughly J-shaped.

PRINCIPAL PARTS AND OPENINGS OF THE STOMACH. The principal parts and openings of the stomach include the cardiac orifice, body, fundus or fornix, pyloric orifice, and pyloric portion.

Cardiac Orifice. This is the opening from the esophagus. It is guarded by the cardiac sphincter. At this point the stratified epithelium of the esophagus undergoes transition into the columnar epithelium of the stomach.

Body of the Stomach. This is the main portion of the stomach.

Fundus or Fornix. This is the dilated portion lying to the left of the cardiac orifice.

Pyloric Orifice. This is the opening into the duodenum of the small intestine. It is guarded by the *pyloric sphincter.*

Pyloric Portion. The region nearest the pyloric orifice consists of a dilated portion, the *pyloric antrum,* and a narrower *pyloric canal.* The antrum is set off from the body of the stomach by a slightly constricted area, the *incisura angularis.* At this point there is a band of circular muscle fibers, the *prepyloric sphincter.*

SURFACES AND BORDERS OF THE STOMACH. The anterior surface of the stomach faces forward and upward; the posterior surface faces backward and downward. The *lesser curvature* forms the concave medial border and is directed to the right; the *greater curvature* forms the convex lateral border and is directed to the left.

HISTOLOGIC STRUCTURE OF THE STOMACH. In an empty contracted stomach, the innermost coat, the *mucosa,* is thrown into numerous longitudinal folds called *rugae,* which disappear when the stomach is full. The surface epithelium, consisting of simple columnar epithelial cells, contains many invaginations, the *gastric pits* or *foveolae,* into the bottom of which the gastric glands open. Goblet cells and villi are lacking. The *submucosa* of the stomach consists of areolar connective tissue containing blood and lymph vessels and nerves. It is situated between the mucosa and the muscular coat, and it binds them together. The *muscular coat* comprises three layers: an outer longitudinal, a middle circular, and an inner oblique. The *serosa,* the outermost coat of the stomach wall, consists of a thin layer of connective tissue covered with mesothelium. It is continuous with the peritoneum of the greater and lesser omenta.

GASTRIC GLANDS. These are simple or branched tubular glands, each with a narrow neck that opens into a gastric pit. They are confined entirely to the mucosa. It is estimated that there are 3 to 4 million pits and 15 to 20 million glands in an average stomach. Four types of cells found in gastric glands are mucous neck cells, chief or zymogenic cells,

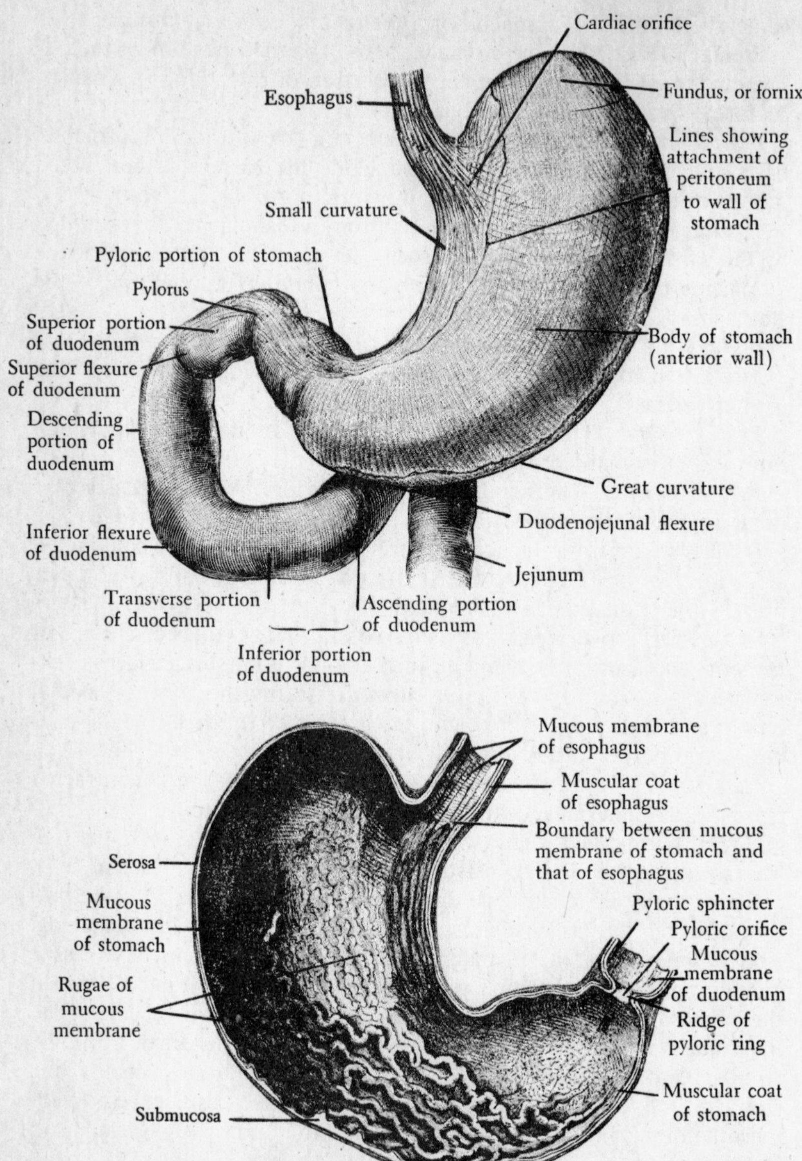

Fig. 7-8. Two views of the stomach. *Above,* ventral view, including part of the small intestine. *Below,* anterior half seen from inside. (Reprinted with permission of the Macmillan Company from C. V. Toldt and A. D. Rosa, *Atlas of Human Anatomy for Students and Physicians,* 1948.)

parietal or oxyntic cells, and argentaffin cells.

Mucous neck cells, found in the neck portion of the glands, are few in number and lodged between the parietal cells; *chief* or *zymogenic cells* secrete pepsinogen, the precursor of pepsin, and small amounts of gastric lipase; *parietal* or *oxyntic cells,* large pyramidal cells wedged between the chief cells, secrete hydrochloric acid and an intrinsic or anti-pernicious anemia factor; *argentaffin cells,* also called *enterochromaffin cells,* synthesize and store serotonin (5-hydroxytryptamine) and are possibly involved in the production of the hormone gastrin.

Gastric glands are named according to the region in which they occur. *Cardiac glands* occupy a limited area adjacent to the cardia. *Fundic glands,* shorter and more branched than the cardiac, occupy the superior two-thirds of the stomach. *Pyloric glands* occupy the pyloric region. These last-named glands contain only one type of cell, which is similar to the mucous neck cells. The lumen of pyloric glands is larger, and their secretory portion is more coiled.

There is no sharp demarcation between the glands of one region of the stomach and those of another region. The glands of one type mix freely with those of another at the borders.

Small Intestine. The small intestine is a much-coiled tube that extends from the stomach to the large intestine. It averages about 7 m in length and consists of three parts: the *duodenum,* about 25 cm long; the *jejunum,* about 2.9 m long; and the *ileum,* about 3.8 m long. These figures apply to the intestine immediately after death and before embalming. In a living subject, however, it is much shorter; in fact, in radiographic examinations, if a flexible radiopaque tube is passed from the mouth to the anus, the part of the tube in the small intestines averages only 1.5 to 1.8 m in length.

The *duodenum,* the widest and shortest portion, consists of four regions (Fig. 7-8). The first or superior part, the *duodenal bulb,* passes upward and backward from the pylorus and to the right. Turning downward, it continues as the second or *descending portion,* which receives the common bile duct. The third or *horizontal portion* crosses to the left and turns upward as the fourth or *ascending portion,* which joins the jejunum at the *duodenojejunal flexure.* The C-shaped duodenum encloses the head of the pancreas. It receives the common bile duct, which transports bile from the liver and pancreatic juice from the pancreas.

The remaining portion of the small intestine (Fig. 7-1) consists of the jejunum and the ileum. The *jejunum* forms approximately the upper two-fifths, and its coils are situated above and to the left; the *ileum* comprises the lower three-fifths, and its coils are to the right. It terminates at its junction with the large intestine, guarded by the *ileocecal (ileocolic) valve.*

HISTOLOGIC STRUCTURE OF THE SMALL INTESTINE (Figs. 7-9 and 7-10). The wall of the small intestine consists of four layers: mucosa, submucosa, muscular coat, and serosa.

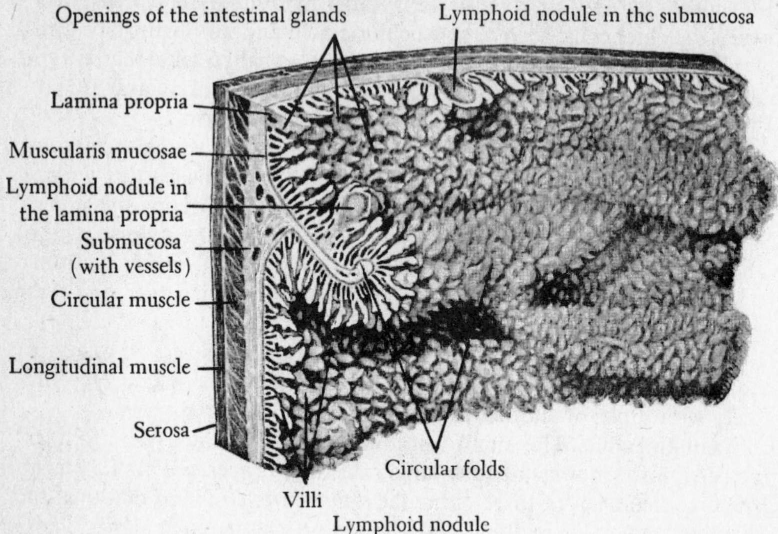

Fig. 7-9. Portion of small intestine. (After Braus.) (Reprinted with permission of W. B. Saunders Co., Philadelphia, from W. Bloom and D. W. Fawcett, *Histology,* 9th ed., 1968.)

Mucosa. The mucosa, the innermost layer of the intestine, consists of three layers: the *epithelium* that covers the innermost surface, a *lamina propria* of connective tissue, and a thin *muscularis mucosa.*

Cells of the epithelium include *columnar absorptive cells,* most of which possess microvilli, *goblet cells* of two types that secrete mucus, *cells of Paneth* with argentaffin granules present in the crypts of Lieberkühn, *enteroendocrine cells* that produce serotonin and intestinal hormones (secretin, cholecystokinin), and *caveolated cells,* peculiar cells bearing minute, tubular invaginations (*caveolae*) that extend from the free surface into the cytoplasm.

The absorbing surface of the mucosa is increased by *circular folds (plicae circulares)* or *valvulae conniventes*, constant, well-developed structures that disappear about the midregion, and by *villi.* Villi are fingerlike outgrowths of the mucosa averaging 0.5 to 1.5 mm in length. In the duodenum the villi are broad; in the jejunum they are elongated and narrow; in the ileum they are short and club-shaped. In the core of a villus are argyrophil fibers, a capillary net, a lacteal, strands of smooth muscle cells and free cells such as lymphocytes, plasma cells, and granu-

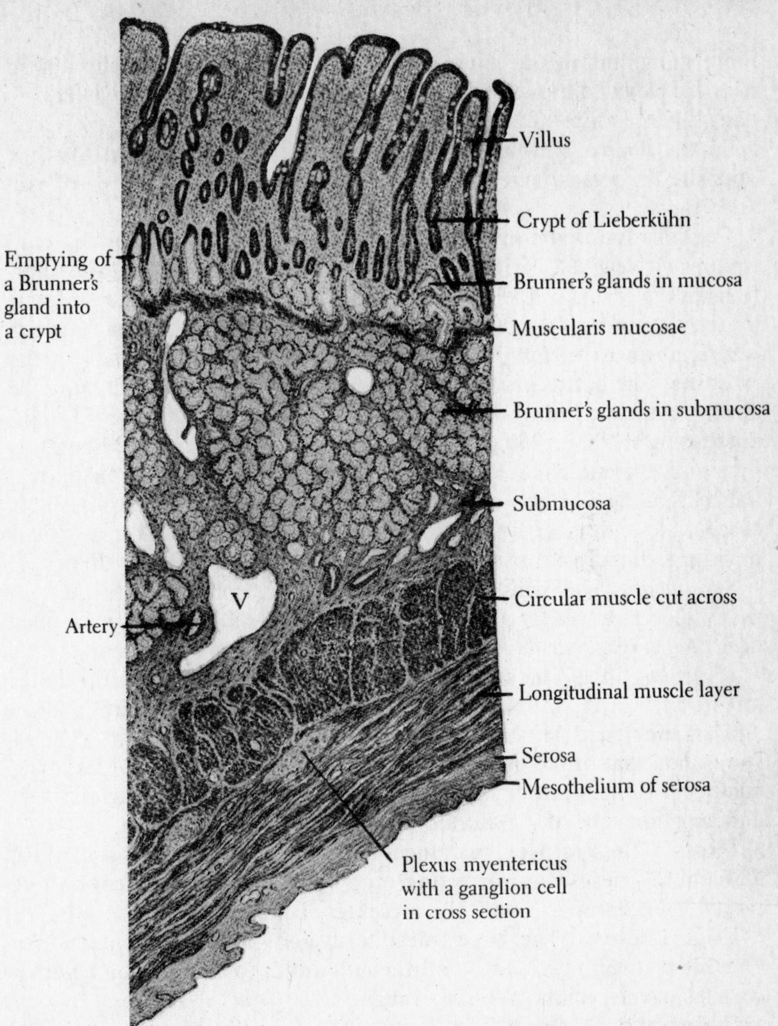

Fig. 7-10. Longitudinal section of the duodenum. (After Schaffer.) (Reprinted with permission of W. B. Saunders Co., Philadelphia, from W. Bloom and D. W. Fawcett, *Histology,* 9th ed., 1968.)

lar leucocytes, especially eosinophils. Villi are most numerous in the jejunum, sparer in the lower portion of the ileum. A lamina propria supports the epithelium and glands and forms the core of a villus.

The mucosa also contains *lymphatic tissue* in the form of nodules located in the lamina propria. These nodules are of two types, solitary and aggregated. The *solitary nodules*, varying in size from 0.6 mm to 3 mm, are fewer in number in the duodenum but become progressively more numerous in the jejunum, with villi and glands usually absent over the elevated area. The *aggregated nodules* (better known as *Peyer's patches*) are groups of 20 to 30 solitary nodules, found almost exclusively in the ileum. They appear as slightly elevated masses on the side opposite the attachment of the mesentery. Their surface is free of villi and crypts.

The cells lining the intestine have a limited life, being replaced continuously. New cells originate by mitosis from cells in the lower portions of the crypts of Lieberkühn. As new cells are formed, they migrate up the side of a gland and continue along the side of a villus to its tip, where, at an *extrusion zone,* they are expelled into the lumen of the intestine. The entire process takes two to four days, in which time the entire epithelial lining of the intestine is replaced. Discharged cells, amounting to about 250 g daily, disintegrate in the intestine, liberating intestinal enzymes. The continuous mitotic activity of intestinal epithelial cells, along with those of the bone marrow, makes these cells highly susceptible to damage from excessive radiation such as that from x-rays or from radioactive substances such as cobalt used in cancer therapy.

Submucosa. This consists of compact areolar connective tissue with many blood vessels. In the duodenum, Brunner's glands form a thick layer. A nerve plexus (*submucous plexus of Meissner*) is present.

Muscular Coat. This consists of two well-developed layers: an outer longitudinal layer, with fibers running longitudinally in a spiral course, and an inner circular layer, with fibers running circularly in a closer spiral than that of the outer layer. Between these two layers of the muscular coat is the *myenteric plexus of Auerbach,* consisting of nerve fibers and ganglion cells of the autonomic nervous system.

Serosa. The serosa or peritoneum consists of a layer of squamous epithelium (mesothelium) resting on a thin layer of loose connective tissue. It is continuous with the mesentery supporting the intestine.

Large Intestine. The large intestine extends from the ileum to the anus. It averages about 150 cm in length and 3 to 8 cm in diameter. It consists of the cecum, colon, and rectum.

FUNCTIONS. The large intestine receives the fluid by-products of digestion from the ileum and slowly converts them through the absorption of water into the more solid *feces* preparatory to their excretion. In the process, it serves as the avenue of excretion of certain minerals (iron, calcium, potassium). It incidentally serves as a breeding place for bacteria which, through their ability to synthesize vitamins, contribute to the well-being of the organism. Bacterial action also results in the production of intestinal gases (hydrogen, methane, and carbon dioxide) as a result of the fermentation and putrefaction of food residues. The gases are absorbed or expelled externally.

GENERAL CHARACTERISTICS. The mucous membrane of the large intestine is not thrown into folds, as is that of the small intestine, except in the rectum. Villi are lacking. The longitudinal layer of muscles consists of three bands, called *taeniae,* which are visible macroscopically; only two such bands are present in the sigmoid colon and the rectum. Each band is about 1 cm wide. Large sacculations called *haustra* protrude from the wall of the large intestine, and hanging from these are small peritoneal pouches containing fat (*appendices epiploicae*). Lymphatic tissue is present in the form of solitary nodules.

CECUM. The cecum, the first portion of the large intestine, is a large, blind pouch situated on the right side below the ileocecal valve. It averages about 6.5 cm in length and 7.5 cm in width. Its blind end is directed downward, and projecting from it is a narrow, worm-shaped tube, the *vermiform process* or *appendix,* which averages about 8.5 cm in length. Histologically, the cecum and the appendix possess the same four coats as does the colon, with which they are continuous. A characteristic feature of the appendix, however, is the large amount of lymphoid tissue present in the mucous layer. This is in the form of a ring of solitary nodules projecting into the submucosa. The lumen is small and irregular.

The lower portion of the ileum ends at an opening at the junction of the cecum and the colon. It is guarded by a valve, the *ileocecal valve* (*ileocolic valve*), consisting of two folds or lips that project slightly into the lumen of the large intestine. The valve prevents the reverse passage of food from the cecum into the ileum and also acts as a sphincter controlling passage of food into the large intestine.

COLON. The colon has four parts: ascending, transverse, descending, and sigmoid. It passes superiorly from the cecum as the *ascending colon* to the region of the liver, where it turns sharply to the left as the *transverse colon.* This bend is called the *right colic* or *hepatic flexure.* The transverse colon passes to the left, then turns downward (*left colic* or *splenic flexure*) and continues as the *descending colon* to the brim of the pelvis. Within the cavity of the true pelvis it forms an S-shaped *sigmoid* or *pelvic colon,* which, at the level of the 3rd sacral segment, becomes continuous with the rectum.

RECTUM (Fig. 7-11). The rectum consists of the *rectum proper* and the *anal canal,* which terminates at the *anus.*

Gross Structure of the Rectum. The *rectum proper,* about 12 cm long, follows the curve of the sacrum and coccyx. Internally the mucosa projects into the lumen, forming three *transverse folds.* In its lower portion is a dilatation, the *rectal ampulla,* located at the junction of the rectum with the anal canal. The rectum is normally empty except immediately prior to defecation.

The *anal canal* continues downward and posteriorly to its exterior opening, the *anus.* On the inner surface of the canal are 5 to 10 vertical folds, the *anal columns,* whose lower ends are joined together, forming shallow, pocketlike folds called *anal valves,* behind which are spaces, the

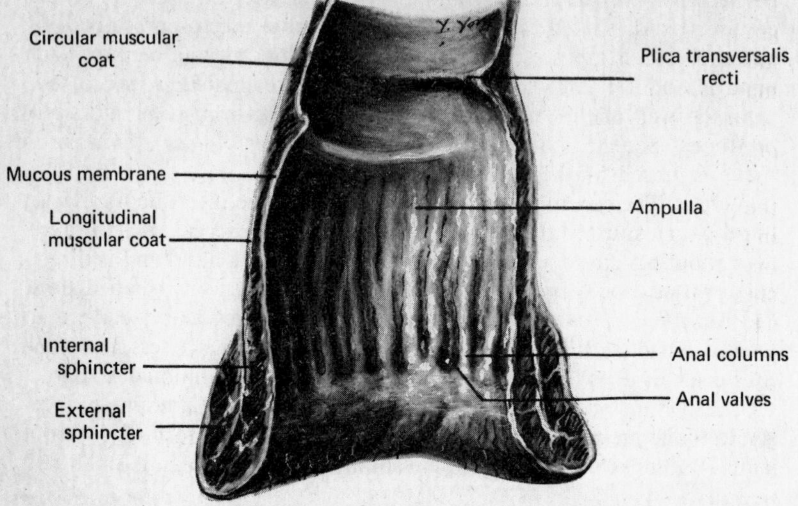

Fig. 7-11. Interior of rectum and anal canal. (After Spalteholz.) (Reprinted with permission of J. B. Lippincott Co. from E. M. Greisheimer and M. P. Wiedeman, *Physiology and Anatomy,* 9th ed., 1972.)

anal sinuses. Distension and inflammation of veins in the anal columns constitutes *hemorrhoids.* The relationship of the rectum and anal canal to other organs of the pelvic cavity is shown in Figs. 7-1 and 7-6, Vol. 2.

Muscles of the Anal Canal. The circular muscle layer of the anal canal is much thickened and forms the *internal sphincter,* which surrounds the anal canal for a distance of about 1 in.; this sphincter is innervated by autonomic nerves. The *external sphincter,* which is amenable to voluntary control, is a layer of striated muscle tissue surrounding the terminal portion of the anal canal. Its fibers are attached to the coccyx and the surrounding skin. The function of these muscles is to keep the anal canal and anal opening closed. On each side of the anal canal are fibers of the *pubococcygeus muscles,* a division of the *levator ani muscle.* Its fibers, which are voluntary, act to compress the anal canal. Some fibers enter the wall of the canal and merge with those of the external sphincter. It assists the sphincters in keeping the anal canal and opening closed. They relax voluntarily during defecation.

Histologic Structure of the Large Intestine. The same four layers that are found in the small intestine are present in the large intestine, with the following differences: The *mucosa* is relatively smooth and lacking in villi; tubular pits or glands are present. The epithelium is of the simple columnar type, containing many goblet cells that secrete mucus. Lymphatic tissue occurs only in the form of solitary nodules in the mucosa and the submucosa. The *muscular coat* includes an inner circular layer and an outer longitudinal layer, the fibers of which are grouped in three bands, the *taeniae coli.* In the serosa are found pendulous *appendices epiploicae,* extensions of the peritoneum containing masses of fat.

Intestinal Glands. There are two types of intestinal glands, the *crypts of Lieberkühn* and the *duodenal glands (Brunner's glands).*

CRYPTS OF LIEBERKÜHN. These are simple tubular glands located in the mucosa and opening between the bases of the villi. They are found in all portions of the intestine. They extend downward through the lamina propria to the muscularis mucosa. Large *cells of Paneth* containing chromophil substance and enteroendocrine cells occur in these glands, and goblet cells are numerous in the epithelium. These glands do not produce intestinal enzymes as formerly postulated but serve as a source of cells for the continual renewal of the intestinal epithelium.

DUODENAL GLANDS. Brunner's glands are abundant near the pylorus, decreasing in number and disappearing in the lower portion of the duodenum. The terminal portions of these glands consist of copiously branched and coiled tubes, each with a conspicuous lumen. The glands are most numerous in the submucosa. Their excretory ducts open into the bottom or the side of a crypt of Lieberkühn. They secrete a mucus with a high bicarbonate content.

ACCESSORY DIGESTIVE GLANDS

The accessory digestive glands include the salivary glands, the liver, and the pancreas.

Salivary Glands. Numerous glands contribute to the secretion of saliva. Many are small glands located in the mucosa or submucosa of the mouth cavity. These are named according to their location (lingual, buccal, labial). The major portion of the saliva, however, is produced by the *salivary glands proper,* the parotid, submaxillary (submandibular), and sublingual glands, which are located outside the mouth cavity (Fig. 7-1). All three are compound tubuloalveolar glands. Salivary secretion by these glands is under nervous control, occurring reflexly when (1) mechanical, chemical, or thermal stimuli act on sensory receptors in the mouth and other parts of the digestive tract, (2) olfactory stimuli act on olfactory receptors, or (3) psychic stimuli arise within the higher nervous centers.

SALIVA. A viscous, colorless, opalescent fluid, saliva is about 99 percent water. The balance is made up of mucin, inorganic salts, enzymes, organic compounds, and miscellaneous cells. Foreign substances, such as iodides or bromides, which have been introduced by mouth or injected into the blood, may appear in the saliva; this accounts for the taste of morphine following its injection. The normal total daily production of saliva is 2 to 3 pints (1000 to 1500 ml).

Constituents of Saliva. Mucin is a complex glycoprotein of high viscosity, which gives saliva its sticky consistency, a property that causes food particles to stick together and makes them easier to swallow. Mucin also lubricates the mouth cavity. The *inorganic salts* include chlorides, carbonates, phosphates, and sulfates of sodium, potassium, calcium, and magnesium. Evidence of their presence in saliva is the tartar that forms on teeth as a result of precipitation of insoluble calcium salts with associated substances when carbon dioxide arises from carbonates. The *enzymes* include *ptylalin,* an amylase that converts starches to dextrins and maltose, and *maltase,* which converts maltose to dextrose. However, digestion of starches is generally incomplete because of the limited time food remains in the mouth. Starch digestion may continue a short time in the stomach but ceases when the stomach contents become acidified by hydrochloric acid. Another enzyme in saliva is *lysozyme* (muramidase), which has a bactericidal action on various pathogenic organisms. The *organic compounds* consist of mucin, traces of urea, and other substances. The "miscellaneous cells" are salivary corpuscles (modified leucocytes from lymphatic tissue) and epithelial cells from the oral mucosa.

Reaction of Saliva. The reaction of saliva when exposed to air is normally slightly alkaline (pH 7.2). When loss of carbon dioxide is prevented, it is slightly acid (pH 6.6).

Functions of Saliva. Saliva (1) moistens and lubricates mucous surfaces, making speech and swallowing possible; (2) moistens and lubricates food, facilitating swallowing; (3) acts as a solvent whereby food particles go into solution and are thus enabled to stimulate taste buds; (4) helps food particles to stick together and form a *bolus* so that they can be swallowed as a mass; (5) initiates the process of digestion through the action of ferments or enzymes; (6) acts as a cleansing agent for the mouth and a neutralizing agent that tends to counteract tooth decay; and (7) aids in the elimination of foreign substances from the blood.

STRUCTURE OF THE SALIVARY GLANDS PROPER. The largest of the salivary glands, the *parotids,* weigh 14 to 30 g each (Fig. 6-13). They lie anterior and ventral to the ears, partly covering the rami of the mandible.

A portion of each extends forward over the masseter muscle. Each is tightly encased in a sheath that is continuous with the *cervical fascia.* The superficial layer of this fascia, the *parotid fascia,* is dense and closely adherent to the gland, being attached to the zygomatic process, the styloid process, and the angle of the mandible, the latter two attachments forming the stylomandibular ligament. The *parotid duct* passes anteriorly across the masseter muscle to empty into the vestibule of the mouth opposite the upper second molar. The parotids are pure serous (albuminous) glands. Their secretion is watery and lacks mucus. Each of the *submaxillary (submandibular) glands* weighs 7 to 10 g. They lie just within the lower jaw anterior to the angle, occupying a fossa on the inner side of the mandible. The major portion of the submaxillary gland consists of serous alveoli; some mucous alveoli are present. A few *demilunes* (crescent-shaped groups of serous cells) are present. A *submaxillary duct* extends from the anterior end of each gland and opens into the floor of the mouth on each side of the frenulum.The *sublingual glands,* each weighing about 3 g, lie in the loose areolar tissue in the floor of the mouth. They consist of several separate lobes or aggregations of tissue and do not possess a capsule. The sublingual gland is a mixed gland. Mucous cells are more abundant than in the submaxillary gland. Serous cells are in the minority and are generally arranged in the form of crescents. The *sublingual ducts,* which vary in number from 4 to 20, open in the floor of the mouth. Some of them may join the duct of the submaxillary gland.

The Liver and Biliary Apparatus (Fig. 7-12). The liver, the largest gland of the body, is located in the upper right portion of the abdominal cavity, immediately below the diaphragm. Its weight averages about 1.5 kg.

LOBES AND SURFACES OF THE LIVER. The liver comprises a large right lobe and a smaller left lobe. The right lobe is further subdivided into the

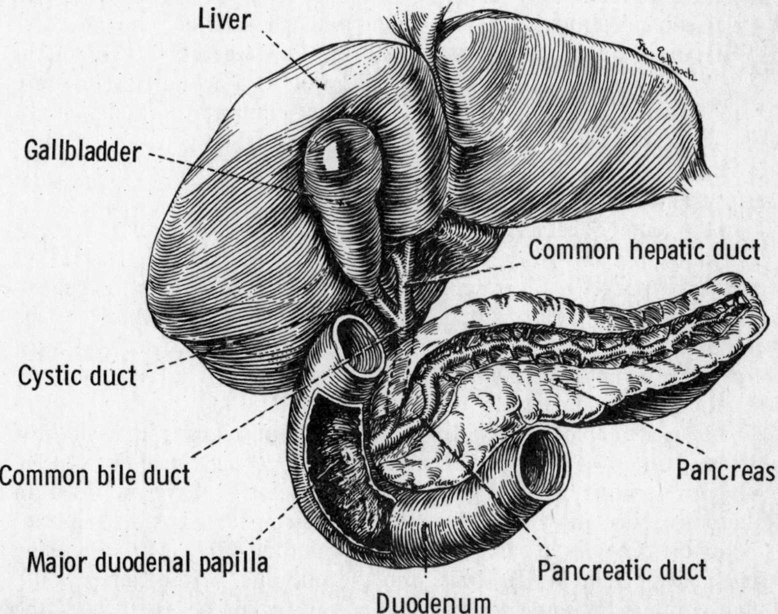

Fig. 7-12. The liver, gallbladder, pancreas, and their ducts. (Reprinted with permission of J. B. Lippincott Co. from E. M. Greisheimer and M. P. Wiedeman, *Physiology and Anatomy,* 9th ed., 1972.)

right lobe proper, the quadrate lobe, and the caudate lobe.

The *surfaces* of the liver are right, posterior, and inferior. On the convex *anterior* and *superior surfaces* is the *umbilical notch,* which separates the right from the left lobe. In this notch is a curved fold of the peritoneum, the *falciform ligament,* which extends superiorly to the diaphragm and the anterior body wall. This ligament is a rudiment of the ventral mesentery. Its free edge, known as the *round ligament (ligamentum teres)* represents the occluded umbilical vein. The *inferior surface* is divided by a number of fissures that separate the lobes. The most important of these is the *portal* or *transverse fissure (porta hepatis).* This constitutes the *hilum,* a wide cleft about 2 in. long, which encloses the portal vein, the hepatic artery, the common bile duct and its divisions, lymphatics, and nerves. All are enclosed in a membrane of connective tissue called *Glisson's capsule (capsula fibrosa perivascularis),* which continues over the entire surface of the liver except that adjacent to the diaphragm. Adjacent to Glisson's capsule, the peritoneum forms the outermost covering.

Other ligaments that attach to the liver are the *lesser omentum,* divided into the hepatogastric and hepatoduodenal ligaments; the *coronary ligaments;* the right and left *triangular ligaments;* and the *ligamentum venosum,* a fibrous cord that represents the *ductus venosus,* an embryonic vessel connecting the umbilical vein with the inferior vena cava.

HISTOLOGIC STRUCTURE OF THE LIVER. The structural and functional unit of the liver is the *lobule.* Each lobule is a cylindrical structure with flattened sides, separated from adjacent lobules by a small amount of connective tissue. At the points of junction between two or three lobules are found the terminal branches of the portal vein, the hepatic artery, and the hepatic ducts (Fig. 7-13).

In the center of each lobule is a *central* or *intralobular vein* (Fig. 7-14). Extending radially from the central vein are anastomosing *cords* or *plates of liver cells.* Liver cells, or *hepatocytes,* are unique in that, although the liver performs many functions, only one morphologic type can be identified. Between these plates of cells are blood spaces (*sinusoids*) that communicate with the branches of the portal vein and the hepatic artery and centrally with the central vein.

The hexagonal lobule just described, with a portal canal at each of its corners, is known as the *traditional* or *classic lobule.* Since this type of structure is not typical of glands, some histologists have proposed an alternative view in which the portal canal is considered as the center of the lobule. The portal canal and the surrounding liver cells, comprising several liver *acini* which drain into the bile duct of the canal, would then comprise a *portal lobule.* In this unit, called a liver *acinus,* the central core contains the terminal branches of the hepatic artery, portal vein, and bile vessels. Although this new concept of liver structure has

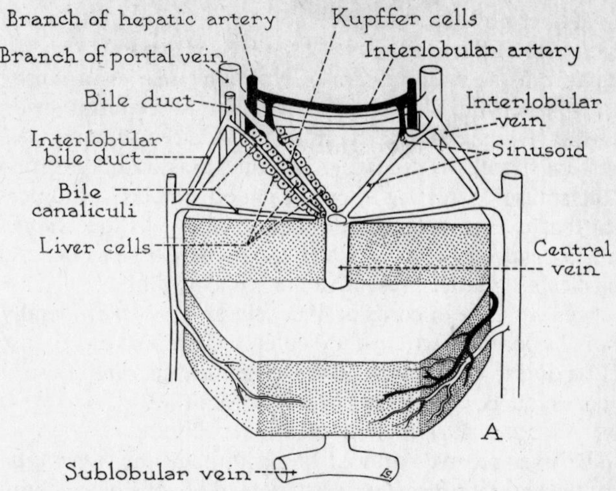

Fig. 7-13. Diagram of a liver lobule, showing the hepatic circulation and the relation of liver cells to sinusoids and bile canaliculi. (Reprinted with permission of W. B. Saunders Co., Philadelphia, from B. G. King and M. J. Showers, *Human Anatomy and Physiology*, 6th ed., 1969.)

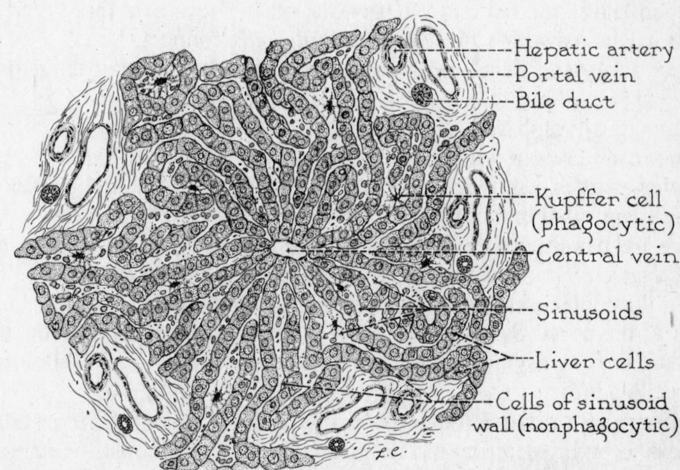

Fig. 7-14. Semidiagrammatic cross section of a liver lobule. (Reprinted with permission of W. B. Saunders Co., Philadelphia, from B. G. King and M. J. Showers, *Human Anatomy and Physiology*, 6th ed., 1969.)

not been generally accepted, it provides, in some cases, a better understanding of liver physiology and pathology.

The *sinusoids* of the liver are irregular spaces lined with endothelium that differs from the endothelium of typical capillaries in that some of its cells are phagocytic. The cells do not constitute a continuous epithelium, and the cells are separated from the adjacent hepatic cells by a *perisinusoidal space of Disse,* which contains fat-storing cells (*lipocytes*). Within the sinusoids, resting upon the endothelial cells, and possessing processes that extend between the endothelial cells are *Kupffer cells.* These reticuloendothelial cells, which are markedly phagocytic, actively engulf particulate matter present in the bloodstream.

Bile capillaries arise in cords of liver cells and pass peripherally to the margin of the lobules, where they enter the primary divisions of the interlobular ducts, which lead to the larger bile ducts and these, in turn, to the hepatic ducts.

BLOOD VESSELS, LYMPH VESSELS, AND NERVE SUPPLY OF THE LIVER. The liver receives arterial blood through the *hepatic artery,* a branch of the superior mesenteric artery, and venous blood through the *portal vein,* which drains the digestive tract, pancreas, and spleen. These vessels enter the liver at the porta. Portal blood contains the products of digestion, except fatty acids, from the intestine and the products of red cell destruction from the spleen. The hepatic artery provides less blood than the portal vein, but this oxygenated blood is received at higher pressure.

Blood leaves the liver through two short *hepatic veins* that enter the inferior vena cava. Within the liver, *sinusoids* connect the ends of the hepatic artery and portal vein with the terminal branches of the hepatic vein.

Numerous lymph vessels accompany the branches of the portal vein, but none are present within the lobules. Lymph from the liver is rich in proteins synthesized by liver cells. The liver provides the major portion of the lymph that enters the thoracic duct.

The liver is innervated by parasympathetic fibers from the left and right vagus nerves and sympathetic fibers from the coeliac plexus. The fibers form plexuses along the hepatic artery and portal vein, which enter at the porta. Sympathetic fibers end in hepatic vessels; both sympathetic and parasympathetic fibers pass to the walls of the bile ducts and gallbladder.

FUNCTIONS OF THE LIVER. The principal functions of the liver fall in these categories: secretion of bile, formation of blood constituents, metabolic functions, protective functions, and miscellaneous functions.

Secretion of Bile. The secretion of bile and the role of its constituents in various digestive and excretory activities are discussed on page 247.

Formation of Blood Constituents (Hemopoietic function). These activities include (1) red cell formation in the embryo; (2) synthesis of blood constituents, including such plasma proteins as albumin, fibrinogen, various globulins, glycoproteins, and lipoproteins; (3) synthesis of prothrombin and other factors involved in clotting; and (4) synthesis of heparin, an anticoagulant.

Metabolic Functions. Metabolic functions of the liver consist of (1) *carbohydrate metabolism* (glycogenesis, or formation and storage of glycogen, and glycogenolysis, or conversion of glycogen to glucose and liberation into the bloodstream); (2) *fat metabolism* (synthesis of fatty acids from carbohydrates, desaturation of fatty acids, production of ketones as intermediary products in fat metabolism, synthesis and storage of fat, oxidation of fat, synthesis of cholesterol); (3) *protein metabolism* (deamination of amino acids and formation of urea, synthesis of amino acids, synthesis of hippuric acid and uric acid); (4) *mineral metabolism* (storage of iron and copper); (5) *vitamin metabolism* (formation of vitamin A, storage of vitamins, especially fat-soluble vitamins); (6) *metabolism of hormones,* especially steroid hormones, which may include their activation or inactivation, their binding to blood proteins for transportation, or their conversion to various substances for utilization by cells or for excretion; and (7) *metabolism of drugs,* such as the degradation of alcohol and barbiturates.

Protective Functions. These include (1) *detoxication* of poisonous substances, such as indol and skatol, that are absorbed by the intestine; (2) *conversion* of harmful substances, such as ammonia, to useful substances or innocuous substances, as urea, which are utilized or excreted; and (3) *phagocytic action* of Kupffer cells.

Miscellaneous Functions. The liver serves as (1) a primary source of body heat, which results from the numerous chemical reactions occurring within its cells; (2) a reservoir for blood, which may be quickly released into general circulation; and (3) a source of a vasodepressor material (VDM), which is produced in shock resulting from severe hemorrhage.

THE DUCT SYSTEM OF THE LIVER. This system transports bile from its place of origin, the liver cells, to the duodenum or the gallbladder. The smallest branches are small *bile capillaries.* These lead to *intralobular ducts,* which lead to *interlobular ducts.* The last join others and form still larger ducts, which lead to the right and left tributaries of the *hepatic duct.* The hepatic duct unites with the *cystic duct* from the gallbladder to form the *common bile duct,* which enters the duodenum a short distance below the pylorus.

THE GALLBLADDER. The gallbladder is a reservoir for concentrating and storing bile. It is pear-shaped and lies in a fossa in the inferior surface of the right lobe of the liver. It is about 8 to 10 cm in length and has a capacity of about 50 ml. The constricted portion, or *neck,* is bent and attached closely to the peritoneal covering of the organ. Its expanded portion, or *fundus,* is directed anteriorly and lies near the end of the

9th costal cartilage. The *cystic duct,* 3 to 4 cm long, leads from the gallbladder to the hepatic duct, with which it unites to form the common bile duct. It contains a *spiral valve,* a fold that serves to keep the duct open.

Histologic Structure of the Gallbladder. The wall of the gallbladder has four layers:

1. *Mucosa:* the innermost layer, composed of simple columnar epithelium. Goblet cells are lacking. The mucosa is thrown into folds (primary and secondary) that unite to form polygonal, pocketlike areas of variable size. Glandlike evaginations may extend into the muscular layer. The mucosa rests on the *tunica propria,* a thin layer of connective tissue.
2. *Muscular layer:* composed of smooth muscles arranged in small bundles with connective tissue between them.
3. *Connective tissue layer:* lying outside the muscular layer.
4. *Serosa:* the outermost or peritoneal layer.

THE COMMON BILE DUCT. This duct, about 7 cm long, extends from the union of the cystic and hepatic ducts to the duodenum. It passes obliquely through the intestinal wall and opens on the *papilla of Vater,* which is located about 8 to 10 cm from the pylorus. The common bile duct is dilated slightly within the papilla, that portion forming the *ampulla of Vater.* This duct receives the main pancreatic duct just before it terminates. At this point, a sphincter muscle, the *sphincter of Oddi,* controls the flow of bile and pancreatic juice.

Pancreas (Figs. 7-12, 7-15). The pancreas is a compound tubuloacinar gland lying below the liver and the stomach, at the level of the 2nd and 3rd lumbar vertebrae. Its larger portion, or *head,* lies adjacent to the middle portion of the duodenum. The remainder (a neck, a body, and a tail) extends transversely to the left, terminating near the spleen.

The pancreas is a finely lobulated, straw-colored organ, averaging 20 to 25 cm in length and weighing from 65 to 160 g. It lacks a definite capsule but is enclosed in a thin covering of connective tissue. The pancreas consists of two types of secreting tissue: an *exocrine portion,* which secretes pancreatic juice containing enzymes (see p. 246) and sodium bicarbonate, and an *endocrine portion,* which secretes hormones. Pancreatic juice is discharged through the pancreatic ducts into the duodenum; pancreatic hormones, *insulin* and *glucagon,* are absorbed into the bloodstream.

EXOCRINE PORTION OF THE PANCREAS. The *acini,* terminal secreting units, are grouped in lobules that are bound together by connective tissue. Some acini are flask-shaped; others are tubular. Each acinus consists of a single row of pyramidal cells surrounding a lumen that is small in the inactive state but distended when the gland is active. The lobules are separated by interlobular septa. The exocrine portion of the pancreas contains two excretory ducts:

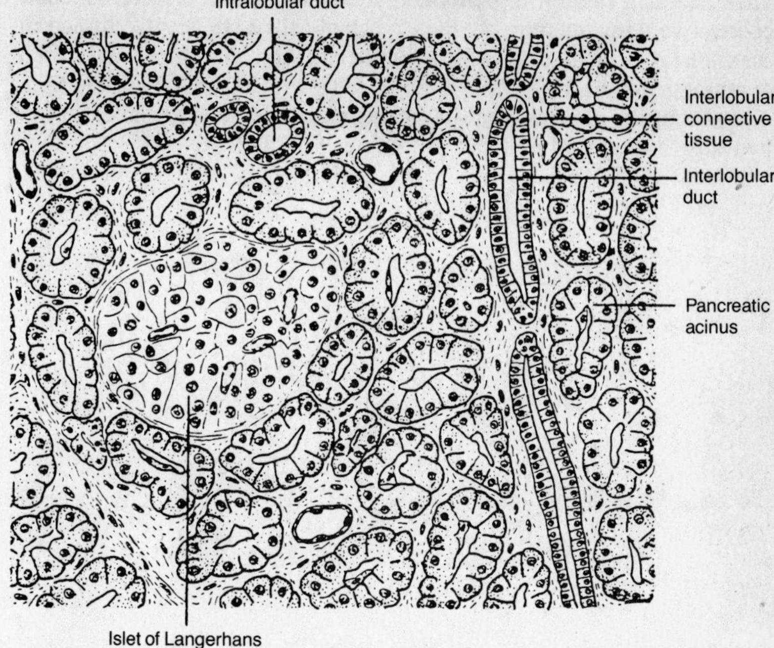

Fig. 7-15. Section of human pancreas. (Reprinted with permission of Oxford University Press from *Cunningham's Textbook of Anatomy,* edited by G. J. Romanes, 11th ed., 1972.)

1. The main excretory canal, or *pancreatic duct,* passes through the middle of the pancreas from left to right. At about the midpoint it bends sharply downward and continues to its termination in the common bile duct at the *ampulla of Vater,* where its secretion enters the duodenum.
2. The *accessory pancreatic duct* is variable. Typically it opens into the duodenum about 2.5 cm above the opening of the common bile duct. It usually has two other branches, one draining the head of the pancreas, the other connecting with the main duct.

Both main and accessory ducts receive branches, the *interlobular ducts,* which in turn receive *intralobular ducts,* one from each primary lobule. The intralobular ducts receive *intercalated ducts,* one from each tubule or acinus.

There is also a system of anastomosing tubules that connect with both the larger ducts and the islets of Langerhans. They do not, however, carry any secretion.

Endocrine Portion of the Pancreas. The *islets of Langerhans* comprise the endocrine portion of the pancreas. These small masses of cells scattered throughout the pancreas number 200,000 to 2,000,000.

Each islet is a small, irregular structure enclosed in a delicate reticular connective tissue membrane. Four types of cells have been identified in the islets: *A* or *alpha cells,* which secrete glucagon, *B* or *beta cells,* which secrete insulin, and *C cells* and *D* or *delta cells,* of unknown function. The role of the pancreas and its hormones in the metabolism of carbohydrates is discussed in detail in Chapter 6, Vol. 2.

8: THE PHYSIOLOGY OF DIGESTION

Digestion is the process of converting food substances to a state in which they can be absorbed through the lining of the digestive tract. It is accomplished by *mechanical processes,* such as changes in form brought about by the action of the teeth or the movements of the muscular walls of the various portions of the alimentary canal, and by *chemical processes,* such as the changes induced by enzymes produced by digestive glands.

FOODS

A thorough understanding of the physiology of the processes of digestion requires for its foundation a knowledge of the chemical nature of foodstuffs, their classification, and the role of food accessories (water, vitamins, inorganic salts, and other substances).

Functions of Foods. Foods are substances taken into the body that serve the following functions:

1. To provide the chemical elements or compounds essential for the *synthesis of protoplasm* (that is, for the building of new tissue or the repair of injured or worn-out tissue). They also provide the materials for growth and those from which secretions of glands are synthesized.
2. To serve as sources of *energy* for metabolic activities and *heat* for the maintenance of body temperature.
3. To provide essential substances for *regulation* of cellular or bodily functions.

Classification of Foodstuffs. Foodstuffs are grouped in two general classes: foods proper and food accessories. The main divisions within these classes follow:

Foods proper (*organic compounds serving as sources of energy*)

Carbohydrates	Proteins	Lipids (fats)
monosaccharides	simple proteins	simple lipids
disaccharides	conjugated proteins	compound lipids
polysaccharides	derived proteins	derived lipids

Food accessories (*substances needed for life activities but providing no energy*)

Water	Inorganic salts	Vitamins

Carbohydrates. In this class of foodstuffs are found compounds of carbon, hydrogen, and oxygen, the two last being in the ratio of 2:1. A carbohydrate molecule usually consists of six, or a multiple of six, carbon atoms. The three principal kinds are monosaccharides, disaccharides, and polysaccharides.

MONOSACCHARIDES. These are *simple sugars.* Their empirical formula is $C_6H_{12}O_6$. Examples are grape sugar (*glucose* or *dextrose*), fruit sugar (*fructose* or *levulose*), and *galactose.* These sugars are soluble in water and readily absorbed by the mucosa of the digestive tract. In Fehling's or Benedict's solution, copper hydroxide is reduced to cuprous oxide, with an accompanying color change from blue to orange or brick red; for this reason these solutions are used in testing substances such as urine for presence of sugar. Monosaccharides are readily oxidized in the tissues; consequently, they are readily available sources of energy.

DISACCHARIDES. These are *double sugars.* Upon hydrolysis they yield two molecules of a simple sugar. Their empirical formula is $C_{12}H_{22}O_{11}$. Examples are cane sugar (*sucrose*), malt sugar (*maltose*), and milk sugar (*lactose*). Sucrose is the common form found in the stem of the sugar cane and the root of the sugar beet. Maltose is found in germinating grains and in malt; lactose is found in the milk of all mammals. Disaccharides are formed in the hydrolysis of starch.

POLYSACCHARIDES. These are *multiple sugars,* which are insoluble in water. Their empirical formula is $(C_6H_{10}O_5)_n$. Examples are *starch* (in plant cells), *glycogen* (animal starch, abundant in liver and muscle cells), and *cellulose* (in walls of plant cells). They are a common form of storage for carbohydrates.

FUNCTION OF CARBOHYDRATES. Carbohydrates serve the following functions in the body: (1) They are a primary source of energy; (2) they are readily stored and serve as a quick reserve supply of energy; (3) they exert a regulatory effect on lipid and protein metabolism; (4) they are essential for the normal functioning of nervous tissue; and (5) they provide nondigestible substances that constitute the major portion of the fiber of foods that gives bulk to the feces and is essential for the normal muscular tone of the intestine.

Proteins. Proteins are compounds containing carbon, hydrogen, nitrogen, and oxygen and sometimes sulfur and phosphorus. They are classified as follows:

SIMPLE PROTEINS. These include *albumins* (serum albumin, egg albumin), *globulins* (serum globulin, fibrinogen, vitellin), *glutelins* (gluten), *prolamines* (gliadin, zein), *albuminoids* or *scleroproteins* (keratin, collagen, elastin, gelatin), *histones* (globin), and *protamines* (salmine).

CONJUGATED PROTEINS. These are compounds of simple proteins combined with a nonprotein substance. Included are *nucleoproteins* (in cell nuclei), *chromoproteins* (hemoglobin, hemocyanin), *glycoproteins* and *mucoproteins* (mucin, chondromucoid), *lipoproteins* (of blood plasma and cell membranes), *phosphoproteins* (casein, vitellin), and *metaloproteins* (heme).

DERIVED PROTEINS. These are substances resulting from the action of acids, alkalies, or proteolytic enzymes on simple or conjugated proteins. They include *primary derivatives* (proteans, metaproteins, and coagulated proteins) and *secondary derivatives* (proteoses, peptones, polypeptides, and peptides).

CHEMICAL NATURE OF PROTEINS. The protein molecule is composed of a number of chemical units linked together. These units are *amino acids,* the so-called "building blocks" of proteins. Some 21 or more different amino acids have been identified. Some proteins contain all or most of the amino acids; others, such as gelatin, contain only 14 or 15, while some simple proteins may have only 3 or 4. Protein molecules are extremely complex and have a very high molecular weight. They vary in shape, the *fibrous* proteins having an elongated fiberlike form, the *globular* proteins a more or less compact, globular form. Owing to their structural and chemical properties, proteins constitute the most important component in the structural organization of the protoplasm of cells.

Proteins that enable an animal to grow and carry on fundamental life activities are *complete proteins;* that is, they contain *all* the essential amino acids. Examples are proteins of animal origin: meat, milk, milk products, and eggs. Proteins lacking some or most of the essential amino acids are *incomplete proteins.* Most proteins of plant origin such as those in grains, legumes, and nuts are incomplete proteins.

AMINO ACIDS. An amino acid is an organic compound possessing both an *amino* (NH_2) and a *carboxyl* (—COOH) radical. Their basic formula is NH_2—R—COOH, in which R stands for an aromatic or aliphatic radical. Amino acids are amphoteric; that is, they can act either as acids, because of the carboxyl group, or as bases, because of the amino group, the reaction depending upon the pH of the solution. This enables them to play an important role as buffering agents.

Essential amino acids are those that are necessary for normal growth and development and are not synthesized in the body. These must be obtained from proteins in food. *Nonessential amino acids* are those the body can synthesize. Listed here are the important amino acids.

Essential	*Nonessential*
Arginine (Arg)*	Alanine (Ala)
Histidine (His)*	Asparagine (Asn)
Isoleucine (Ileu)	Aspartic acid (Asp)
Leucine (Leu)	Cysteine (Cys)
Lysine (Lys)	Glutamic acid (Glu)
Methionine (Met)	Glutamine (Gln)
Phenylalanine (Phe)	Glycine (Gly)
Threonine (Thr)	Norleucine (Nor)
Tryptophane (Try)	Proline (Pro)
Valine (Val)	Serine (Ser)
	Tyrosine (Tyr)

*Essential only in growing children.

FUNCTION OF PROTEINS. Proteins, or amino acids derived from them, serve the following functions in the body: (1) They are the primary constituents of all cells; consequently, they are the structural base of membranes, muscles, blood, and all tissue and are essential for the building of any new tissue; (2) they are essential for the synthesis of all enzymes, some hormones, the secretions of many glands, and the formation of antibodies; (3) they serve as transport agents for various substances such as oxygen, some hormones, and certain minerals; (4) they are of importance in the maintenance of fluid balance between the blood and the tissues; (5) they are a component of nucleoproteins; hence, they are of importance in the structure of all cell nuclei and function in the transmission of genetic traits; and (6) they are sources of energy.

Lipids or Fats. Most lipids are compounds of carbon, hydrogen, and oxygen, but some also contain phosphorus and nitrogen. They are compounds that yield fatty acids when hydrolyzed or complex alcohols capable of combining with fatty acids to form esters. Lipids are soluble in organic liquids such as ether or benzene but are nearly or completely insoluble in water. Lipids are classified as follows.

SIMPLE LIPIDS. These include the neutral fats and waxes. Neutral fats are triglycerides (esters of fatty acids and glycerol). The principal fatty acids are oleic, stearic, palmitic, and linoleic. Stearic and palmitic acids are *saturated* (they contain as many hydrogen atoms as the carbon chain will hold); oleic and linoleic acids are *unsaturated.* Animal fats contain a larger percentage of saturated fatty acids than most vegetable fats or oils.

COMPOUND LIPIDS. These include phospholipids, glycolipids, and lipoproteins. *Phospholipids* are compounds of fatty acids, phosphoric acid, and a nitrogenous base. Examples are lecithin, cephalin, and sphingomyelin. *Glycolipids* are compounds of fatty acid combined with a carbohydrate and a nitrogenous base. Examples are cerebrosides and gangliosides. *Lipoproteins* are lipids combined with a protein.

DERIVED LIPIDS. These are substances derived from simple or compound lipids by hydrolysis or enzymatic action. Included are the *fatty acids* (mono- and diglycerides), *glycerol, sterols* (cholesterol, ergosterol, steroid hormones, vitamin D, and bile salts), and the *fat-soluble vitamins* A, E, and K.

FUNCTIONS OF FATS. Fats are high-energy foods, each gram providing twice the amount of energy provided by 1 g of carbohydrate or protein. Fat in adipose tissue is the principal stored form of energy. Fats support and protect organs, form an insulating layer under the skin, aid in the absorption and transport of fat-soluble vitamins, form an integral part of cell membranes, provide fatty acids essential to growth, and are components or precursors of many substances, such as hormones and vitamins, that are essential for normal physiological activities. In foods they contribute much to palatability and flavor.

Food Accessories. A number of food substances, some organic, others inorganic, do not serve as sources of energy. These, called *food accessories,* include water, inorganic salts, vitamins, and pseudovitamins.

WATER. Water is the most important single compound in the body, comprising about 70 percent of body weight. Its role in bodily activities is summarized on pages 27 and 28.

INORGANIC SALTS. These include the mineral compounds or the ash constituents of the body. The principal minerals of the body and the percentage of total ash each comprises as follows:

Calcium (Ca)	39.0%	Chlorine (Cl)	3.0%
Phosphorus (P)	22.0%	Sodium (Na)	2.0%
Potassium (K)	5.0%	Magnesium (Mg)	0.7%
Sulfur (S)	4.0%	Iron (Fe)	0.15%

The following appear in only minute amounts:

Iodine (I)	Zinc (Zn)	Cobalt (Co)
Manganese (Mn)	Aluminum (Al)	Molybdenum (Mo)
Copper (Cu)	Bromine (Br)	Vanadium (V)
Nickel (Ni)	Silicon (Si)	Chromium (Cr)
Fluorine (Fl)	Selenium (Se)	(traces of others)

The principal minerals (*macrominerals*) in the body have the following specific roles:

Calcium—essential for development of bone and teeth, coagulation of blood, normal heart action, and normal neuromuscular irritability. It also plays a role in the permeability of cell membranes and in the activities of certain enzymes.

Phosphorus—functions principally with calcium as calcium phosphate; consequently, it is also essential for development of bone and teeth. Phosphorus plays an important role in the liberation of energy for muscle contraction and in other bodily processes involving energy transfer. Combined with lipids, it plays an important role in the structure of cell membranes and, through its phosphoric acid buffer system, helps maintain acid-base balance.

Potassium—an essential constituent of all cells and of importance in the maintenance of normal osmotic pressure and acid-base balance of the body fluids; essential for normal nerve conduction and muscle contractility and for protein and carbohydrate metabolism. It is the principal intracellular cation.

Sodium—an important extracellular ion. It functions in acid-base and water balance, osmotic pressure, carbon dioxide transport, cell membrane permeability, and muscle irritability.

Chlorine—one of the principal ions in extracellular fluids and of primary importance in maintaining osmolarity of body fluids and acid-base balance.

Sulfur—an important constituent of nearly all proteins. It is present in all cells and plays a role in the activation of enzymes and in certain detoxication reactions.

Magnesium—an important cation in intracellular fluid; a constituent of bones and teeth, and an essential component of some enzyme systems.

Iron—a component of hemoglobin, hence essential in the formation of red blood cells; a component of certain enzyme systems involved in energy release.

The following minerals (*microminerals*), which are present only in trace amounts, play the following roles in bodily metabolism.

Iodine—essential in the formation of thyroid hormones and in the regulation of cellular metabolism.

Manganese—an essential cofactor in certain enzyme systems.

Copper—essential for absorption and transport of iron with which it functions in certain enzyme systems and in the formation of hemoglobin. It is also involved in bone formation and in the maintenance of myelin in the nervous system.

Fluorine—essential for normal bone and tooth structure; increases resistance of dentine to dental caries.

Zinc—an important component in certain enzyme systems, especially the carbonic anhydrase system; essential in protein biosynthesis and utilization.

Cobalt—an essential constituent of vitamin B_{12}.

Molybdenum—a constituent of certain enzymes.

Chromium—essential for glucose utilization.

Selenium—essential in tissue respiration, inhibits lipid peroxidation.

The following trace materials (nickel, aluminum, silicon, vanadium, and possibly others) have not been shown to be essential to human health.

VITAMINS. The term *vitamin* is applied to a number of unrelated substances found in many foods in minute quantities and essential for the normal metabolic processes. All are complex compounds. Most are produced by animal or plant cells; a few have been synthesized.

In the body most vitamins are obtained from ingested organic foods. In some cases they may be synthesized within the tissues (e.g., vitamins A and D); in others, symbiotic bacteria residing in the intestines are the source (e.g., vitamin K).

There are two general classes of vitamins: *fat-soluble vitamins,* which include vitamins A, D, E, and K, and *water-soluble vitamins,* which include those of the B complex and vitamin C.

Functions. Vitamins are essential for normal growth and development. Vitamins provide no energy, but they participate as coenzymes or prosthetic groups for enzymes that catalyze the chemical reactions involved in various metabolic activities and the release of energy. With the exception of vitamins A and D, vitamins are not stored in significant quantities in the body. Water-soluble vitamins are excreted principally in the urine. Vitamins are also excreted in the feces, sweat, and milk.

Hypovitaminosis is the condition in which there is an absence or a deficiency of one or more vitamins. This may be due to inadequate intake, defective absorption or utilization, faulty bacterial synthesis, excessive elimination, the action of antivitamins, or other factors.

Hypervitaminosis is the condition in which there is an excessive intake of one or more vitamins. This may result from excessive vitamin content of food ingested, excessive ingestion of vitamins by mistake, or overzealous vitamin therapy, often the result of food faddism. Excessive intake of vitamins A, D, K or thiamine (B_1) may give rise to toxic manifestations.

FAT-SOLUBLE VITAMINS. The following tables give information about the principal fat-soluble vitamins (A, D, E, and K). (IU is the abbreviation for *International Unit*).

VITAMIN A
(Retinol)

Chemical Nature

An alcohol ($C_{20}H_{20}OH$) formed in the body from precursors, yellow pigments of plants (alpha, beta, and gamma carotene, and cryptoxanthin). It is stored in the liver of vertebrates.

Function

Normal cell growth and development; integrity of epithelial tissues; activity of osteoblasts and odontoblasts; formation of visual purple (*rhodopsin*), which is essential for vision in dim light.

Source

Yellow and dark green vegetables, especially carrots, spinach, sweet potatoes; yellow fruits; egg yolk; butter, whole milk, cheese; fresh liver and fish oils (cod and halibut liver oil, in particular).

Daily Requirement

5000 IU for average adult; pregnant women need 6000 to 8000 IU.

Effects of Deficiency

Dryness, hardness, and roughening of skin; night blindness or lessened visual acuity in dim light; degenerative changes in epithelia of mucous membranes, with tendency to cornification of stratified squamous epithelia; failure of normal growth and development; increased susceptibility to infections; alterations in glandular secretions.

Effects of Excess

Acute hypervitaminosis A may result from eating the liver of polar bears or other Arctic animals. Severe headaches, nausea, and vomiting result. *Chronic hypervitaminosis A* may result from therapeutic administration of excessive amounts of vitamin A (50,000 IU or more) daily for long periods of time. It is characterized by anorexia, weight loss, loss of hair, and premature calcification of bone.

VITAMIN D
(Calciferol)

Chemical Nature

Mixture of D_2 (calciferol, $C_{28}H_{43}OH$) and D_3 (activated dehydro-

Source

Fish liver oils, especially cod liver oil; butter; cream; egg yolk. Many

VITAMIN D (CONT.)

cholesterol). Ergosterol, precursor of calciferol, is transformed to vitamin D by action of ultraviolet rays. In the body, sunlight or ultraviolet irradiation via the skin bring about production of vitamin D from its precursors in the dermis (hence, "sunshine vitamin"). It is stored in the liver.

foods, especially milk, bread, and cereals, are fortified with vitamin D. Because it is synthesized in the body from precursors, it is now regarded as a hormone.

Function

Calcium and phosphorus metabolism; promotes their absorption from intestine and maintains their proper concentrations in the blood, especially in bone-forming regions. Essential for normal eruption of teeth and in preventing dental caries.

Daily Requirement

Normal adults need no supplementary intake. Infants and children, and women during the latter half of pregnancy and in lactation, require 400 IU. For elderly persons or persons receiving inadequate exposure to sunlight, a small amount per day is recommended.

Effects of Deficiency

In children, lack of vitamin D is evidenced by irritability, abnormal tooth and bone development, and malaise. Severe deficiency causes *rickets.* In adults, severe deficiency is manifested by *osteomalacia,* which predominates in women. It is usually accompanied by tetany.

Effects of Excess

Hypervitaminosis D may result from excessive intake. Excessive calcification of bone and calcification of soft tissues, especially the kidney, heart, and lungs, may occur. Gastrointestinal disturbances develop, and renal function is impaired.

VITAMIN E
(Tocopherol)

Chemical Nature

Consists of a mixture of tocopherols (alpha, beta, gamma, delta), of which alpha is the most effective biologically. Stable to acids and to heat; unstable in alkalis; oxidizes slowly.

Source

Vegetable oils, especially those rich in polyunsaturated fatty acids; fish; liver; whole-wheat products; unpolished rice; oatmeal.

Function

Specific role in human nutrition has not been established, but in laboratory animals it is essential for normal reproduction and liver function and the integrity of muscle tissue and red blood cells. It may function as a coenzyme in cell respiration.

Daily requirement

For adults, 25 to 30 IU.

Effects of Deficiency

Not definitely established; possibly impaired utilization of vitamins A and D, hemolysis of red blood cells, cardiovascular disorders, impaired functioning of reproductive systems.

VITAMIN K
(Phylloquinone, Farnoquinone)

Chemical Nature

Occurs naturally in two forms: K_1 or phylloquinone ($C_{31}H_{45}O_2$) and K_2 or farnoquinone ($C_{41}H_{56}O_2$). The former is a light yellow oil, the latter a yellow crystalline compound. A synthetic compound, menadione, possesses similar properties. It is not stored in the body.

Source

Abundant in many foods, but especially in leafy green vegetables. Synthesized in the intestines by bacteria.

Function

Formation by the liver of prothrombin, a substance necessary for normal clotting of blood. In its absence coagulation time is prolonged. Vitamin K appears to be an essential part of the enzyme system concerned with the synthesis of prothrombin.

Daily Requirement

Not definitely known, but 2 mg is considered adequate.

Effects of Deficiency

This is uncommon in man because bacteria in the intestine synthesize vitamin K. In infants, deficiency causes excessive hemorrhaging. This may result from lack of prenatal stores or from the time required for bacterial flora to become established in the intestine. In adults, deficiency may result from biliary obstruction, because the natural vitamin is not absorbed in the absence of bile. Diarrhea or excessive use of lubricants may cause loss of vitamin K in feces.

Effects of Excess

Hypervitaminosis K sometimes occurs in children, resulting in hemolytic anemia, excess bilirubin in the blood, and deposition of a yellow pigment in parts of the brain.

WATER-SOLUBLE VITAMINS. These include vitamins of the *B complex* and *vitamin C.* B vitamins include thiamine (B_1), riboflavin (B_2), pyridoxine (B_6), niacin, pantothenic acid (B_5), folic acid (B_9 or B_{10}), biotin (B_7 or H), cyanocobalamine (B_{12}), para-aminobenzoic acid (PABA), inositol, choline, and lipoic acid. The following tables summarize information on each of these vitamins.

Thiamine
VITAMIN B_1

Chemical Nature

A white crystalline compound with the empiric formula $C_{12}H_{17}N_4OS$ produced synthetically as thiamine hydrochloride. It is stable in a dry state and not rapidly destroyed by heat.

Source

Widely distributed in various animal and plant foods. Dry yeast and wheat germ are its richest natural sources. Occurs in outer layers of seeds and in nuts, legumes, most vegetables, some meats (especially pork, liver, and muscle). Has been synthesized.

Function

Carbohydrate metabolism. Acts as coenzyme of carboxylases in decarboxylation of pyruvic acid, an intermediate product in conversion of glucose to carbon dioxide and water. Thiamine is essential for liberation of energy and disposal of pyruvic acid.

Daily Requirement

For normal adults, 0.5 mg for each 1000 cal, or 1.5 to 3 mg, depending on activity and carbohydrate intake. It is not stored to any extent and must be supplied regularly in the diet. Excess is eliminated in urine and through skin.

Effects of Deficiency

Thiamine deficiency is the commonest vitamin lack. It is also one of the easiest to recognize and to correct. When thiamine is deficient, pyruvic and lactic acids accumulate in the blood and brain, with marked impairment of nervous, circulatory, digestive, and endocrine functions. Moderate deficiency is manifested by irritability, loss of appetite, muscle tenderness, and fatigue. Pronounced deficiency causes *beriberi,* whose symptoms are neuritis, paralysis, nerve lesions, digestive disorders, cardiac disturbances, and neurasthenia (*fatigue syndrome*). Symptoms disappear quickly upon administration of thiamine.

Effects of Excess

Thiamine in excessive amounts may give rise to nervousness, tremor, dyspnea, tachycardia, and sweating. Sometimes severe allergic-type anaphylactic reactions occur.

Riboflavin
VITAMIN B_2

Chemical Nature

An orange-yellow crystalline powder ($C_{17}H_{20}N_4O_6$), comparatively stable to heat and air but unstable to light. It is absorbed readily by the small intestine but stored only to limited extent in the liver and kidneys. It is excreted in the urine and the feces.

Source

Milk and milk products, leafy green vegetables, liver, beef, fish, and dry yeast.

VITAMIN B_2 (CONT.)

Function

Tissue respiration. In combination with phosphoric acid, it is a component of flavoprotein coenzymes, which, acting with nicotinamide-containing enzymes, are essential in oxidation reactions in the normal metabolism of cells. Riboflavin is essential for normal health, vigor, and resistance to disease. It is necessary for tissue repair. It is especially important in the metabolism of proteins.

Daily Requirement

Depending on activity, 1.5 to 2.0 mg for normal adults, 2.5 to 3.0 mg for pregnant and lactating women. Intake should be increased after tissue injury, since growth of new tissues is dependent on supply of riboflavin.

Effects of Deficiency

Owing to the wide distribution of riboflavin and to its bacterial synthesis in the body, deficiency of this vitamin is not common in man, and there is no distinct, general clinical picture. The most common signs are eye disorders (vascularization of cornea, conjunctivitis, increased lacrimal secretion, photophobia); cracking and splitting of lips at angles (cheilosis); magenta-colored tongue resulting from inflammation (glossitis), and seborrheic dermatitis, especially of face and scalp.

Pyridoxine
VITAMIN B_6
(Pyridoxal, Pyridoxamine)

Chemical Nature

A white, crystalline, odorless compound ($C_8H_{10}0_3N$), soluble in water and alcohol. Stable to heat and acids; unstable in alkaline solutions and to light.

Source

Whole-wheat products, yeast, pork, liver, egg yolk.

Functions

Plays an important role in the absorption and interconversions of amino acids; essential in utilization of fatty acids and in tryptophane metabolism.

Daily Requirement

For adults, 1.5 to 2 mg.

Effects of Deficiency

Anemia resulting from a defect in the synthesis of heme; nervous system disorders, hyperirritability, convulsions, mental retardation; cheilosis, seborrheic dermatitis about eyes and mouth; impaired growth. Isoniazid, an antituberculosis drug, is an antagonist.

Niacin
(Nicotinic acid, Niacinamide)

Chemical Nature

A white, crystalline substance produced from tryptophane, readily converted into nicotinamide. Stable in heat, air, light, and alkalies.

Functions

Niacin is a component of coenzymes NAD and NADP, pyridine nucleotides involved in hydrogen transport. It plays an essential role in cellular respiration, being involved in the release of energy from all classes of foods, especially carbohydrates. It plays a role in glycolysis and the synthesis of fats.

Source

Same foods that provide riboflavin, plus mushrooms, whole-wheat products, poultry.

Daily Requirement

12 to 20 mg for men and 11 to 15 mg for women; this is approximately 5 mg per 1000 cal. Increased amounts with unusual energy expenditure and in stress situations.

Effects of Deficiency

Same as those of pellagra: inflammation of mouth cavity and tongue, gastrointestinal disturbances, bilateral dermatitis (especially of arms, legs, neck, and submammary regions), and nervous symptoms (insomnia, irritability, neurasthenia). Degeneration of spinal pathways may occur. In severest form may be fatal.

Pantothenic Acid
Vitamin B_5

Chemical Nature

A white, crystalline compound ($C_9H_{17}NO_5$). Stable in dry form but easily destroyed by acids, alkalies, and heat.

Functions

Forms a part of coenzyme A, which is involved in intermediate metabolism of carbohydrates, proteins, and fats.

Source

Widely distributed in foods, such as yeast, liver, eggs, milk, many vegetables, molasses. Synthesized by intestinal bacteria.

Daily Requirement

For adults, 5 to 10 mg.

Effects of Deficiency

Deficiencies are rare. Experimentally induced deficiency results in nervous disorders, insomnia, fatigue, and gastrointestinal disturbances.

Folic Acid
(Folacin, Pteroylglutamic Acid)

Chemical Nature

The folic acids include a large and poorly defined group of B complex vitamins, the best-known of which is pteroylglutamic acid (PGA), a bright yellow, crystalline substance slightly soluble in water. Its sodium salts are readily soluble.

Source

Widespread in animal and plant foods, especially abundant in leafy dark-green vegetables, liver, kidney, and yeast. Synthesized by bacteria in the intestine. A synthetic form has been prepared.

Functions

Serve as coenzymes in the transfer of 1-carbon units in various metabolic reactions. Important resulting products are thymine and other purines and pyrimidines, which are components of nucleoproteins. They are essential for cellular growth and reproduction, especially in the formation of red blood cells.

Daily Requirements

4 mg daily; 8 mg during pregnancy; 5 mg during lactation.

Effects of Deficiency

Results in macrocytic anemia, leucopenia, and hyperplasia of the bone marrow. It has proved to be effective in the treatment of *sprue*, a tropical disease, frequently fatal, characterized by glossitis, gastrointestinal disorders, loss of appetite, emaciation, and marked macrocytic anemia. It is used in the treatment of *pernicious anemia* but does not prevent degenerative neurologic changes associated with it. Folic acid antagonists (e.g., *aminopterin*) or the use of antibiotics that inhibit intestinal bacterial growth may induce some of these symptoms.

Biotin
(Vitamin H, Vitamin B_7)

Chemical Nature

A substance ($C_{10}H_{16}N_2O_3S$) present in every living cell. Fine, long needles insoluble in organic solvents. Stable in air and to heat.

Source

Liver, kidney, yeast, many vegetables. Synthesized by intestinal bacteria.

Function

Serves as a coenzyme in carboxylation, deamination, and oxidation reactions. Essential for synthesis and oxidation of fatty acids and synthesis of amino acids.

Daily Requirement

Has not been established. Only a minute amount is required.

BIOTIN (CONT.)

Effects of Deficiency

Deficiencies are rare. May occur in individuals or laboratory animals who are fed large amounts of raw egg white, which contains a substance, *avidin,* that interferes with the absorption of biotin.

Cyanocobalamin VITAMIN B_{12}

Chemical Nature

Consists of deep red, needlelike, cobalt-containing crystals soluble in water and alcohol. Stable to heat but inactivated by acids and alkalies. Stored in the liver.

Functions

When combined with cyanide, it becomes an active coenzyme that functions in the synthesis of nucleoproteins and in the utilization of carbohydrates and fats. With folic acid, it stimulates red blood cell formation. It is essential in the metabolic activities of all cells.

Source

Animal tissues, especially liver, kidney, muscle, and milk products. Synthesized by intestinal bacteria. Not found in plants.

Daily Requirement

For adults, 5 μg; 5 to 8 μg during pregnancy and lactation; 6 μg in old age. Two gastric factors, *hydrochloric acid* and an *intrinsic factor,* plus calcium are essential for its absorption by the intestinal mucosa.

Effects of Deficiency

Pernicious anemia due to failure of red blood cells to mature. This may result from inadequate intake of B_{12} (extrinsic factor) or failure of the stomach to produce the intrinsic factor necessary for absorption. Demyelination of nerve fibers in the spinal cord may occur. People living on a meatless diet have a low dietary intake of B_{12}.

Ascorbic Acid VITAMIN C

Chemical Nature

A white crystalline compound ($C_6H_8O_6$). Soluble in water. Stable in air when dry, but in solution it is readily oxidized. Light, alkalies, and certain metals (iron and copper) accelerate its oxidation. Inactivated by prolonged application of heat.

Source

Citrus fruits, other raw fruits, tomatoes, vegetables, greens (especially cabbage and peppers). Synthesized by most plants and by many animals (but human, monkey, guinea pig *do not* synthesize it). Fruits and vegetables lose their vitamin C activity rapidly when stored at room temperature.

VITAMIN C (CONT.)

Function

Exact mode of action not well understood. Plays a role in metabolism of certain amino acids. Necessary for formation of collagen, a protein forming an essential part of the white fibers of connective tissue and the organic substance of bone. Maintains integrity of intercellular cement in many tissues, especially capillary walls. Involved in formation and utilization of adrenal cortical hormones. Enhances absorption of iron.

Daily Requirement

60 mg for men, 55 mg for women. During pregnancy and lactation, 60 mg is recommended. Growing children need relatively more than do adults. Intake should be increased during convalescence from fevers, infections, or injuries such as burns and fractures. Not stored in any one organ, but widely distributed in body tissues. Excreted in urine. The value of massive doses is questionable.

Effects of Deficiency

Pronounced deficiency results in *scurvy*. Some symptoms are redness, swelling, and bleeding of gums, with resultant loss of teeth; subcutaneous hemorrhages due to increased fragility of capillaries; delayed healing of soft tissues and injured bones; edema; secondary anemia; pain and tenderness in extremities. In *subclinical* cases the symptoms are less pronounced but may also include irritability, loss of appetite, and weight loss.

PSEUDOVITAMINS. The following factors of the B complex are not regarded as true vitamins, but since they play vitaminlike roles and are important in human nutrition, essential data will be given for each. They are called *pseudovitamins* or *growth factors.*

Para-aminobenzoic acid (PABA) is a growth-promoting factor in chicks and anti–gray-hair factor in rats and mice. It is an essential growth factor also for certain microorganisms, especially those inhibited by the sulfonamides and related drugs. It is theorized that the sulfa drugs, which are related to PABA, produce their antibacterial effects by replacing this factor.

Inositol, in the diet of laboratory animals, is a necessary factor for normal growth and development. In mice, rats, and guinea pigs, deficiency results in retarded growth, loss of hair, dermatitis, and deposition of fat in the liver. Since inositol is present in living tissues in greater quantities than true vitamins, it probably serves as a structural component rather than as a catalyst.

The specific role of inositol in human nutrition has not been established. It is widely distributed in foods, being present in meats, milk, most fruits (especially citrus fruits), and in various grains, nuts, and legumes.

Choline ($C_5H_{15}NO_2$) is an essential factor in human nutrition, playing an important role in protein and fat metabolism. It is a lipotropic agent that promotes the utilization of fats. It is a precursor of *acetylcholine,* an important neurotransmitter, and a basic component of two lipids, lecithin and sphingomyelin.

Choline is widely distributed in animal foods but sparse in most fruits and vegetables. A daily requirement for humans has not been established. Deficiency results in excessive accumulation of fat in the liver and degenerative changes in other parts of the body, especially the kidneys.

Lipoic acid, also called *pyruvate oxidation factor* (POF), is a fatty acid, $C_7H_{13}S_2$(—COOH), capable of being synthesized in the body. It functions, with thiamine, as a coenzyme in the oxidative decarboxylation of pyruvate. It is involved with other vitamins in various reactions in the Krebs cycle that result in the release of energy. It is widespread in animal and plant tissues, liver and yeast being important sources. No dietary requirements for lipoic acid have been established, and no effects of deficiencies have been noted.

ANTIVITAMINS. An increasing number of cases are being reported in which a substance interferes with the absorption of a vitamin or its utilization within the body. Such a substance is called an *antivitamin, vitamin antagonist,* or *antimetabolite.* An antivitamin may act in either of two ways: (1) By being structurally similar to a vitamin, it may become involved in reactions without being able to produce the effect of the vitamin; or (2) by becoming attached to an enzyme, it may block the incorporation of an essential vitamin into an enzyme system.

Among substances known to act as an antivitamin are *avidin,* a substance in raw egg white that prevents the absorption of biotin; *isoniazid,* an antituberculosis drug that is an antagonist for pyridoxine; *aminopteran,* which is an antagonist for folic acid; and *dicoumarol,* an anticoagulant, which is an antagonist for vitamin K. Some antibiotics may cause vitamin deficiencies by their adverse effects on intestinal bacteria that synthesize vitamins.

MOVEMENTS OF THE ALIMENTARY CANAL

The muscles of the lips, jaws, tongue, cheeks, walls of the pharynx, and upper portion of the esophagus are striated; accordingly, the movements of these parts, with the exception of the pharynx and the upper portion of the esophagus, are *voluntary.* Throughout the remainder of the digestive tract, all muscles are of the smooth type and are innervated by the autonomic nervous system; therefore, they are *involuntary* in action. The following description of the movements of the alimentary canal is confined to a consideration of these smooth muscles.

Characteristics of Smooth Muscles of the Alimentary Canal. The smooth muscles of the viscera are generally arranged in two layers: an outer *longitudinal* layer, which, upon contraction, brings about a shortening of the digestive tube and dilation of its lumen, and an inner *circular* layer, which, upon contraction, decreases the size of the lumen and increases the length of the tube.

Smooth muscles differ functionally from striated muscles in the following respects: (1) They have a much longer latent period (3 sec or more); (2) the threshold stimulus is higher; (3) they are much less excitable; and (4) their tonus does not depend on impulses from the central nervous system; thus, when the stomach or intestines are empty, their cavities are practically obliterated, owing to the muscle tone of the smooth muscles in their walls.

A fifth, and perhaps the most noteworthy, difference between the action of smooth muscles and that of striated muscles is that *smooth muscles exhibit slow contraction and relaxation repeated rhythmically at regular intervals.* Such rhythmic activity occurs in the smooth muscles of the arterial walls and even in a loop or section of the intestine removed from the body.

Muscles of the alimentary canal, in general, are under nervous and hormonal control. Nervous control is accomplished through two nerve plexuses, a *myenteric plexus* located between the layers of circular and longitudinal muscles and a *submucous plexus* located in the submucosa. These plexuses are present in the digestive tract from esophagus to anus. They contain nerve cells whose axons synapse with other neurons in the same plexus, or their axons may pass to smooth muscles or glands. Axons of a myenteric plexus may synapse with neurons in a submucous plexus or vice versa. Stimulation of a plexus may initiate impulses that can travel both up and down the intestine, affecting muscles or glands some distance from the point of stimulation.

The digestive tract receives fibers from both divisions of the autonomic nervous system. Sympathetic fibers from cell bodies in sympathetic ganglia may pass directly to glands or muscle fibers, or they may end in a plexus. The principal autonomic nerve innervating the intestine is the *vagus,* a parasympathetic nerve that contains both efferent (motor) fibers and afferent (sensory) fibers. Some of its fibers end in the plexuses that serve as terminal ganglia. Dendrites of cells in the plexuses may act as sensory receptors, in which case local reflexes in which the impulses do not reach the spinal cord may be effected. This could account for muscular movement or glandular secretion in isolated sections. Other receptors are the terminations of sensory neurons whose cell bodies lie in spinal or cranial ganglia.

The smooth muscles of the intestine, being of the single-unit type, are also capable of spontaneous, rhythmic contractions independent of nervous or hormonal input. Epinephrine, an adrenal medullary hormone, inhibits intestinal motility.

Movements of Various Portions of the Alimentary Canal. The functions of the various portions of the alimentary canal are accomplished by a variety of movements that are peculiar to the local structure and to the immediate objectives.

Movements of the Mouth and Pharynx. The first two actions of the digestive tract, performed in the mouth and the pharynx, are *mastication* and *deglutition.*

Mastication. In mastication the lower jaw is raised and lowered, protracted and retracted, and moved laterally. Through combinations of these movements, the particles of food ingested are ground by the teeth into very small particles. The tongue and cheeks act to keep the food between the teeth; for soft foods, the tongue alone may be used to break up the larger particles. During mastication, saliva, a product of the salivary glands, is released into the mouth cavity. It moistens the food, causing soluble materials to go into solution. The mucin in saliva makes the solid particles stick together to form a *bolus.*

Deglutition. In deglutition, the process of *swallowing,* there are three stages. The first is under voluntary control, but after food has entered the pharynx all the successive movements are of an involuntary, reflex nature.

In the *first stage,* food passes from the mouth to the pharynx. The food mass is placed on the dorsal depressed surface of the tongue, and the tip of the tongue is raised and placed against the teeth and hard palate. Elevation and backward movement of the tongue then forces the food posteriorly through the isthmus of the fauces into the pharynx.

In the *second stage,* food passes through the pharynx and into the esophagus. The pharynx is the common passageway for food and air. To conduct food to the proper channel (so that it will not enter the air passages), as the food is pushed backward by the tongue, the soft palate and uvula are elevated, closing the *posterior choanae* (openings into the nasal cavity). The larynx is now elevated, causing the epiglottis, which is directed upward, to move posteriorly to serve as a shield that prevents food from entering the laryngeal opening and directs it into the esophagus. Simultaneously, the vocal cords approach each other, closing the glottis. During this stage, all inspiratory and expiratory movements are reflexly inhibited. Fluid substances pass quickly through the pharynx to the esophagus. Semisolid or solid substances are acted on by the constrictor muscles of the pharynx, which press on the bolus, forcing it into the esophagus. Should the respiratory passageways be opened while food is in the pharynx (as sometimes occurs from a cough, sudden laughter, or an attempt to speak), food may enter the larynx or the posterior portion of the nasal cavity.

In the *third stage* of deglutition, food passes through the esophagus and into the stomach. Liquids are expelled by the pharynx and travel to the lowest part of the esophagus without any assistance from the esophagus. Semisolid or solid foods must be forced through this organ by peristaltic contractions of the circular muscle in its wall. Relaxation of the muscles immediately below the bolus, combined with contraction of the muscles directly above it, causes the bolus to be moved downward slowly. As it progresses downward, the muscles of the area just passed contract, and by such successive muscular actions the bolus is propelled toward the stomach. The foregoing process is called *peristalsis.* The cardiac sphincter, at the opening of the stomach, is normally in a state of relaxation when the stomach is empty, but it is contracted when food is present. When the bolus reaches this sphincter, the sphincter relaxes and the food enters the stomach.

Except for the first stage, deglutition is a *reflex act.* The sensory nerve endings of the pharynx are stimulated by the presence of food or saliva, and the impulses are transmitted over afferent nerves (trigeminal, glossopharyngeal, and vagus) to the medulla of the brain, in which there is a *deglutition center.* Here are initiated motor impulses that pass over efferent nerves (trigeminal, glossopharyngeal, vagus, spinal accessory, and hypoglossal) and bring about contraction of the muscles of the pharynx and the esophagus. Although the muscles are striated, they are not under voluntary control.

The *time required* for deglutition varies. Fluids reach the cardiac sphincter in about 1 sec, but they remain above the contracted sphincter for 4 or 5 sec until it relaxes from the peristaltic wave. For most semisolid substances, the time averages 6 sec, although it may be longer.

MOVEMENTS OF THE STOMACH. The first food to enter the stomach fills the pyloric end and the peripheral portion of the body. As food continues to enter, it assumes a more central position and is almost completely enclosed by the food that preceded it. In this way the continuation of salivary digestion is facilitated.

The *muscles* of the stomach differ from those of the other parts of the digestive tract by having, in addition to the layers of circular and longitudinal muscles, a third layer in which the fibers run obliquely. The muscle layers are thin, and contractions are weak in the body and fundus; in the antrum, the layers are thicker and contractions are stronger.

In an empty stomach, the walls are almost in apposition, there being little or no cavity except in the region of the fundus, which may be distended with gas. As food enters and fills the stomach, the muscles relax, and the volume of the cavity is adjusted to its contents with little or no increase in intragastric pressure. The muscles in the upper half of the stomach remain in a state of steady tonic contraction and show little movement, but at about the middle of the stomach active peristaltic contractions occur at intervals of 10 to 20 sec. They pass down the stomach to the pyloric region, about 20 to 30 sec being required for their passage. The waves increase in strength in the antrum, this region serving as a mixing chamber.

Peristaltic contractions are initiated by changes in membrane potential originating in a *pacemaker center* located near the esophageal opening. This center discharges impulses continuously at a rate of about 3 per minute even when the stomach is empty. After filling, stomach contractions are weak for 20 to 30 min, then increase in intensity and continue until stomach contents are digested and expelled. The strong stomach contractions that occur during fasting are apparently the result of parasympathetic stimulation. These were thought to be the cause of *hunger pangs,* but a definite cause-and-effect relationship has not been established.

EMPTYING THE STOMACH. This involves the interrelated actions of three structures: the *pyloric antrum,* the *pylorus,* and the *duodenal bulb.* When the duodenum is empty, the pyloric sphincter is relaxed and the passageway from the stomach is open. As the stomach fills, peristaltic waves are initiated. When a wave arrives at the antrum, pressure there is increased, and some of the semifluid *chyme* is passed through the open pylorus. Antral pressure, however, causes the pyloric sphincter to contract, and solid material within the antrum is forced back into the body of the stomach and subjected to further churning action and the action of digestive enzymes.

The rate at which chyme enters the duodenum depends principally upon the amount of food in the stomach; the greater the amount of food, the greater the frequency and strength of peristaltic and antral contractions. These are reflex activities mediated through the internal nerve plexuses or through spinal reflex pathways. Stimuli are distention of the stomach wall and the presence of food or products of digestion in the stomach.

Gastric motility is markedly affected by the contents of the duodenum. Distention of the duodenum, the presence of fats or acids, and the presence of hypotonic solutions or the products of digestion markedly reduce the activity of gastric musculature. Control is exerted through two mechanisms, nervous and endocrine. Nervous control is accomplished through the *enterogastric reflex* involving the vagus nerve; humoral factors involve the hormones *secretin* and *cholecystokinin,* which are liberated when acid chyme enters the duodenum. The duodenal contents are acted upon by bile and pancreatic enzymes and are moved on to the jejunum; this removes the inhibiting influences, and stomach activity is resumed until the stomach is completely emptied. Gastric activity is stimulated by a hormone, *gastrin,* released by the pyloric antrum.

Emotional states affect gastric motility, their effects being mediated through the autonomic nervous system. Grief, sadness, or depression generally decrease motility; excitement, anger, or fear tend to increase it. Decreased activity generally results from parasympathetic stimulation, increased activity from sympathetic stimulation.

As to the *length of time food remains* in the stomach, foods *begin* to leave the stomach a few minutes after ingestion. The food consumed in an ordinary meal requires 3 to 5 hr to pass through this portion of the digestive system. Water leaves the stomach almost immediately after it has been swallowed. The first of the organic foods to leave are carbohydrates; they are followed by the proteins and their derivatives and, finally, the fats.

MOVEMENTS OF THE SMALL INTESTINE. Two distinct types of movements occur in the small intestine: rhythmic segmentation and diastalsis (peristalsis).

In *rhythmic segmentation* (a "dividing" motion), the intestine is observed to divide itself into a number of segments by constriction of the circular muscles. After 2 or 3 sec, the muscle fibers in the center of each of these segments contract, dividing the segment into two parts and uniting the halves of adjoining segments. Then the fibers in the center of these segments contract. This rhythmic activity produces a churning movement that thoroughly mixes the intestinal contents with glandular secretions and facilitates absorption by bringing the contents into contact with the villi. It also stimulates the flow of blood and lymph in the vessels in the intestinal wall. Food is *not propelled onward* by this movement.

The *peristaltic wave* (or diastalsis) comes about when the intestines are stimulated by the presence of a mass of food. Circular fibers at or above the food mass contract, while those below, over a considerable distance, relax. This is known as the *myenteric reflex* and is thought to be dependent upon local reflexes. The contraction wave moves along the small intestine at a variable rate (1 to 21 cm/min).

In addition to the slow peristaltic wave, there may occur at times a "peristaltic rush," in which the wave travels much faster (from 2.5 to 25 cm/sec). Such a movement takes place over a much longer portion of the intestine than does the peristaltic wave.

It requires 2 to 4 hr for chyme to traverse the length of the small intestine and reach the ileocecal sphincter, which then relaxes, permitting food to enter the upper part of the cecum. If the cecum is full, the sphincter remains closed.

Intestinal movements are affected by various factors. Distention of the intestinal wall is the principal stimulus for normal activity. Motility is increased during periods when the stomach is discharging chyme into the duodenum. This is known as the *gastroileal reflex*. Conversely, gastric motility is reduced when the ileum is distended, the *ileogastric reflex*. Excessive distention, injury, or bacterial inflammation of the intestine may result in cessation of intestinal movements. This is known as the *intestinointestinal reflex*.

MOVEMENTS OF INTESTINAL VILLI. The villi of the intestines are almost constantly in motion—constricting, shortening, elongating, and swaying from side to side. Such movements are due to contractions of smooth muscle fibers within each villus. In this manner the intestinal contents that bathe the villi are stirred, and the chyle within the lacteals is forced out into the larger lymph vessels.

MOVEMENTS OF THE LARGE INTESTINE. In the large intestine, segmentation movements and weak peristaltic movements serve principally to mix the intestinal contents. The main propulsive movement is a strong wave of contraction (*mass movement* or *peristalsis*) that sweeps along the colon a considerable distance, moving the bowel contents ahead of it into the sigmoid colon or rectum. This wave is frequently initiated by the entrance of food into the stomach or duodenum, providing the stimulus for this *gastrocolic* or *duodenocolic reflex.* This accounts for the desire for defecation often experienced after a meal, especially breakfast. Mass movements occur three or four times a day. If defecation does not occur, the intestinal contents, now called *feces,* are withdrawn back into the sigmoid colon to await the next wave.

Defecation. This is the process of emptying the rectum. It is a reflex act initiated by the presence of feces in the rectum, which is generally empty until just before defecation. Mass peristalsis causes the feces to enter the rectum. When intrarectal pressure reaches 30 to 40 mm Hg, the *desire* to defecate is initiated; pressure of 40 to 50 mm Hg initiates the defecation reflex. This results automatically in a bowel movement in an infant (or in an adult suffering from spinal cord injury). In normal children or adults, however, the reflex can be inhibited or assisted voluntarily.

The muscular mechanism involved in intestinal activity during defecation is as follows: Sensory fibers carry impulses to the lumbosacral region of the spinal cord, and the musculature of the colon and rectum is reflexly stimulated to contract. A strong wave of mass peristaltic contractions takes place in the colon. At the same time, the internal and external sphincters are inhibited and relaxed. Contraction of the abdominal wall muscles brings about compression of the abdomen and thus increases the pressure on the lower portion of the colon and the rectum. To keep the abdominal contents from moving upward, the contraction of abdominal muscles is usually preceded by a deep inspiration, and the glottis is kept closed. As a result, pressure in the rectum is greatly increased, leading to the explusion of feces.

The levator ani and other muscles of the pelvic floor play an important role in defecation. They support the pelvic floor against increased intra-abdominal pressure. Fibers of the levator ani are inserted into the walls of the anal canal and, on contraction, tend to enlarge the anal opening and pull the anus over the feces. At the end of the act of defecation, they tend to bring about expulsion of any material that may remain.

The number of bowel movements normally required in one day varies with individuals. Most people have one or more a day; others have bowel movements only every other day; some have even less frequent movements.

Composition of Feces. Feces consists of (1) undigested food such as muscle fibers, starch grains, and fat droplets; (2) indigestible materials such as cellulose and mucin; (3) products of fermentation and putrefaction, such as skatole and indole; (4) bacteria, both living and dead, which make up about 25 percent of the feces; (5) inorganic salts, especially compounds of calcium, iron, magnesium, and phosphorus; and (6) bile pigments or their derivatives. The brown color of normal feces is due principally to the presence of *stercobilin*, which is derived from bile pigments.

Foreign substances may appear in feces under pathologic conditions. For example, blood cells may be present in hookworm infestation, in ulceration, or in a late stage of carcinoma. Animal parasites and their eggs or larvae may be found.

Amount of Feces. The amount of feces formed each 24 hr depends on the amount and nature of the food that has been ingested. It is larger on a vegetable diet. The average daily amount of fecal discharge for a person in normal health is about 150 g.

THE ROLE OF ENZYMES IN DIGESTION

Most of the foods taken into the body are in the form of large complex molecules that are incapable of being absorbed by the epithelium of the intestinal walls and of being drawn into the blood and lymph. Digestion, which makes the necessary changes, is brought about principally by enzymes produced by the glands of the digestive system.

Enzymes as Catalysts. An *enzyme* is a catalytic agent produced by a living cell. (A catalyst is a substance that increases the velocity of a chemical reaction.) Catalysts are not reacting agents in the reactions they modify, for they can be recovered unaltered after the reaction has taken place. They are effective in very small quantities and apparently can exert their effects again and again. Catalysts may be either inorganic or organic.

ACTION OF AN INORGANIC CATALYST. If *sucrose* is dissolved in water, the following reaction takes place:

$$C_{12}H_{22}O_{11} + H_2O \rightarrow C_6H_{12}O_6 + C_6H_{12}O_6$$
$$\text{(sucrose)} + \text{(water)} \rightarrow \text{(glucose)} + \text{(fructose)}$$

This reaction (*hydrolysis*) takes place at an extremely slow rate, even though the temperature may be increased to boiling. If, however, a little hydrochloric acid is added, the hydrolysis of sucrose is greatly speeded up. Other acids may produce the same effect; apparently the presence of hydrogen ions is responsible for the effect. In this case, hydrochloric acid serves as the catalyst.

ACTION OF AN ORGANIC CATALYST. *Sucrase* is an enzyme that may be obtained by extraction from the mucosa of the intestine. If a small quantity of sucrase is added to a solution of sucrose, hydrolysis (as in the case of adding hydrochloric acid) will occur speedily, even at low temperatures. A small quantity will bring about the hydrolysis of a large amount of sucrose.

In the reaction described, the enzyme reacted with the substrate with the result that the substrate was modified and converted into one or more products. The reaction, which is generally reversible, can be expressed thus:

$$\text{Enzyme + substrate} \rightarrow \begin{matrix}\text{Enzyme-}\\ \text{substrate}\\ \text{complex}\end{matrix} \rightarrow \text{Enzyme + product(s)}$$

Characteristics of Enzymes. Enzymes have the following nine characteristics:

1. They are *organic in origin,* being produced only by living cells.
2. They are *catalysts,* producing their effects only when the *substrate* (the substance acted upon) is in solution.
3. They are *soluble,* being extracted from animal or plant sources by various solvents, such as water, salt solutions, dilute alcohol, and glycerol.
4. They are *precipitated by alcohol or metallic salts,* such as copper sulfate or mercuric chloride, in this respect resembling proteins.
5. They *behave as colloids* and do not diffuse through semipermeable membranes, such as parchment paper and collodion.
6. They are *responsive to temperature changes,* each enzyme having an optimum temperature of activity. Enzymes are inactivated, but not destroyed by freezing (0°C). Most are destroyed by temperatures of 70°C; all are destroyed by boiling (100°C). The optimum temperature for most enzyme activity is body temperature (38° to 40°C).
7. They are *sensitive to changes in hydrogen ion concentration.* Each enzyme produces its effects only within a limited pH range. Some act in acid solutions (for example, pepsin, which acts in the stomach at a pH of 1 to 3); others act in an alkaline medium (for example, the protease of pancreatic juice, which acts in the intestine at a pH of 6.8 to 9.5). Most enzymes produce their effects in solutions having a pH of around 7 (neutrality).
8. They are *specific in their action.* A given enzyme will act on only one type of chemical substance or one very closely related to it. For instance, sucrase acts on the double sugar sucrose, but it will not act on proteins or fats or even on other double sugars, such as lactose or maltose. Some enzymes possess *absolute specificity* and will act only on a specific substance; others possess *relative specificity* and will act on compounds related to the substrate.
9. They are *capable of reversing the reactions catalyzed.* Lipase, an enzyme that hydrolyzes fats to fatty acids and glycerine during digestion can also act to reverse this process, wherein resynthesis of fat occurs.

Theory of Enzyme Action. The current theory of enzyme action is that the substrate attaches itself to the protein component of the enzyme, forming a specific *enzyme-substrate complex* at which the proper reaction catalyzed (hydrolysis, dehydration, oxidation, reduction, or ion transfer) may take place. Only a small portion of the protein molecule is engaged in the catalytic activity. It is postulated that the enzyme molecule bears *active sites,* with a specific configuration, at which the substrate, with a complementary configuration, may combine. The analogy between this action and the action of a lock and key has given rise to the *lock-and-key theory* of enzyme action. The substrate (key) fits the configuration of a particular site on the enzyme molecule (lock), permitting an interaction to occur by which parts of the key-lock complex either separate or unite. This concept explains the specificity of enzymes, the active site of the molecule accepting only those substrate molecules with a configuration that fits into it.

Chemical Composition of Enzymes. All enzymes are proteins or a protein combined with a nonprotein called the *prosthetic group.* Over a thousand enzymes have been identified, and some have been isolated as pure substances. Some enzymes require the action of low-molecular-weight substances called cofactors or coenzymes for their activation. *Cofactors* include such metallic ions as copper, iron, zinc, cobalt, or magnesium to bind the substrate to the active site and activate the enzyme. *Coenzymes* are nonprotein molecules that during the reaction, become bound to the enzyme and activate it. The inactive protein (*apoenzyme*) combined with the activator (*coenzyme*) forms a complete enzyme (*holoenzyme*). Important coenzymes are nicotinamide adenine dinucleotide (NAD) and NAD phosphate (NADP), essential in the conversion of lactic acid to pyruvate. An important function of vitamins is their role as cofactors in various enzyme reactions. Vitamins of the B group (thiamine, riboflavin, niacin, and others) are coenzymes or parts of enzymes or act as cofactors.

Enzyme Inhibition. Enzyme activity may be inhibited in a number of ways. Excessive heat, unfavorable changes in pH, the addition of certain protein precipitants, the action of certain oxidizing agents, or the action of substances like cyanides that combine with metals essential for enzyme action may stop or prevent enzyme activity. Some compounds may compete with the substrate for the active site on the enzyme molecule. These are *competitive inhibitors. Noncompetitive inhibitors* combine with a group at the active site and cannot be displaced even by an increase in the concentration of the substrate. The effects of various drugs in the body are often the result of their inhibitory action on enzymes. Such antibiotic drugs as sulfanilamides and penicillin produce their effects by inhibiting enzyme and coenzyme reactions in pathogenic organisms.

Inactive Form of Enzymes. Some enzymes are secreted in an inactive form and require activation to exert their catalytic effect. The inactive form of an enzyme is called a *zymogen.* For example, when pepsin is obtained from gastric glands it is not proteolytic, but upon reaching the lumen of the stomach it is activated by the hydrochloric acid and becomes actively proteolytic. The precursor of pepsin (that is, its zymogen) is called *pepsinogen.* Similarly, trypsinogen, secreted by the pancreas, is the zymogen of trypsin. Trypsin is activated by *enterokinase* liberated from cells of the intestinal mucosa.

Names and Classification of Enzymes. Enzymes are usually named by adding the suffix *-ase* to the root of the substrate acted upon or the type of reaction catalyzed. The principal groups of enzymes are as follows:

Groups of enzymes based on type of reaction catalyzed.

Oxidoreductases, which catalyze oxidation-reduction reactions
Transferases, which participate in the transfer of a chemical group from one substrate to another
Hydrolases, which catalyze hydrolytic reactions
Lyases, which permit the removal of chemical groups from substrate without hydrolysis
Isomerases, which catalyze isomerization reactions
Ligases, which permit the linking of two molecules by the breaking of a phosphate bond

Groups of enzymes based on substrate acted upon.

Amylases, starch-splitting enzymes (ptyalin, amylopsin)
Disaccharases, which hydrolyze double sugars (sucrase, lactase, maltase)
Lipases, which hydrolyze fats (gastric lipase, steapsin)
Proteases, proteolytic enzymes, which hydrolyze proteins (pepsin, trypsin)

Types of enzymes based on site where action takes place.

Intracellular, within cells. Most of the enzyme action occurring within the body takes place within cells. All the chemical reactions involving energy release or synthetic activities occur within cells and are catalyzed by enzymes.
Extracellular, outside of cells. All digestive enzymes act outside of cells within the lumen of the alimentary tract. Chemical reactions occurring in the blood plasma or interstitial fluid are also extracellular.

Digestive Enzymes, Sources and Action. The following table lists the principal digestive enzymes, their source, the substrate, and the final products. All enzymes listed are hydrolases.

Glands and Enzymes Produced	*Substrate*	*Final Product(s)*
Salivary glands		
Ptyalin (salivary amylase)	Starches	Disaccharides (maltose)

Glands and Enzymes Produced	Substrate	Final Product(s)
Gastric glands		
Pepsinogen (pepsin)	Proteins	Proteoses and peptones
Gastric lipase	Emulsified fats	Fatty acids and glycerol
Pancreas		
Trypsin (chymotrypsin)	Proteins (natural or derived)	Proteoses, peptones, peptids, amino acids
Carboxypeptidase	Proteins	Amino acids
Lipase (steapsin)	Fats (emulsified by bile)	Fatty acids and glycerol
Amylopsin (pancreatic amylase)	Starches	Maltose
Intestinal mucosa		
Aminopeptidase	Peptides	Amino acids
Sucrase	Sucrose	Fructose and glucose
Lactase	Lactose	Galactose and glucose
Maltase	Maltose	Glucose

CHANGES IN FOODSTUFFS DURING DIGESTION

Through the action of the teeth and the effects of secretions of the digestive glands, the ingested foodstuffs are changed to their absorbable forms.

Changes that Occur in the Mouth. In the oral cavity, food is ground by the teeth (mastication) and moistened by saliva. The saliva helps it to form a *bolus* (ball) and makes it slippery, thus facilitating its passage through the pharynx and esophagus. Food remains in the mouth such a short time that salivary digestion is limited, but enzyme action continues in the stomach until the salivary enzymes, which require an alkaline medium, are inactivated by the acidity of gastric secretions.

Changes that Occur in the Stomach. Gastric juice is the secretion of the gastric glands. It contains *hydrochloric acid,* which is produced by the parietal cells, and *pepsin,* produced by the chief cells, which are also the source of *gastric lipase* and an *intrinsic anti-pernicious-anemia factor* essential for the absorption of vitamin B_{12}. In addition, gastric juice contains water, mucin, mucinlike proteins, and inorganic salts. The amount of hydrochloric acid varies from 0.2 to 0.5 percent by volume. Gastric juice has a pH of 0.9 to 1.5. The volume secreted after an average meal is about 700 ml; total daily volume is about 2000 ml. The volume varies with the content of a meal, the most abundant secretion occurring after the ingestion of meat.

Pepsin hydrolyzes proteins through successive stages into metaproteins, proteoses, and peptones. It also acts on the protein of cow's milk *(caseinogen)*, transforming it into casein (paracasein). In the presence of calcium ions, insoluble calcium caseinate is formed. This brings about coagulation (or *curdling*) of milk, a semigelatinous mass being formed. By such action milk is prevented from passing directly to the intestine and is held in the stomach long enough for pepsin to act on its protein constituents. This coagulation of milk has been attributed to an enzyme, *rennin,* but careful studies have shown that rennin is absent from the gastric juice of adults. Pepsin requires the presence of hydrochloric acid for its action.

The action of *gastric lipase* is limited because only traces of it are present. It acts on finely emulsified fats, such as egg yolk, butter, and cream, hydrolyzing them into fatty acids and glycerol.

Through muscular and enzymatic action, the food in the stomach is reduced to *chyme,* a substance of semiliquid consistency in which all food substances are in solution or in an emulsified state. Chyme is distinctly acid. On its discharge into the duodenum, the acidic chyme initiates the secretion of *cholecystokinin* and *secretin,* hormones that play an important role in the regulation of gastric motility and the discharge of bile.

The *time required* for gastric digestion of an average meal is 3 to 5 hr.

Changes that Occur in the Small Intestine. Three juices are of importance in intestinal digestion: pancreatic juice, bile, and intestinal juice.

PANCREATIC JUICE. The exocrine secretion of the pancreas is an alkaline fluid resembling saliva in consistency. From 500 to 800 ml are secreted daily. Two types of secretion are produced. One, initiated by secretin, has a high sodium bicarbonate content but is poor in enzymes; the other, initiated by vagal stimulation, is scanty in amount but rich in enzymes. The principal constituents of pancreatic juice are water, inorganic salts, and enzymes. Enzymes secreted include trypsin, chymotrypsin, carboxypeptidase, an amylase, maltase, steapsin, and ribonuclease.

Within the intestine, pancreatic juice acts on all classes of foods. In conjunction with bile, its alkaline nature counteracts the acidity of the chyme from the stomach. The action of enzymes on the various foods is as follows.

Pancreatic Proteases. These include *trypsin,* secreted as *trypsinogen; chymotrypsin,* secreted as *chymotrpysinogen;* and *carboxypeptidase.* In the intestine, trypsinogen is converted to trypsin by the action of an intestinal enzyme, *enterokinase;* trypsin in turn converts chymotrypsinogen to chymotrypsin. These enzymes act on proteins or partially digested proteins, converting them to proteoses, peptones, polypeptides, peptides, and amino acids, the last-named being the final products of protein digestion.

Pancreatic Amylases. Pancreatic *amylase* continues the digestion of starches that is initiated in the mouth by ptyalin. The starches are hydrolyzed to *dextrins,* which are then hydrolyzed to maltose. *Maltase* hydrolyzes maltose to *glucose,* a simple sugar.

Pancreatic Lipase. Steapsin is the principal enzyme involved in the digestion of fats. It acts on fats emulsified by bile salts, hydrolyzing them to *diglycerides, monoglycerides, fatty acids,* and *glycerol,* the end products of fat digestion. Since digestion occurs in a stepwise fashion, probably several lipases are involved.

Pancreatic Ribonuclease. This enzyme acts on nucleic acids (DNA and RNA), converting them to nucleotides.

BILE. Bile is secreted continuously by the liver at the rate of about 1000 ml a day. It is golden brown in color and alkaline in reaction. Bile passes from the liver through the hepatic ducts into the common bile duct, which enters the duodenum at the *ampulla of Vater.* At this point the bile duct is surrounded by a ring of smooth muscle, the *sphincter of Oddi.* When the duodenum is empty, the sphincter is usually closed. This causes the bile to back up into the gallbladder, where it is stored and concentrated. The organic constituents of bile are 5 to 10 times more concentrated in gallbladder bile.

COMPOSITION OF BILE. The principal constituents of bile are *bile salts, bile pigments, cholesterol,* and *lecithin.*

Bile Salts. The principal bile salts are sodium salts of glycocholic and taurocholic acids. They are alkaline and have the property of lowering the surface tension of water to a marked degree. This enables them to emulsify fats, that is, to break the larger fat particles into smaller particles. This permits more effective access of pancreatic lipase to the surfaces of fats and results in a more complete hydrolysis of these substances. Bile salts are also capable of dissolving fatty acids and water-insoluble soaps, which enhance the emulsifying power of bile.

Bile Pigments. These substances, which give bile its color, are *bilirubin* (reddish) and *biliverdin* (greenish), the latter excreted only in small amounts. Bile pigments originate from products resulting from the breakdown of hemoglobin, which is accomplished by phagocytes in the bone marrow, liver, and spleen.

The first pigment to appear is biliverdin. It is then converted to bilirubin. In the liver, bilirubin is conjugated with glucuronic acid to form mono- and diglucuronides, which, along with free bilirubin, are excreted in bile. In the intestine, bile pigments may be converted by bacterial action into *stercobilin* and *urobilin,* pigments excreted in the feces and urine. Stercobilin is responsible for the brown color of feces. Some bile pigments may be reabsorbed by the intestine and returned to the liver. In pathologic conditions or bile duct obstruction, bile pigments may appear in the blood, giving the skin and mucous membranes a yellowish color. This condition is called *icterus* or *jaundice.*

Cholesterol. Cholesterol ($C_{27}H_{45}OH$) is a sterol present in many foods, especially meats, dairy products, and egg yolk. Animal fats and oils have a high cholesterol content. It is present in all tissues and especially abundant in nervous and glandular tissue. The brain and adrenal glands show a high content.

Cholesterol is synthesized in the body in many tissues, but the liver is the principal source of plasma cholesterol. Its presence in bile is the result of its ready solubility in the presence of bile salts and lecithin. If either is deficient, cholesterol in bile may precipitate, causing the formation of *gallstones,* which may accumulate in the gallbladder. They may also obstruct the common bile duct, giving rise to jaundice.

Cholesterol content of the blood plasma varies from 100 to 280 mg/100 ml. In the blood it occurs free or combined with fatty acids as esters. Cholesterol plays an important role in the biosynthesis of a number of steroids, including bile acids, adrenocortical hormones, sex hormones (androgens, estrogens, progesterone), and vitamin D. Excessive amounts of cholesterol in the body are associated with a number of pathologic conditions, especially the formation of atheromatous plaques in blood vessels (atherosclerosis).

Lecithin. Lecithin, a phospholipid, is widely distributed in the body. It is a component of all body cells and is present in bile. It is essential for the absorption and utilization of cholesterol in the tissues. A decrease in lecithin results in a rise in blood cholesterol. Lecithin is synthesized in the liver.

FUNCTIONS OF BILE. Bile (1) reduces the acidity of chyme in the upper portion of the small intestine; (2) emulsifies fats, ensuring more complete digestion and absorption; (3) increases the solubility of fatty acids, thus aiding in their utilization; (4) serves as a vehicle for excretion of waste substances, such as those resulting from destruction of hemoglobin in the liver; (5) lessens fermentation and putrefaction in the intestine by aiding in more complete utilization of proteins and carbohydrates; (6) favors the absorption of vitamins, especially the fat-soluble vitamins A, D, and K; and (7) stimulates intestinal motility.

INTESTINAL JUICE. Intestinal juice *(succus entericus),* secreted by intestinal glands, the crypts of Lieberkühn, consists of water, salts, and mucus. Approximately 2000 ml is secreted each day. Intestinal enzymes present in the juice are not secreted by the intestinal glands but are produced in mucosal cells. These cells migrate along the sides of villi and are discharged at their tips. When these cells disintegrate, enzymes are released into the lumen of the intestine. Intestinal enzymes have also been identified in the brush borders of intestinal absorptive cells, and some of their catalytic activity may occur intracellularly.

Enzymes acting in the small intestine include (1) *aminopeptidase* and *dipeptidase,* which complete the hydrolysis of proteins to amino acids; (2) *inverting enzymes* (*sucrase, lactase,* and *maltase*), which hydrolyze disaccharides (sucrose, lactose, and maltose) to monosaccharides (glucose, levulose, galactose); (3) *enterokinin,* which activates trypsinogen; and (4) a miscellany of enzymes that includes an amylase, lipase, and nucleases that hydrolyze, respectively, starches, fats, and nucleic acids.

By the time chyme has reached the large intestine, the digestible materials have been acted on by enzymes and reduced to their end products (amino acids, fatty acids, glycerol, and simple sugars), and these substances have been absorbed into the blood and lymph. The remaining chyme, consisting principally of undigestible material and water, enters the colon through the ileocecal valve.

Changes that Occur in the Large Intestine. The principal processes that take place in the large intestine involve (1) absorption of water, (2) elimination of waste products, and (3) synthetic activities, especially the formation of vitamins, accomplished through the action of symbiotic organisms, principally bacteria, that inhabit the large intestine.

ABSORPTION OF WATER. Fluid chyme passing through the ileocecal valve is of about the same consistency as that entering at the pylorus. Although much water with substances in solution is absorbed by the small intestine, an equivalent amount is added in the secretions of the mucosa and the various glands. Feces leaving the rectum are semisolid in consistency, indicating that a very considerable amount of water is absorbed by the large intestine.

ELIMINATION OF WASTE PRODUCTS. In addition to the elimination of undigested food, the large intestine is involved in the excretion of calcium, magnesium, iron, and phosphates. Foreign substances, such as bismuth and mercury, are also eliminated by this organ. The fluid secreted by the large intestine is thick and alkaline (pH 8.4) and contains much mucus but no enzymes.

Bacterial Action in the Intestines. Large numbers of bacteria are usually consumed along with food. Most of these are destroyed by the action of the acid contents of the stomach, though some acid-resistant bacteria may pass through unaffected.

The contents of the first portion of the small intestine are sterile, or nearly so. The bile has a mild antiseptic action. As the chyme moves downward through the intestines, the floral content increases both in number and in species. It consists principally of bacteria, although yeasts and filamentous fungi may be present; the types of bacteria vary among individuals, but in one individual they remain relatively constant. These organisms live *symbiotically*.* For most, the relationship between them and their host is that of *commensalism*. Sometimes a *mutualism* exists in that these organisms, through fermentation that aids in the digestion of foods or through synthetic activities that result

* *Symbiosis* is the condition in which two organisms of different species live in intimate association with each other. It is designated *commensalism* if one organism benefits, but the other does not; *mutualism* if the relationship benefits both; *parasitism* when one organism, the *host*, provides one or more essential nutrient factors for the other, the *parasite*. Parasites often, but not always, affect the host adversely.

in substances, such as vitamins (B, K, and pantothenic acid) and amino acids, which are absorbed and utilized by the body. When their activities are harmful, the organisms are regarded as *parasites.* Included in this group are the pathogenic forms that cause disease (for example, the dysentery, cholera, and typhoid bacilli) and bacteria that cause undesirable fermentations.

The principal changes brought about by bacteria in the intestines are fermentations. Carbohydrates are decomposed, with alcohol and acid products resulting. Indeed, this action helps to account for the increased acidity of chyme as it moves through the intestine. Proteins are acted upon, giving rise to putrefactive products that include amines (histamine, thylamine, thyramine), phenols (skatole, indole, creosole), volatile acids, and various gases (hydrogen sulfide, hydrogen, carbon dioxide, and methane). These products are responsible for the typical odor of feces. Some of the amines are potentially toxic, but they have little or no effect on the body, for most are eliminated through the feces and, if absorbed, are rendered innocuous through the detoxifying action of the liver.

SECRETION BY DIGESTIVE GLANDS

The initiation and regulation of secretion by digestive glands are accomplished by neural, humoral, and mechanical factors. In *neural control,* nerve impulses are initiated through the action of stimuli on receptors (ends of afferent or sensory nerves). This gives rise to impulses that are transmitted to the reflex centers in the brain or the spinal cord. From these centers impulses travel to the gland involved, and secretory activity is initiated or inhibited. In *humoral control,* hormones or other chemical substances are transmitted to glands by way of the bloodstream. These are specific in their action and may initiate or inhibit secretion by direct effect upon gland cells. *Mechanical factors* (for example, the presence of food in the stomach or intestines) are believed to play a role in secretion, but the extent of their influence is not well established.

Secretion by the Salivary Glands. The salivary glands are entirely under neural control. They receive branches from both the sympathetic and parasympathetic divisions of the autonomic nervous system. Parasympathetic stimulation produces copious secretion of a thin and watery fluid of low enzyme content; sympathetic stimulation produces a thick, viscid fluid, low in volume but high in enzyme content. If the nerves to a gland are severed, secretion ceases. Impulses initiating salivary secretion originate in chemoreceptors, pressure receptors, and the higher nerve centers.

IMPULSES FROM CHEMORECEPTORS. Food or other substances, such as acetic acid, stimulate chemoreceptors. The impulses give rise to sensations of taste or smell and reflexly initiate gland secretion.

IMPULSES FROM PRESSURE RECEPTORS. These sense organs respond to physical factors such as the presence of food or other substances in the mouth or to the movement of mouth parts.

IMPULSES FROM HIGHER NERVE CENTERS. Impulses may arise in these centers when the individual sees food, hears the name of a food or the sounds associated with food preparation (the frying of a steak), or thinks of food and situations associated with it. Secretion resulting from such stimuli is called *psychic secretion.* The basis for such action is the conditioned reflex. As in other cases, it depends on previous experience and training, which have established conditioned reflex paths.

Secretion of Gastric Juice.

PSYCHIC (CEPHALIC) PHASE. The presence of food in the mouth, the sight, taste, and smell of foods, or memories associated with food usually initiate the secretion of gastric juice. This is accomplished through the nervous system, impulses reaching the stomach from a reflex center in the medulla by way of the vagus nerve. Stimulation of gastric glands results. Emotional states influence gastric secretion. Anger and hostility increase secretion; fear, anxiety, and depression decrease it.

GASTRIC PHASE. Food, upon entering the stomach, initiates gastric secretion. This results from the distention of the stomach wall and the action of chemical substances in the food. Certain foods are more effective than others in initiating secretory activities. Meat broths, meat juices, meat extracts, and products of protein digestion in the stomach, especially proteoses and peptones, all have a stimulating effect on gastric secretion. This is accomplished through the production of a hormone, *gastrin,* secreted by glands in the pyloric region of the stomach. This hormone is secreted, absorbed into the bloodstream, and then transported back to the gastric glands, which it stimulates.

Alcohol and caffeine also stimulate the production of gastrin. The secretion of a highly acidic stomach fluid in the absence of food may have a highly irritating effect on the mucosa of the stomach and adjoining esophagus and duodenum. Excess stomach acidity is thought to be a significant factor in the causation of peptic ulcer.

INTESTINAL PHASE. The presence in the duodenum of fats and various products of gastric digestion reduces gastric secretion and motility and delays emptying of the stomach. This is accomplished in two ways: (1) *reflexly,* through the nervous system, and (2) *humorally,* through the production by the duodenal mucosa of three hormones, *secretin, cholecystokinin,* and a *gastric inhibitory peptide,* which are transported to the stomach via the bloodstream. Their production is initiated by the presence of chyme in the duodenum.

Secretion of Pancreatic Juice. The secretion of pancreatic juice, like that of gastric juice, involves three phases: cephalic, gastric, and intestinal. The *cephalic phase,* involving neural control, plays a minor role.

There is a flow of pancreatic juice a few minutes after ingestion of food. This is the result of a reflex arising from stimulation of receptors in the mouth. Stimulation of the vagus nerve increases the rate of secretion; severing this nerve inhibits secretion. Psychic stimulation may occur to a limited extent.

In the *gastric phase,* the release of gastrin stimulates bicarbonate secretion.

The principal control of pancreatic secretion is exerted in the *intestinal phase* by two intestinal hormones, *secretin* and *cholecystokinin.* Secretin is a potent stimulus for the secretion of pancreatic juice that is rich in enzymes or their zymogens; cholesystokinin stimulates the secretion of a pancreatic juice rich in sodium bicarbonate. The latter action was formerly ascribed to a specific hormone, *pancreozymin,* but careful studies have shown that pancreozymin and cholecystokinin are identical in chemical structure.

Secretion of Bile. The secretory activity of liver cells is practically continuous and apparently depends largely on blood supply. It is independent of nervous activity, although stimulation of the vagus nerve increases bile secretion, and stimulation of sympathetic nerves inhibits it. The presence of bile or of certain foods in the intestine stimulates secretion by the liver cells. Substances that increase the secretion of bile are called *choleretics;* bile salts and bile acids are the most potent. Among foodstuffs that are effective choleretics are fats or the products of their digestion. Production of bile is stimulated by the hormones *secretin* and *cholecystokinin,* whose secretion is increased by the presence of fats in the duodenum. Secretin is the most potent stimulus. *Gastrin* from the stomach also stimulates bile production.

The concentration of bile salts in bile is determined primarily by the concentration of bile salts in blood plasma. An increase in the amount of bile salts in the blood increases the amount of bile salts excreted in bile.

DISCHARGE OF BILE FROM THE GALLBLADDER. As acid chyme enters the intestine, the tonus of the muscular layer of the gallbladder wall is increased, forcing bile through the relaxed sphincter muscle at the ampulla of Vater. The contraction of the gallbladder is brought about primarily by cholecystokinin.

Secretion of Intestinal Juice. The secretion of intestinal juice (succus entericus) is principally under neural control. The presence of food or of undigested food residues in the intestine acts as a mechanical stimulus reflexly initiating secretory activity. Inasmuch as this occurs in a denervated intestine, it is assumed that these reflexes are local, involving the nerve plexus (Meissner's plexus) within the intestinal wall. Chemical factors may also be involved. Secretin stimulates the secretion of intestinal juice as well as pancreatic juice and bile. Pancreatic juice locally stimulates the intestinal glands. In addition, certain substances that have been isolated from the intestinal mucosa act as hormones, stimulating secretion by the intestinal glands. The secretion of mucus by the goblet cells of the intestinal epithelium is brought about by the direct stimulation of the goblet cells by mechanical and chemical stimuli.

ABSORPTION BY THE DIGESTIVE SYSTEM

Absorption is the process in which the final products of digestion pass through the intestinal mucosa and are taken into the blood or lymph. It depends on physical and chemical processes utilizing diffusion, osmosis, active transport, pinocytosis, and phagocytosis.

Pathways of Absorption. Absorbed substances, especially amino acids and simple sugars, are picked up by the capillaries of the mucosa. These lead to veins (mesenteric), which lead to the *portal vein,* which enters the liver. After passing through the liver sinusoids, the blood enters the *inferior vena cava* via the hepatic veins. This is the pathway of absorption into the bloodstream. As for absorption into the lymphatic system, absorbed substances, particularly fats, are taken into the *lacteals,* which are blind lymph capillaries within the villi. The lacteals lead to larger lymph vessels, which in turn lead to the main lymph vessel, the *thoracic duct.* This duct empties into the bloodstream at the junction of the left subclavian and left innominate veins. The material carried by lymphatic vessels is called *chyle.*

ABSORPTION IN THE STOMACH. Absorption of foods in the stomach is extremely limited. Although there is some evidence that water, some salts, and glucose may be absorbed in small amounts, the quantity absorbed is negligible. Alcohol and weak acids, such as acetylsalicylic acid (aspirin), may be absorbed, but the usual products of digestion are not.

ABSORPTION IN THE SMALL INTESTINE. The small intestine is the organ in which major absorption of nutrients occurs. Absorption is facilitated in the small intestine by the presence of *circular folds* and *villi,* both of which greatly increase the absorptive surface. The *total absorptive surface* of the small intestine is estimated to be about 10 m^2. The processes by which the various food substances pass through the mucosa into the blood or lymph are as follows.

Absorption of Carbohydrates. Carbohydrates—with the exception of cellulose, which passes through the intestine unchanged except for fermentative changes—are digested to simple sugars (glucose galactose, fructose). These sugars differ in their rate of absorption, galactose being taken in most readily, followed by glucose and then fructose. Galactose seems to be absorbed through the process of facilitated diffusion; glucose and fructose require a carrier for active transport. The presence of sodium in the intestinal fluid is essential for this process, sodium apparently assisting in the combination of the sugar with the carrier. On passing through the mucosa, sugars enter the capillaries of the villi.

Absorption of Proteins. Through the action of proteolytic enzymes, proteins in the small intestine are completely hydrolyzed to free amino acids. These amino acids are actively transported by a number of specific carriers across the mucosa of the intestine and are taken up by the blood in the capillaries of the villi. This blood passes by way of the portal vein to the liver.

Absorption of Fats (Lipids). Most digestion of lipids occurs in the small intestine. Through the action of pancreatic lipase on lipids emulsified by bile salts, the lipids are digested to fatty acids, mono- and diglycerides, and glycerol. Through further action by bile salts, some free fatty acids and monoglycerides form combinations called *micelles,* which are soluble in water.

Some free fatty acids and monoglycerides, as well as unhydrolyzed lipids, can enter the mucosal cells directly by diffusion. The lipid nature of the cell membrane facilitates this action.

After the fatty acids, monoglycerides, and diglycerides enter the mucosal cells, they are immediately resynthesized to triglycerides (neutral fats) or phospholipids. The glycerol involved in this reaction is not the glycerol that results from lipid hydrolysis but that of a metabolic pool present in intestinal cells. The triglycerides then combine with free fatty acids, phospholipids, and proteins to form low-density lipoprotein particles called *chylomicrons,* most of which are absorbed into the lymph of the lacteals and transported by way of the thoracic duct to the bloodstream. Some short-chain fatty acids may enter the portal vein directly.

Absorption of Miscellaneous Substances. The small intestine also absorbs the accessory food substances comprising water, salts, and vitamins. From 8 to 10 l of water enter the digestive tract each day. This includes water ingested, water in the secretions of glands, and water entering by osmosis. Most of this water is reabsorbed at an estimated rate of 200 to 400 ml/hr. Digestive changes and absorption usually bring about a hypotonic condition of the intestinal contents, which causes water to pass into the blood and lymph. If the intestinal contents become hypertonic, water will pass from the blood into the lumen.

Minerals are absorbed in various ways. Sodium is actively transported, often in combination with monosaccharides or amino acids. Other inorganic ions, such as chlorine, potassium, magnesium, and calcium ions, are absorbed by active transport or sometimes by simple diffusion. Calcium ions require vitamin D as a cofactor; iron absorption depends upon its ability to form a compound, *ferritin,* within mucosal cells.

Some *vitamins* are absorbed by simple diffusion; others have special conditions for absorption. The absorption of vitamin D is associated with absorption of calcium and phosphorus. All fat-soluble vitamins require the presence of bile salts and, in their absorption, accompany fats. Vitamin B_{12} requires the presence of an *intrinsic factor* produced by the stomach for its absorption.

ABSORPTION IN THE LARGE INTESTINE. Because most of the ingested food has been digested and absorbed before reaching the large intestine, the principal task remaining is that of taking up water. Water, some amino acids, vitamins (especially those synthesized by colonic microorganisms), minerals (sodium, calcium, iron), and some phosphates and carbonates are absorbed by the colon. The remaining material is eliminated as feces.

METABOLISM OF FOODS

The term *metabolism,* in its general sense, refers to the processes that take place in the utilization of foods after they have been absorbed into the circulatory fluids. Constructive synthetic activities—such as the building of protoplasm, which involves the synthesis of substances within the cells—constitute *anabolism;* destructive or degradative activities, in which larger molecules are broken down to simpler substances, constitute *catabolism.* Normally these processes are in balance (a state of *dynamic equilibrium*). Anabolic processes predominate in early life when growth and development with increase in size occur; catabolic processes may predominate in old age, pathologic conditions, or in such abnormal states as starvation or poor nutrition.

Energy. All bodily activities require *energy.* Energy may be defined as the capacity to perform work or the ability to bring about change. Energy exists in two forms, kinetic and potential. *Kinetic energy* results from activity, such as a change in structure or position. *Movement,* as in muscle contraction, *heat* (which is the result of molecular movement), light, sound, and electricity are forms of kinetic energy. *Potential energy* is that which is due to internal structure or position. A stretched spring, molecules of glucose, and stored fat in the body are examples of potential energy. These forms of energy are interchangeable, one form being convertable to the other.

Caloric Value of Foods. When foods are oxidized in the body, heat is produced. The standard unit for the measurement of heat is the *calorie.* It is possible to determine the amount of heat produced by foods by burning them in a *bomb calorimeter.* A *small calorie* (cal) is the amount of heat required to raise the temperature of 1 gram of water 1 degree Celsius; a *kilocalorie* (kcal, also designated Calorie, Cal, or kg cal) is that amount of heat required to raise the temperature of 1 kilogram (1 liter) of water 1 degree Celsius. In studies of human metabolism, the kilocalorie is used. In terms of other units of energy, a kilocalorie equals 426.85 kilogram-meters or 3087.4 foot-pounds or 1000 calories.

Calorimetry. Two methods are employed in measuring the amount of heat produced by an organism: direct calorimetry and indirect calorimetry.

Direct Calorimetry. In this method, the test animal is placed in an apparatus in which the amount of heat liberated can be measured. Because such an apparatus is expensive to build and difficult to operate, direct calorimetric methods are used only in research laboratories.

Indirect Calorimetry. By this method, heat production of an animal is calculated from the amount of oxygen consumed, the carbon dioxide liberated, or both. The results from indirect calorimetry agree within less than 1 percent with those from direct calorimetry.

A number of types of apparatus have been developed for use in indirect calorimetry. The type most widely used for clinical purposes in the determination of the human metabolic rate is the *Benedict-Ross closed-circuit apparatus.* The heat produced is calculated from the oxygen consumption alone. In the determination of the basal metabolic rate, the subject lies on a couch and breathes through a mouthpiece connected to the apparatus. Pure oxygen contained within the spirometer bell is inspired; carbon dioxide expired passes through a canister of soda lime and is absorbed. As oxygen is consumed, the spirometer bell slowly falls, its movements being recorded on a sheet of paper attached to a slowly revolving drum. The heat production is calculated from the quantity of oxygen used during a 6-min period.

Respiratory Quotient (RQ). This is the ratio of the volume of CO_2 produced and oxygen consumed, expressed as follows:

$$\frac{\text{Volume of } CO_2 \text{ expired}}{\text{Volume of } O_2 \text{ inspired}} = \text{Respiratory quotient}$$

In the oxidation of a carbohydrate, the following reaction takes place:

$$C_6H_{12}O_6 + 6O_2 \rightarrow 6CO_2 + 6H_2O \textit{ plus } \text{energy}$$

Then,

$$\frac{6 \text{ volumes of } CO_2}{6 \text{ volumes of } O_2} = 1.0 \text{ (RQ of a carbohydrate)}$$

Similarly, the RQ for a fat is 0.71; for a protein, 0.80. For a person on an average mixed diet, it is 0.85. For a person who has fasted for 12 hours, the RQ is 0.82.

Total Caloric Requirements. The total bodily requirement of kilocalories varies with the activity. The general requirement each 24 hours for a person weighing 70 kg engaged in various activities is shown in the following table.

Activity	*Energy Expenditure (kcal)*
Sleeping	1560
Lying in bed	1850
Sedentary occupation	2400
Moderate activity	3200
Above-average activity	4400
Heavy work	6000 to 8000

For the average person, proteins contribute 10 to 15 percent of the caloric value, carbohydrates 50 to 70 percent, and fats 20 to 30 percent.

Basal Metabolism. The heat production of the body, determined 14 to 18 hr after the last meal, at a room temperature of approximately 20°C and with the body at complete rest and the subject lying quietly and free from environmental distractions, is called the *basal metabolism* or the *basal metabolic rate* (BMR). It represents the minimum expenditure of energy compatible with life, that is, the energy utilized in all activities involved in vital functions, such as heartbeat, respiratory movements, digestive activities (such as glandular secretion and peristaltic contractions), maintenance of body temperature, and maintenance of muscle tonus. For a young adult male, the BMR is:

In terms of kilocalories	1500 to 1800 kcal/day
In terms of body weight	1 kcal/kg/hr
In terms of body surface	40 kcal/m^2/hr

NORMAL RANGE OF BMR. The basal metabolic rate is commonly expressed as the percentage deviation from the normal figure established for individuals of corresponding age and sex. For example, a person with a BMR of +20 would have a rate 20 percent above normal, while a person with a BMR of −15 would have a rate 15 percent below normal. Values within 15 percent of the mean are considered to be in the normal range; values above or below these variations are regarded as indicative of some abnormal condition.

FACTORS INFLUENCING BASAL METABOLISM. The basal metabolism of an individual is influenced by the following factors:

Size of the Individual. The number of kilocalories produced in a given unit of time per unit of surface area is fairly constant for an individual. Because a small person has proportionately a much larger surface exposed to the environment and radiates more heat per unit of body weight than does a larger person, the small person must produce more heat to maintain body temperature and as a consequence usually has a higher basal metabolic rate than that of the larger person.

Age. The basal metabolic rate varies with age. It is lowest in a newborn child; it increases rapidly, reaching its peak between the ages of 5 and 10, after which it decreases steadily thereafter.

Sex. Males have a slightly higher (by 6 to 7 percent) BMR than females. In females it rises during pregnancy and lactation.

Weight. In terms of body weight, thin people have a much higher rate than stout people—at least 50 percent higher.

External Temperature. Lowering of the external temperature greatly increases metabolism, owing to increased muscle tone and increased bodily activity in general.

Internal Temperature. For each degree Celsius of increase in body temperature, the basal metabolic rate increases about 5 percent.

Activity. Approximately 75 percent of the energy liberated by muscular activity is in the form of heat. Exercise, work, or any other form of muscular activity is therefore the greatest single factor in energy output and has a profound effect on the metabolic rate.

Ingestion of Food. The ingestion of food brings about an increase in heat production over and above that which can be accounted for by increased activities of the digestive tract. Proteins especially exhibit this characteristic. For example, if a person ingests 100 kcal of protein, 130 kcal of heat are produced—an increase of 30 percent. The stimulating effect food intake has on metabolism is referred to as *specific dynamic action.* All three classes of foods show this specific dynamic effect in the approximate rates of 30 percent for proteins, 7 percent for carbohydrates, and 4 percent for fats.

Endocrine Secretions. The hormone *thyroxine* is a primary factor in regulating basal metabolism. An excess causes *hypermetabolism,* which has many effects including increased oxygen consumption and heat production *(calorigenic effect).* Undersecretion causes *hypometabolism,* in which there is a slowing of all body functions. Several other hormones may influence the metabolic rate. These include epinephrine, norepinephrine, androgens, somatotrophin, cortisol, and insulin.

Pathologic Conditions. The metabolic rate is *decreased* in starvation, malnutrition, obesity due to pituitary disorders, hypothyroidism, adrenal cortical insufficiency (Addison's disease), and lipoid nephrosis. It is *increased* in hyperthyroidism, fever, diabetes insipidus, cardiorenal disease, dyspnea, leukemia, and polycythemia.

Metabolism of Specific Foods. The mechanisms of metabolism of the three classes of foods are specific for each class, as described in the following paragraphs.

METABOLISM OF PROTEINS. Proteins form the framework of all cells and comprise about 15 percent of human body weight. All contain carbon, hydrogen, oxygen, and nitrogen; most also contain sulfur and phosphorus. They are composed of *amino acids* which, in a protein molecule, are joined together by peptide bonds. There are 20 or more amino acids, all different in structure; all follow independent metabolic pathways in the synthesis and degradation of proteins. Protein molecules are extremely complex, having molecular weights ranging from 10,000 to over 1,000,000.

In the process of digestion, proteins are hydrolyzed to polypeptides, peptides, and amino acids. The amino acids are absorbed from the small intestine and enter circulation principally by way of the portal vein. Small amounts may enter the lymph and be transported to the bloodstream through the thoracic duct.

On passing through the liver, most of the amino acids, about 80 percent, are deaminized. The remainder continue through the bloodstream and are carried to all parts of the body, where they enter into the cells of the tissues. Amino acids in the blood plasma and in tissue cells constitute an *amino acid pool* that provides amino acids for use in a variety of processes, especially the synthesis of proteins. Wherever and whenever new cells are being produced, as in the bone marrow, in the reproductive organs, in the formation of new tissue as in the skin, or in the repair of injured tissues, amino acids are required. Amino acids are also utilized in the synthesis of substances present in glandular secretions, such as mucus, enzymes, and hormones.

Neither amino acids nor proteins, as such, are stored in the body, except temporarily in the liver. Amino acids not immediately used may undergo *deamination, transamination,* or *denaturation.* Some are excreted by the kidneys.

Deamination. This process, carried on principally by the liver, involves the removal of the amino (NH_2) group with the formation of ammonia (NH_3) and a keto acid. It is accomplished principally by oxidation. The ammonia formed combines with carbon dioxide to form *urea,* which is excreted by the kidneys. The nonnitrogenous keto acid residue may be oxidized to carbon dioxide and water or converted to glucose. The glucose may be burned or stored as glycogen or converted to fat and stored in adipose tissue.

Transamination. Following deamination, the keto acid formed may accept another amino group. In this process, called *transamination,* dietary amino acids can be readily exchanged with tissue amino acids.

Denaturation. This is the process in which a substance reacts with a protein in such a way as to alter its structure, resulting in the loss of solubility and biological activity. The poisonous effects of the heavy metal salts of mercury, silver, and lead are the result of their denaturing action on proteins, which is usually followed by their coagulation.

In the catabolism of proteins, a number of nitrogenous products may be formed. These include urea, uric acid, creatine, creatinine, hippuric acid, and ammonium salts. All are excreted in the urine.

Urea. This white crystalline substance is soluble in water. It is a diamide of carbonic acid ($NH_2CO \cdot NH_2$). The amount excreted daily averages 20 to 30 g, depending on the nature of the food consumed. Urea contains the major portion of the nitrogen excreted by the urine. The urea in urine is derived not only from deamination of amino acids from the proteins of food but also from the breakdown of proteins in various tissues of the body.

Uric Acid. This crystallizable acid ($C_5H_4N_4O_3$) is a normal constituent of blood and urine. Practically insoluble in water, it is soluble in alkaline salts. Uric acid is a product of the metabolism of purines, substances forming a part of the nucleoproteins present in all cells. When uric acid is present in excess in the blood, it may be deposited at joints in the form of urates, causing *gout.* The amount of uric acid excreted daily ranges from 0.1 to 1.0 g.

Creatine. This compound is present in most tissues, especially muscle tissues, in which it is combined with phosphates to form phosphocreatine. Creatine ($C_4H_9N_3O_2$) plays an important role in muscle contraction.

Creatinine. This compound, a creatine anhydride ($C_6H_9N_3O_2$), is excreted in the urine in amounts of 1 to 2 g daily. It is derived from phosphocreatine.

Hippuric Acid. A crystallizable aminoacetic acid, it is synthesized in the kidneys and the liver from glycine and benzoic acid. This is a detoxication reaction that renders the toxic benzoic acid harmless. *Ammonium salts,* chiefly salts of inorganic acids, are present to a limited extent in the blood. Most are converted to urea in the liver, but the reverse of this process may occur in the kidneys, with the salts being eliminated in the urine.

Nitrogen Equilibrium. In the utilization of proteins, except in those processes involving synthesis of body proteins, nitrogenous waste products are formed. When proteins are converted to fats and carbohydrates or utilized for energy, nitrogenous end products are produced and eliminated—the greater portion in the urine and small amounts in the feces and sweat. Under normal conditions, the intake and output of nitrogen are equal, and the body is said to be in *nitrogen equilibrium.* When the intake exceeds the output, as occurs during growth or pregnancy or in convalescence following a wasting disease, *positive nitrogen balance* exists. When the nitrogen intake is less than the output, *negative nitrogen balance* exists; this is seen in malnutrition and in certain disease states, especially wasting diseases.

The amount of protein necessary to maintain nitrogen equilibrium depends not only on nitrogen intake but also on the amount of fats and carboyhdrates consumed. In their absence, tissue protein is utilized for energy, with a consequent increase in nitrogen output. Even on a diet consisting exclusively of proteins, the excretion of nitrogen always exceeds the intake. The ingestion of fats and carbohydrates, especially the latter, along with proteins, makes possible a reduced protein intake to meet the needs of the body. In this role these foods serve as *protein sparers;* their presence permits proteins to be used for other purposes.

Metabolism of Carbohydrates. Carbohydrates are absorbed by the intestine in the form of monosaccharides, principally *glucose.* Glucose enters the blood of the portal vein, which carries it to the liver. In the liver, excess glucose is resynthesized into *glycogen,* or animal starch, and stored in the liver cells. This process is called *glycogenesis.* Glycogen can also be formed from noncarbohydrate sources (proteins and fats), a process called *gluconeogenesis,* which takes place in the liver and kidneys.

The amount of sugar in the blood ranges from 70 to 120 mg/100 ml (average 100 mg, or 0.1 percent). The blood sugar level is the difference between glucose entering the blood and that leaving it. Glucose *enters* the bloodstream (1) from the digestive tract by absorption, (2) from the tissue fluids, and (3) from the liver as a result of the breakdown of glycogen to glucose *(glycogenolysis).* Glucose *leaves* the bloodstream (1) by diffusion into the tissue fluids, (2) by conversion to glycogen in the liver and muscle cells, (3) by conversion to fat, (4) by being oxidized in the tissues, and (5) in cases of excess, by renal excretion.

Endocrine and Neural Control. The formation and utilization of glucose are under the control of endocrine and neural factors. Endocrine control is effected through hormones of the pancreas, the anterior lobe of the pituitary, the adrenal gland, the thyroid gland, and the gonads. Hormones of these glands regulate the storage and liberation of glucose by the liver and its utilization by the tissues. Neural control is effected primarily through nervous regulation of the endocrine glands.

Insulin. The maintenance of a normal blood sugar level is regulated principally by a hormone, *insulin,* secreted by the islets of Langerhans in the pancreas. If the pancreas is removed from an animal experimentally, or if the islets fail to secrete an adequate supply of insulin, the percentage of sugar in the blood rises. If the amount is greater than 0.17 percent, a condition of excess blood sugar *(hyperglycemia)* exists, and the kidneys strive to offset this by excreting sugar in the urine *(glycosuria).* As a result, glycogen stored in the liver becomes depleted, and the body converts fats and tissue proteins into glucose.

Other Hormones. Although insulin acts to lower the blood sugar level, a number of other hormones act to raise it. These include *glucagon* from the pancreatic islets, *cortisol* and *sex steroids* from the adrenal cortex, *epinephrine* from the adrenal medulla, *somatotrophic* or *growth hormone* and the *adrenocorticotrophic hormone* (ACTH) from the anterior pituitary, and *thyroxine* from the thyroid gland. All are insulin antagonists in that they raise the blood sugar level. Details on how their effects are produced are given in Chapter 6, Vol. 2.

RELEASE OF ENERGY FROM GLUCOSE. Glucose, the end product of carbohydrate digestion, is transported to all cells of the body, where it serves as the primary source of energy. Within the cell, it undergoes a series of reactions in which it is ultimately oxidized to carbon dioxide and water with the release of energy, the major portion of which is stored in high-energy bonds of a phosphorus compound, adenosine triphosphate (ATP). The chemical reactions that take place within cells are catalyzed by intracellular enzymes; they constitute *intermediary metabolism.*

Two primary pathways are utilized in the breakdown of glucose to carbon dioxide and water. The first is the *anaerobic* or *Embden-Meyerhof glycolytic pathway,* which involves the phosphorylation of glucose, its conversion to glycogen, reconversion to glucose, and final breakdown to pyruvic acid and lactic acid with the formation of energy-rich ATP. A number of reactions involving a number of enzymes and coenzymes, especially coenzyme A (CoA), are involved in the process.

The second and more important pathway is the *aerobic* pathway or *citric acid cycle,* also called the *Krebs cycle* or *tricarboxylic acid (TCA) cycle.* In this cycle, which requires free oxygen, pyruvic and lactic acids are oxidized to carbon dioxide and water with the release of energy. In this series of reactions, through the process of oxidative phosphorylation, the energy released is incorporated into molecules of ATP, which then becomes available for muscular work or for other activities requiring energy.

The Krebs cycle is the final common pathway through which nutrients (carbohydrates, proteins, fats) can be utilized for the release of energy. It is a highly efficient process by which energy can be liberated and stored. The reactions involved are extremely complicated and require a large number of enzymes. For details of the reactions occurring, intermediate compounds produced, enzymes and coenzymes participating, and final products, the reader is referred to texts in biochemistry or advanced physiology.

METABOLISM OF FATS. In the process of digestion, fats are hydrolyzed to fatty acids and glycerol, in which form they pass into epithelial cells of the intestinal mucosa, where they are resynthesized into neutral fats, glycerides, and phosphatides, which enter the blood of the portal vein or the lymph. The major portion (70 percent, on the average) passes into the lymph contained within the lacteals of the villi and thence through lymphatic vessels into the thoracic duct. The thoracic duct opens into the left subclavian vein near its junction with the internal jugular vein, at which point the fat molecules enter the bloodstream. The milklike fluid of the lymphatic vessels draining the intestine is called *chyle.* Following the ingestion of large quantities of fat, a distinct milkiness of blood plasma may be observed, owing to the presence of microscopic fat particles, or *chylomicrons.* A marked rise in blood lipids (fats) is called *lipemia.*

In the body, fats are (1) stored in adipose tissue or incorporated in cells as constituent fats of tissues, (2) stored and metabolized in the liver, (3) oxidized to carbon dioxide and water with release of energy, (4) converted to carbohydrates, or (5) excreted in the secretions of certain glands (mammary and sebaceous) or in the feces.

The utilization of fats involves two processes that occur simultaneously, principally in the liver and adipose tissue. They are *lipogenesis,* which involves the synthesis and deposition of fat, and *lipolysis,* the degradation and oxidation of fat.

Adipose tissue, in which fat is stored, is not a static reserve of stored food but a dynamic tissue in which there is a constant turnover of fat molecules. Within it fats are being continuously synthesized from fatty acids and glycerol. *Fatty acids* are obtained from dietary fats and degraded body fats or synthesized from carbohydrate precursors. *Glycerol,* utilized in lipogenesis, is produced as a product of carbohydrate metabolism. It is converted into active glycerol by an enzyme, *glycerokinase.* The deposition of fat takes place in the fat cells (*adipocytes* or *lipocytes*) of adipose tissue, which are widely distributed in the body. It is abundant in subcutaneous tissue, especially over the buttocks, in the greater omentum, between muscles, and around organs, especially the kidneys. It also fills depressions and crevices of the body.

As fats are being deposited in adipose tissue, they are also being withdrawn and transported to organs and tissues where they are broken down to glycerol and fatty acids, which follow independent metabolic pathways.

The liver plays a primary role in the metabolism of fats. It functions in the synthesis of fats, desaturation of fatty acids, hydrogenation of unsaturated fatty acids, and oxidation of fatty acids. Fatty acids, which are composed of long chains of carbon atoms, are broken down step by step, with the loss of two carbon atoms at a time, until the entire molecule is converted into acetylcoenzyme A (acetyl-CoA), which enters the citric acid cycle, where it is completely oxidized.

Sometimes acetyl-CoA, instead of being oxidized, is converted into *ketone bodies* (acetoacetic acid, beta-hydroxybutyric acid, and acetone), which are eventually oxidized by way of the citric acid cycle. However, under conditions in which the body depends upon fats as its major source of energy, as in starvation or diabetes mellitus, excessive amounts of these ketone bodies may be produced, and they appear in the blood (*ketonemia*) or in the urine (*ketonuria*). This seriously alters the acid-base balance, resulting in a lowering of the alkaline reserve and the development of acidosis.

Essential Fatty Acids. Like amino acids, some fatty acids play an essential role in body activities. A fatty acid is considered essential (1) if, when absent from foods, a specific pathologic condition develops, or (2) if the body is unable to manufacture it. In human nutrition, one fatty acid, *linoleic acid,* is considered to be essential. In its absence, a severe form of eczema develops in infants. With vitamin B_6, linoleic acid is also essential for the synthesis of linolenic and arachidonic acids, fatty acids essential for growth.

Glycerol. Glycerol, which is formed in the breakdown of neutral fat, enters the citric acid cycle, where, at a number of points, it is involved in the synthesis of ATP and subsequent storage of energy. It is eventually oxidized to carbon dioxide and water.

Regulation of Fat Metabolism. Fat metabolism, like carbohydrate metabolism, is under chemical and neural control. Chemical control involves the same hormones that affect carbohydrate metabolism. The growth hormone (GH), adrenocorticotrophic hormone (ACTH), and thyroid-stimulating hormone (TSH), from the anterior pituitary; cortisone and cortisol, from the adrenal cortex; epinephrine and norepinephrine, from the adrenal medulla; thyroxine, from the thyroid gland; and glucagon, from the pancreatic islets, all have a *lipolytic effect,* releasing fatty acids from fats stored in adipose tissue. Insulin from the pancreas has a *lipogenic effect,* resulting in the deposition of fat in the adipose tissue, the conversion of fatty acids to fats, or both.

Chemical substances other than hormones may have a lipogenic effect. These include choline (a B vitamin synthesized from methionine), inositol, and lipocaic.

Obesity. Obesity is the condition in which there is an excessive deposition of fat. This results in overweight, which may be slight, moderate, or excessive (gross). A weight that is 25 percent above the desirable weight for a normal person constitutes obesity.

Obesity has the following deleterious effects:

1. It decreases efficiency in the performance of work; the greater load interferes with muscular movements.
2. It places a greater burden on the circulatory system (through its heavier demand in the performance of work), with accompanying hypertrophy of the heart.
3. It interferes with the release of body heat; subcutaneous fat acts as insulation.
4. It increases susceptibility to a number of pathologic conditions. These include infectious diseases (pneumonia, nephritis), diabetes, gallstone formation, arthritis, and cardiovascular disorders, especially hypertension and atherosclerosis. The mortality rate for obese persons is higher than for persons of normal weight. Obese persons are poor surgical risks.

The basic cause of obesity is the excess of caloric intake (food) over caloric output (expenditure of energy). Obesity is of two types, simple and glandular. *Simply obesity* may be *exogenous* (due to excessive consumption of food, usually accompanied by reduced physical activity) or *endogenous* (due to some disturbed bodily mechanism, either physiological or psychological). *Glandular obesity* results from the malfunctioning of one or more of the endocrine glands, especially the pituitary gland or the thyroid. Pituitary obesity is frequently associated with hypothalamic disorders in which normal perception of satiety is altered. Excessive secretion of some adrenocortical hormones is characterized by obesity.

Physiological factors include such conditions as hypoglycemia, which causes excess hunger, resulting in high food intake. Psychological factors are of primary importance. Nervous tension, frustration, or unhappiness often result in increased food intake and reduced physical activity. Compulsive eating may compensate for difficult life situations, the pleasure and satisfaction of eating offsetting the loss of physical attractiveness and the reduction in efficiency.

Cultural factors are of importance. Children of obese parents tend to be obese, since they are served the same foods and tend to follow the same eating habits as their parents. Overfeeding in infancy and in early stages of puberty may lead to the development of excessive numbers of cells in adipose tissue. The role of heredity in obesity is not clear, but it is generally recognized that constitutional factors involving basic body structure, such as the nature and structure of bone, amount of muscle mass, and volume of adipose tissue, have a genetic basis and may predispose a person to obesity.

Weight reduction can be accomplished only by a reduction in caloric intake. Reducing diets should be under the direction of a physician so that nutritional needs, especially for minerals and vitamins, are fully met. Diet fads should be avoided, since fat loss is usually only temporary. The use of drugs that stimulate metabolism, increase water loss, or depress the appetite may be dangerous. Sometimes psychotherapy as an adjunct to a dietary regimen is advisable.

Malnutrition. Malnutrition is the condition that results from an inadequacy of one or more of the essential components of one's diet: calories, proteins (including certain essential amino acids), certain fatty acids, vitamins, and minerals. It may be caused by inadequate ingestion of foods, ingestion of foods lacking in essential constituents, defective absorption by the intestine, or defective metabolism within the tissues. The effects of malnutrition are enumerated here.

1. Reduction in body weight. If the malnutrition is mild, the loss is principally from adipose tissue, but if it is continued (as in fasting), loss of protein occurs, with breakdown of essential body tissues, and, ultimately, emaciation.
2. Reduction in the basal metabolic rate, with resultant lower body temperatures and reduced blood pressure.
3. Increased susceptibility to fatigue, accompanied by mental apathy and disinclination for physical exertion.
4. Tendency of organs to become displaced, owing to loss of the fat that normally supports them.
5. Reduction in blood proteins, resulting in edema (swelling from accumulation of fluids).
6. Increased susceptibility to infectious diseases, especially tuberculosis.

A nutritional disorder, *anorexia nervosa,* may result in severe malnutrition. The subjects, usually adolescent girls, have a disturbed sense of body image and excessive anxiety about weight gain; consequently they engage in compulsive and bizarre dieting sometimes interspersed with episodes of excessive eating (*bulimia*) followed by induced vomiting. The illness, which is potentially life-threatening, may have psychological as well as physical causes and frequently requires psychotherapy.

REGULATION OF BODY TEMPERATURE

Because body temperature depends in large part on the various chemical processes that take place in the metabolism of food substances, the subject of its regulation is discussed here.

With respect to body temperature, animals are divided into two classes: warm-blooded, or homoiothermal, and cold-blooded, or poikilothermal.

Homoiothermal (*homeothermal*) animals include the birds and mammals. In this group, except in hibernating mammals, the body temperature remains constant irrespective of environmental temperature. Body temperature depends upon the relationship between heat produced and heat lost. Heat is produced in various chemical reactions, especially in oxidative reactions. Heat loss depends upon various physical rather than chemical factors.

Poikilothermal animals include the fishes, amphibians, and reptiles. In this group the body temperature is variable, changing with the temperature and environment. When the body temperature of a cold-blooded animal falls, metabolic processes are slowed down, and body activities become sluggish. In warm-blooded animals the metabolic processes continue at a fairly constant rate all the time.

Heat Production. Most of the metabolic processes that occur in the body tissues result in the production of heat. Because the muscles constitute the bulk of the body (50 percent of weight of soft tissues and 40 percent of total body weight in males, 36 percent of total body weight in females), most of the heat produced comes from muscular activity. The liver is also an important source of heat. A small amount of heat comes from the ingestion of hot foods and drinks.

Heat Loss. Heat is lost from the body through the skin, the lungs, and urinary and digestive excretions.

Heat Loss Through the Skin. Approximately 85 percent of body heat is lost through the skin by the processes of (1) *convection,* when air in contact with the body has a lower temperature (the skin being reheated and the heat removed by air currents); (2) *conduction,* when the body is in contact with a cooler object (heat being lost to that object); (3) *radiation,* in the giving off of thermal energy; and (4) *evaporation,* the vaporization of water as in sweat, about 0.6 kcal being required for each gram of water vaporized at 37°C.

Heat Loss in Respiration. Heat is lost by the vaporization of water from the lungs. It is also lost in the warm air expired by the lungs and in the process of warming inspired air, which is usually below body temperature.

Heat Loss Through Urinary and Digestive Excretions. Urine and feces are at body temperature when they are passed from the body. This accounts for a small loss of heat.

Heat is also lost when foods taken into the body are below body temperature.

DAILY HEAT LOSS

Process	*Kcal*	*Percent*
Radiation, convection, conduction	2100	70.0
Evaporation from the skin	500	16.9
Vaporization of water from the lungs	300	9.1
Warming of inspired air	80	2.5
Excretion of urine and feces	50	1.5
	3030	

Temperature Control. The balance between heat production and heat loss is maintained primarily through the activity of the nervous system. Numerous reflex centers in the spinal cord and the lower portions of the brain are concerned with the production and loss of heat, principally the latter. They regulate such activities as sweating, vasoconstriction, vasodilatation, muscle tonus, and shivering.

The most important heat-regulating centers are located in the *hypothalamus,* a portion of the diencephalon. These centers are subject to two influences: nerve impulses from *cold* and *warm* receptors in the skin and other parts of the body, and the temperature of the blood flowing near the centers. The hypothalamus acts in two ways. A center located in its anterior portion controls the mechanism involved in heat loss and thus prevents overheating of the body; another center located in the posterior portion regulates heat production and thus prevents chilling of the body.

The hypothalamus acts as a biological thermostat. When heat production is increased in the body, or the outside temperature is increased, activities brought about reflexly increase the heat loss. This is accomplished to a large extent through the circulatory system. Flow of blood to the skin is accelerated, and this is accompanied by dilatation of blood vessels and increased activity of sweat glands, as a result of which heat is lost from the body surface. The rate of respiration tends to increase, and muscular activity is reduced. When outside temperatures are low, peripheral vessels are constricted, and the consequent reduced activity of sweat glands conserves heat. Simultaneously, heat production is increased through muscular activity, sometimes involuntarily, as in shivering, and the metabolic rate is generally increased.

The *normal temperature of the body* averages 37°C (98.6°F), although in healthy individuals it may vary a degree or more. Temperature readings are taken by mouth (oral), under the arm (axillary), or within the rectum. Rectal temperature is usually about 0.6°C higher than by mouth, axillary temperature about the same amount lower. The body temperature may show variations of 0.3° to 0.6°C over the course of 24 hours. It is lowest around 5 A.M., highest in the late afternoon or early evening. The temperature of all parts of the body is not the same. The temperature of the central internal organs (*core temperature*) remains close to average body temperature, while that in the peripheral regions (the skin and underlying tissues, especially in the extremities) varies widely depending upon environmental temperature, amount of clothing worn, and other factors.

Role of Endocrine Glands in Heat Regulation. Two endocrine glands play a major role in heat production and heat loss: the thyroid and the adrenal glands. As the thyroid regulates basal metabolism, hyperthyroidism, by increasing the basal metabolic rate, increases heat production. Conversely, either hypothyroidism or surgical removal of the thyroid gland reduces heat production and reduces body temperature. The adrenal medulla, through its hormone epinephrine (adrenalin), stimulates oxidations in the tissues, with a resulting increase in heat production. At the same time, its vasoconstrictor effect on the blood vessels of the skin reduces heat loss. The adrenocortical hormones also play a role in temperature regulation, since in their absence animals are less resistant to cold.

Two other endocrine glands are involved in temperature regulation: the hypophysis (pituitary) and the ovaries. Anterior hypophyseal insufficiency results in a reduced metabolic rate and a decrease in resistance to cold. Through its effects on the activities of the thyroid and adrenal glands, it exerts a continuous indirect action on heat production.

In women, body temperature fluctuates during the menstrual cycle. It remains normal during the first portion of the cycle, then drops suddenly about 0.6°C at the time of ovulation. Then it rises to about 0.6°C above normal, where it remains during the second half of the cycle, returning to normal at menstruation. The exact mechanism of this temperature change is not known.

Body Temperature Disturbances. Body temperature, under certain conditions, may vary significantly from the normal. The temperature may be above normal (hyperthermia) or below normal (hypothermia).

HYPERTHERMIA (PYREXIA, FEVER). In this condition, the body temperature remains above normal for a protracted period. The highest temperature compatible with life is about 42.5°C. Functionally, fever is due to disorder of the mechanisms involved in heat elimination, particularly those regulating blood volume, vasodilatation, and sweating. The heat-producing mechanisms of the body are usually not accelerated in fever. The causes of fever include (1) *infection by organisms* (bacteria and viruses), as in typhoid fever, yellow fever, septicemia, and viral conditions; (2) *trauma,* as in surgical fever when tissues are injured; (3) *brain injury,* in which the heat-regulating center is affected (neurogenic fever); (4) *dehydration,* in which there is a reduction in the water content of the blood, as following injections of hypertonic solutions of glucose and salt into the bloodstream and in the use of strong cathartics; and (5) *effects of drugs,* such as ergotoxine, dinitrophenol, adrenalin, thyroxine, or foreign proteases.

The fever accompanying infectious diseases is due to the production of endogenous *pyrogens,* substances released from leukocytes following their contact with bacterial endotoxins or other inflammatory stimuli. These pyrogens act directly on the thermoregulatory centers. Fever is thought to be an adaptive mechanism of the body to inhibit the multiplication of infective organisms by interfering with enzymatic activities within bacterial cells. Other defensive activities, such as antibody formation and phagocytosis, are enhanced.

HYPOTHERMIA. This is the condition in which the body temperature remains below normal for a considerable period of time. Core body temperature may fall 10° to 15°F (5° to 8°C) without significant ill effects. It is less common than hyperthermia and occurs usually from *accidental exposure,* as in the case of abandoned infants and alcoholics, or from *immersion* of the body in cold water. A temperature drop to 95° F (35°C) has a stimulating effect; below that temperature, the effect is depressant. At 86°F (30°C) critical effects ensue, death usually resulting when the temperature reaches 70°F (21°C). Sometimes therapeutic hypothermia, in which the body temperature is reduced to 86°F (30°C), is employed in surgery.

Frostbite. Local hypothermia or cold injury results from prolonged exposure of a part of the body (fingers, toes, face, ears) to freezing temperatures. Cells of frozen tissues die as a result of the formation of intracellular ice or the plugging of capillaries. If *superficial,* surface tissues are sloughed off, and regeneration may take place. *Deep frostbite* involves the skin and underlying tissues, and unless circulation can be quickly restored, the tissue dies and gangrene may set in.

Body Heat Disturbances. Conditions in which the thermal balance of the body is upset are heat cramps, heat exhaustion or prostration, and heat stroke.

HEAT CRAMPS. In heat cramps, one encounters muscle spasms, pains, dilated pupils, and a weak pulse. The condition is common among those who work in regions of high temperatures. The fundamental cause is dehydration and reduction in blood chlorides through the elimination of the salts in excessive sweating.

HEAT EXHAUSTION OR PROSTRATION. In heat exhaustion, excessive, long-continued demands upon the body mechanisms regulating heat loss may result in pronounced weakness, dizziness, nausea, and sometimes collapse (prostration). Body temperature usually rises to about 38.5°C.

HEAT STROKE. In heat stroke, including "sunstroke," the body temperature rises markedly, the skin becomes dry, and there are muscular cramps, dizziness, thirst, nausea, and headache. Often, there is loss of consciousness. The heat-regulating center in the hypothalamus is unable to cope effectively with the sudden increase in heat production. Body temperature may rise to 42.5° to 43°C, often producing irreversible changes in the brain. Prolonged body temperature of 42.5°C is usually fatal.

PRACTICAL CONSIDERATIONS

Hunger. When the body has been deprived of food for some time, the sensation of *hunger* is experienced. Hunger is an unpleasant sensation accompanied by dull or acute pains (referred to the epigastric region) and an overwhelming desire to eat. It is distinguished from *appetite* in that the latter is a sensation based on previous experience of ingesting foods that are pleasing to the taste and that provide one with the sensation of well-being. The pains that accompany hunger ("hunger pangs") are due primarily to strong peristaltic contractions that occur in the stomach. However, in prolonged fasting, after the first few days there is a gradual decline in hunger pains, although strong gastric contractions persist.

Vomiting. *Vomiting* (or *emesis*) is the ejection of the stomach contents through the mouth. It is a reflex act controlled by a vomiting center in the medulla. Vomiting is usually preceded by a feeling of *nausea,* at which time the tonus of the stomach decreases and the upper part dilates. The expulsion of the stomach contents is brought about as follows: A deep inspiration occurs, and the diaphragm descends. The soft palate rises and closes the nasopharyngeal opening, and the glottis closes. Vigorous contractions of the abdominal muscles force the abdominal organs upward against the diaphragm, which in turn compresses the stomach. The food or liquid is expelled through the relaxed cardia and esophagus. In the act of vomiting the musculature of the stomach is passive and plays no essential part in the action.

CONTROL OF VOMITING. Vomiting is under neural control effected through a *vomiting center* located in the medulla oblongata. It may occur as the result of a number of disorders, or it may be induced to rid the digestive tract of harmful substances. The vomiting center may respond to (1) afferent impulses from the isthmus of the fauces, pharynx, stomach, intestines, or other parts of the body; (2) impulses from higher cerebral centers; or (3) chemical substances in the bloodstream.

FACTORS THAT INDUCE VOMITING. Some of the conditions that induce vomiting are (1) mechanical stimulation of the fauces or pharynx; (2) inflammation or mechanical disturbances of the digestive tract, as in enteritis, appendicitis, intestinal obstruction, and strangulated hernia; (3) disturbances in other abdominal organs, such as the kidney, urinary bladder, uterus, and gallbladder; and (4) the presence of certain foods, food products, or drugs in the stomach (these substances include tartar emetic, mustard, ipecac, mercuric chloride, and zinc and copper sulfates, which are referred to as *emetics*). Emetics may act reflexly, stimulating nerve endings in the digestive tract, or, if injected intravenously, they may cause vomiting by acting directly on the vomiting center in the medulla.

Vomiting may be initiated by impulses from the higher cerebral centers. Excessive fatigue, severe pain, hysterical states, emotional reactions (as may follow the witnessing of a "nauseating" scene), or psychogenic factors may bring about vomiting. The excitability of the vomiting center varies in different individuals and in the same individual under varying conditions. Blood supply has an influence on the center. A sudden increase in intracranial pressure may bring on a forceful, "projectile" type of vomiting. A reduced oxygen supply, as in anemia or following hemorrhage, may reduce the threshold of the center and induce vomiting. Vomiting often follows the administration of an anesthetic.

The exact cause of *morning sickness* in early pregnancy is unknown, but it is thought that carbohydrate starvation and dehydration accompanied by ketosis are factors. The nausea and vomiting that accompany *seasickness* or other forms of motion sickness result from stimulation of the sensory receptors in the ear, which induce vomiting reflexly through the vomiting center.

Vomiting, when prolonged, may seriously affect water and electrolyte balance. Dehydration and alkalosis with potassium depletion may result from the loss of water and acid secretions from the stomach. Vomiting is especially serious in infants and young children.

Diarrhea. The condition in which there are abnormal *frequency* of bowel movement and increased *liquidity* of the fecal discharge is known as *diarrhea.* It is characterized by abnormally rapid passage of food through the small intestine, which interferes seriously with the digestive processes and reduces the amount of food absorbed.

Types of diarrhea include *osmotic diarrhea,* in which there is an excess of nonabsorbable, water-soluble compounds that draw water into the intestine; *secretory diarrhea,* in which the intestine secretes rather than absorbs electrolytes and water; *malabsorption diarrhea,* as when undigested fats or carbohydrates cause excessive secretion by the colon; *exudative diarrhea,* in which serum proteins, plasma, blood, and mucus enter the bowel through an inflamed mucosa; and *altered-transit diarrhea,* which results from accelerated passage of chyme through the intestine.

Diarrhea is symptomatic of many diseases. It may be caused by (1) enteric infections, such as bacillary and amebic dysentery, cholera, and typhoid and paratyphoid fevers; (2) food poisoning; (3) food sensitivities; (4) poisons, such as mercury and arsenic; (5) avitaminosis, such as pellagra; (6) toxic or septic states accompanying a number of diseases, and (7) emotional states such as fear or grief.

As with vomiting, diarrhea, if persistent, may seriously affect water and electrolyte balance. Dehydration accompanied by acidosis may result from loss of water and bicarbonates.

Constipation. In constipation, bowel movements occur infrequently or with difficulty, and the feces are retained within the rectum. The effects vary in different persons, but general symptoms are headache, a feeling of depression, sluggishness, and general malaise. These effects are due to the mechanical pressure of the foods on the walls of the colon and not to the absorption of toxic products (autointoxication), as was once supposed.

Among the causes of constipation are (1) gross mechanical obstructions, such as tumors, adhesions, strictures, and developmental anomalies; (2) poor toilet habits, that is, failure to evacuate the bowel in response to the sensations aroused by a full rectum ("psychogenic constipation"); (3) poor muscle tone of the intestine ("atonic constipation") which may result from debilitating disease, senility, or obesity; (4) excessive muscle tone ("spastic constipation"), in which the muscles are hypertonic and remain contracted, often the result of emotional or nervous tension; (5) improper diet, especially one lacking in bulk or roughage or containing too little water; (6) excessive use of cathartics ("cathartic constipation"); (7) lesions of the digestive tract, especially in the colon and rectum, such as inflammatory conditions (appendicitis), hemorrhoids, and cancer.

METHODS OF INDUCING BOWEL MOVEMENTS. Various measures are employed to bring about the elimination of the indigestible contents of the digestive tract. These include the use of purgatives, lubricants, and enemas.

Purgatives. Mild purgatives are called *laxatives* or *aperients.* Stronger purgatives are usually called *cathartics.* Common laxatives include *bulk laxatives* (e.g., bran and agar-agar), which act by increasing the bulk of the intestine and thus mechanically stimulate muscular activity (peristalsis); *irritant laxatives* (e.g., castor oil, senna, cascara, and croton oil), which stimulate intestinal contraction; and *saline laxatives* (e.g., Epsom salt, or magnesium sulfate, and citrate of magnesia), which increase the osmotic pressure within the lumen of the intestine, causing water to enter, resulting in an increase in the fluidity of bowel contents.

Lubricants. Normally, the intestinal passage requires no artificial lubricant because the mucus secreted by epithelial cells in its wall provides adequate natural lubrication. It may, however, be desirable to soften the fecal mass to facilitate defecation. Any agent used for this purpose should not be amenable to enzymatic action but should pass through the digestive tract more or less unchanged. *Mineral oil* is such a substance. Continuous use of a lubricant is undesirable, however, since the presence of excessive oils in the intestine interferes with the action of enzymes on food substances and with normal absorption. In the presence of mineral oils, the fat-soluble vitamins A, D, and K may be inadequately absorbed.

Enemas. Injection of a liquid through the anal opening serves (1) to make the contents of the colon more fluid and (2) to stimulate activity in the muscles of the colon by distension of its walls.

Nutrient enemas are occasionally employed when food cannot be taken by mouth. Because no enzymes are produced by glands of the large intestine, food thus injected must be *predigested.* Water is readily absorbed by the large intestine; accordingly, *thirst enemas* are regarded as effective.

DISORDERS AND DISEASES OF THE DIGESTIVE SYSTEM

The digestive system is prey to a wide range of disorders and diseases that may be classified as developmental anomalies and endogenous and exogenous conditions.

Developmental Anomalies. The following developmental anomalies are commonly seen in various portions of the alimentary canal:

ANOMALIES OF THE MOUTH AND PHARYNX. *Harelip* is a condition in which the median nasal processes of the embryo fail to unite; it may be unilateral or bilateral and may involve the fleshy lip, the bony maxilla, or both. *Cleft palate,* in which the lateral palatine processes fail to unite and which may involve the hard palate, the soft palate, or both, is often associated with harelip. *Tongue-tiedness,* or reduced mobility of the tongue, results from an abnormally short frenulum.

ANOMALIES OF THE ESOPHAGUS. *Double esophagus* has been seen, or the esophagus may be *absent* (abnormal development of the communication with the trachea).

ANOMALIES OF THE STOMACH. The stomach may be *transposed* to the right side. There may be *atresia* (closure) or *stenosis* (narrowing) of the esophageal opening. Also seen is *hour-glass stomach,* in which a constriction divides the stomach into two compartments. *Gastroesophageal hernia,* in which the lower end of the esophagus and adjacent part of the stomach herniate into the throax.

Endogenous and Exogenous Conditions. The following are the commonest disorders and diseases that arise from malfunction, circulatory disturbances, invasions of microorganisms, and infestations with parasites. In addition, there are the *neoplasms* (new growths), which occur as either benign or malignant tumors, including various forms of carcinoma (cancer). Neoplasms may develop in any part of the digestive tract, especially where there is an abrupt transition of one type of epithelium to another, as at the lips, cardiac and pyloric regions of the stomach, and anal valves of the rectum. The rectum, colon, and pancreas are common sites of malignant tumors.

CONDITIONS OF THE MOUTH AND PHARYNX. *Caries,* decay of teeth, involves destruction of the enamel and dentine; frequently an opening is made into the pulp cavity that permits infectious organisms to gain access to the alveolus, resulting in abscesses. *Mumps* is a virus disease involving the salivary glands, especially the parotids, and may involve other glands of the body (e.g., the testes). *Gingivitis* is inflammation of the gums. *Pyorrhea* is inflammation of the dental periosteum accompanied by formation of pus; necrosis of the alveoli may concur, with loss of teeth. *Gumboil* is a superficial abscess of the alveolus. *Tonsillitis* is inflammation of the palatine tonsils. *"Adenoids"* refers to hypertrophy of the pharyngeal tonsil.

Conditions of the Esophagus. *Dysphagia,* or difficulty in swallowing, may result from obstruction of the esophagus, neurogenic disorders, or muscular dysfunction (a local spasm or failure of peristalsis). *Obstruction* may result from neoplasms, enlarged adjacent structures, or the presence of foreign bodies, especially in constricted regions where the bronchus crosses the diaphragm or where the esophagus passes through the diaphragm. *Diverticula* (blind pockets) may develop as a result of excessive pressure from within or traction from surrounding tissues. *Stenosis* or *stricture* is a closure or an abnormal narrowing of the esophagus. It may be congenital or follow the ingestion of a caustic chemical.

Conditions of the Stomach. *Gastritis* is inflammation of the gastric mucosa. It may be due to many causes, including various foods and drugs, bacterial infection, or allergenic drugs. *Achlorhydria* is the absence of free hydrochloric acid in gastric secretions; the pH does not fall below 6. *Gastric ulcers,* commonly called *peptic ulcers,* may develop near the pylorus or along the lesser curvature. *Stenosis* or *stricture* of the pyloric opening usually results in dilatation of the stomach. The stomach, or a portion of it, may protrude through the esophagal hiatus of the diaphragm, a condition called *hiatus hernia.*

Conditions of the Intestines. *Enteritis,* or inflammation of the intestine, is a common ailment. It may result from viral, bacterial, or protozoan infections. The intestinal "flu" or "grippe" is of viral origin, cholera and typhoid fever of bacterial origin, amebic dysentery of protozoal origin. Enteritis may also result from allergic reactions to certain foods or food poisoning. *Food poisoning* may be caused by (1) toxins of the botulinus organism (*botulism*) or staphylococci; (2) poisons occurring naturally in certain foods, such as toadstools, fishes, mussels, and rye infected with a certain fungus (ergotism); or (3) poisonous substances such as pesticides, lead, or arsenic in contaminated foods.

Infestations by animal parasites are usually caused by tapeworms or threadworms. *Tapeworms* (*cestodes*) include *Taeniarhynchus saginata* (beef tapeworm), *Taenia solium* (pork tapeworm), and *Diphyllobothrium latum* (fish tapeworm). Tapeworms are acquired by ingesting beef, pork, or fish containing encysted larval forms (*cysticerci*). *Threadworms* (*nematodes*), also called *roundworms,* include the large *Ascaris* and the smaller hookworm, pinworm, and whipworm. The hookworm (*Necator*) is acquired when free-living infective larvae, which live in the soil, penetrate the skin, usually between the toes. The larvae pass via the lymphatics to the bloodstream, thence through the heart to the lungs, from which they are coughed up and swallowed and then pass to the intestines, where they attach themselves to the villi. *Pinworms* or *seatworms* (*Enterobius* or *Oxyurus*), residing in the large intestine, are common in children; they frequent the rectum, causing severe itching about the anus. The *whipworm* (*Trichurus trichiura*), about 5 cm long, with a slender anterior portion, inhabits the large intestine; its presence may cause diarrhea, vomiting, nervous disturbances. Nematodes other than hookworms are acquired by ingesting food or water contaminated with eggs or infective larvae.

Appendicitis is inflammation of the vermiform appendix, rupture of which may permit infectious organisms to gain access to the peritoneal cavity, giving rise to *peritonitis. Colitis,* or inflammation of the colon, is frequently accompanied by excessive secretion of mucus ("mucous colitis"); it may be psychogenic in origin ("irritable colon"). *Intussusception* is a condition in which a portion of the intestine slips over an adjacent region. *Hemorrhoids,* or "piles," result from excessive dilatation of the veins in the wall of the rectum; they are "external" when visible outside the anal sphincter, "internal" when within.

CONDITIONS OF THE LIVER AND GALLBLADDER. *Jaundice* (*icterus*) is a condition in which bile pigments are deposited in the skin and mucous membranes, giving the subject the characteristic yellowish skin coloration. Jaundice may result from (1) obstruction of bile passageways, with resultant absorption of bile into the blood; (2) increased destruction of red blood cells by the spleen (hemolytic jaundice); or (3) toxic or infectious conditions in which liver tissue is damaged. *Cirrhosis* is a chronic progressive disease of the liver characterized by an increase in the amount of connective tissue and a decrease in the parenchymal tissue. Causes include nutritional deficiencies, especially lack of lipotropic factors, poisoning, infections, chronic alcoholism, and biliary obstruction. *Fatty degeneration* or *infiltration* of the liver results when fat is deposited within functional liver cells. Normal fat content (3 to 5 percent) may rise to 40 percent. *Hepatitis* is inflammation of the liver. *Infectious hepatitis* is a common clinical condition caused by a virus. It occurs sporadically or sometimes in small epidemics. *Serum hepatitis* is caused by a virus found only in the blood. It is transmitted by transfusion of blood containing the virus or by use of inadequately sterilized syringes, needles, or surgical equipment. It is commonly seen in drug addicts who inject themselves with dirty needles.

Cholecystitis is inflammation of the gallbladder; *cholelithiasis* is a condition in which biliary concretions or gallstones are present in the gallbladder or bile ducts.

CONDITIONS OF THE PANCREAS. *Pancreatitis* is inflammation of the pancreas. It may be edematous, hemorrhagic, or necrotic. It is often associated with obesity and alcoholism and may be acute or chronic. *Diabetes mellitus* is a systemic disease caused by failure of the islets of Langerhans to secrete an adequate supply of insulin, which is essential for sugar metabolism. *Malignant neoplasms* are common in the pancreas.

9: THE BODY FLUIDS

The *principal body fluids* (found in all body regions) are *blood* (fluid within the heart and in the vessels of the circulatory system), *tissue fluid* (occupying intercellular spaces), and *lymph* (fluid within the lymph vessels and lymphatic organs, such as lymph nodes, tonsils, thymus, spleen). These will be described in this chapter.

Other fluids more specifically located and performing distinct functions are the *cerebrospinal fluid* (filling spaces within and surrounding the brain and spinal cord), the *synovial fluid* (filling cavities of the articulations), the *aqueous humor* (in anterior and posterior chambers of the eye), the *endolymph* (filling the membranous labyrinth of the ear), and the *perilymph* (within the bony labyrinth of the ear).

THE BLOOD

The blood fills the heart and the blood vessels. It is pumped by rhythmic contractions of the heart into the larger arteries, from these into smaller arteries and arterioles, and eventually into the capillaries of all tissues. It passes through capillaries into small vessels (venules), which empty into larger veins that return blood to the heart. The blood is constantly in circulation.

Functions of Blood. Blood is the chief medium of transportation within the body. Its principal functions are as follows:

1. To transport *substances to cells.* Among these are nutrients, water, oxygen, hormones, and other substances required for metabolic or secretory activity.
2. To transport *substances from cells.* These include food substances that have been absorbed by cells in the walls of the intestines, waste products (carbon dioxide, lactic acid, urea), and hormones from glands of internal secretion.
3. To transport *phagocytes and antibodies.* These are defensive cells and immunogenic substances, important in the body's defenses against disease.
4. To transport *excess internal heat to the lungs and body surfaces.* This is important in the regulation of body temperature.

The functions of the blood may also be listed as follows:

1. *Respiratory*—carrying oxygen from the lungs to the tissues and carbon dioxide from the tissues to the lungs.
2. *Nutritive*—carrying food substances (glucose, amino acids, fats) from the intestines or storage "depots" to the tissues.

3. *Excretory*—carrying waste products (urea, lactic acid, creatinine) from the cells to the organs of excretion.

4. *Protective*—carrying defensive cells and antibodies throughout the body to resist disease.

5. *Regulatory*—carrying hormones and other chemical substances that regulate the functioning of many organs; carrying excess internal heat to the lungs and body surfaces, through which it is lost, thus regulating body temperature; and maintaining water balance and a constant internal environment for tissue cells.

Origin of Blood. In the embryo, blood develops from mesenchyme (derived from mesoderm). Subsequently, red and white cells develop as follows: fourth week, wall of yolk sac; fifth week, body mesenchyme and blood vessels; sixth week, liver; seventh to thirteenth weeks, spleen, thymus, lymph nodes; thirteenth week on, red bone marrow and lymph nodes.

General Characteristics of Blood. In systemic arteries, blood is *bright red;* in systemic veins, it is *brownish red* or *purple.* Its *specific gravity* ranges from 1.050 to 1.060 (specific gravity of plasma is 1.027). Its *viscosity* is about five or six times that of water. As to *quantity,* blood constitutes about 7 percent, or one-fifteenth of body weight. For a person weighing 68 kg (150 lb), this would amount to 5 to 6 liters (9 to 11 pints). When the body is at rest, about one-fourth of the blood is in the heart, lung vessels, and large blood vessels, one-fourth in the vessels of liver, one-fourth in the vessels of striated muscles, and one-fourth in the vessels of the other organs.

Composition of Blood. Blood consists of *plasma* (fluid portion) and *formed elements* (corpuscles and platelets) suspended in the plasma. Corpuscles comprise red blood cells and white blood cells.

BLOOD CONSTITUENTS

PLASMA (about 55 percent of blood by volume), consisting of the fluid portion

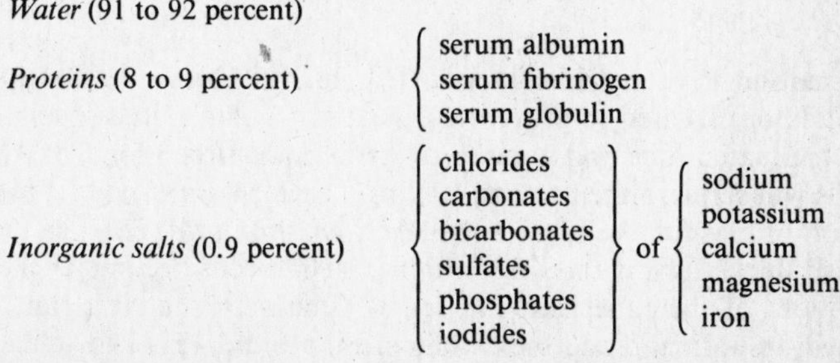

- *Water* (91 to 92 percent)
- *Proteins* (8 to 9 percent)
  - serum albumin
  - serum fibrinogen
  - serum globulin
- *Inorganic salts* (0.9 percent)
  - chlorides, carbonates, bicarbonates, sulfates, phosphates, iodides
  - of sodium, potassium, calcium, magnesium, iron

- *Organic substances other than proteins*
  - nitrogenous substances
    - ammonium salts
    - urea
    - uric acid
    - creatine
    - creatinine
    - amino acids
    - xanthine
    - hypoxanthine
  - nonnitrogenous substances
    - glucose
    - neutral fats
    - phospholipids
    - cholesterol
- *Gases*
  - oxygen
  - carbon dioxide
  - nitrogen
  - gases from the intestines
- *Miscellaneous substances*
  - hormones
  - antibodies (immune substances)
  - enzymes (lipase, catalase, *et al.*)

FORMED ELEMENTS (about 45 percent of blood by volume), consisting of corpuscles and platelets

Erythrocytes, or red blood cells (RBC)

Leukocytes, or white blood cells (WBC)

- Granulocytes
  - neutrophils
  - eosinophils
  - basophils
- Agranulocytes
  - lymphocytes
  - monocytes

Thrombocytes (blood platelets)

Sedimentation Rate and Hematocrit. If a quantity of blood that has been citrated or oxalated to prevent coagulation is placed in a specially prepared graduated tube and allowed to stand undisturbed for 1 hr, the blood cells will settle to the bottom, leaving a clear plasma above. From the amount of plasma, the *erythrocyte sedimentation rate* (ESR) can be determined. If the tube is then centrifuged, the red cells become packed at the bottom of the tube, and the cell volume or *hematocrit* can be determined. In normal blood the hematocrit is around 45 (47 for males, 42 for females), which means that blood cells comprise about 45 percent of the volume of the whole blood. Changes in sedimentation rate and hematocrit are characteristic of certain pathologic conditions, such as anemia, leukemia, rheumatic fever, and rheumatoid arthritis.

PLASMA

The constituents of blood plasma have a variety of functions. The principal ones are listed here.

Water. The importance of water in the body is discussed in Chapter 2 of this volume and in Chapter 1, Vol. 2.

Proteins. The plasma proteins constitute a dynamic pool of amino acids that may be utilized by any of the various tissues of the body. Collectively, they also function in the maintenance of (1) collodial osmotic pressure, (2) acid-base, water, and electrolyte balance, and (3) the viscosity and volume of the blood. In addition to these roles, specific functions are as follows.

Albumin is the most abundant protein, comprising about 60 percent of the total blood proteins. It plays a special role in the transport of substances such as minerals (calcium, iodine), hormones, and vitamins, and in the maintenance of plasma osmotic pressure. Albumin is synthesized in the liver.

Globulins include a group of high-molecular-weight proteins comprising three types, *alpha, beta,* and *gamma* globulins. Gamma globulins constitute the *immunoglobulins,* substances that function as specific antibodies that confer on the body specific immunity to various bacterial and viral infections. There are five classes of immunoglobulins (Ig) in man: IgM, IgG, IgA, IgD, and IgE. Globulins are produced in the plasma cells of the bone marrow and the lymphatic tissue.

Fibrinogin is a protein which, in shed blood, is converted into insoluble fibrin that forms the basis of a clot. (See *Coagulation,* page 287). Fibrogen is formed in the liver.

Inorganic Salts. The inorganic salts, comprising about 0.9 percent of the plasma, play a role in the osmolarity of the blood and in acid-base, water, and electrolyte balance. Their various functions in the body are given in Chapter 2.

Other Substances. Present in the plasma are many products of intermediary metabolism, food substances absorbed from the intestine, waste products eliminated from cells, respiratory and other gases, and miscellaneous substances such as hormones and enzymes. The plasma may also contain abnormal substances, including bacterial toxins and chemical substances ingested, inhaled, or injected, such as drugs of various types.

FORMED ELEMENTS OF THE BLOOD

The blood contains three types of formed elements: erythrocytes, leukocytes, and blood platelets (thrombocytes) (Fig. 9-1).

Erythrocytes. Red blood cells (RBC), or *erythrocytes,* are nonnucleat-

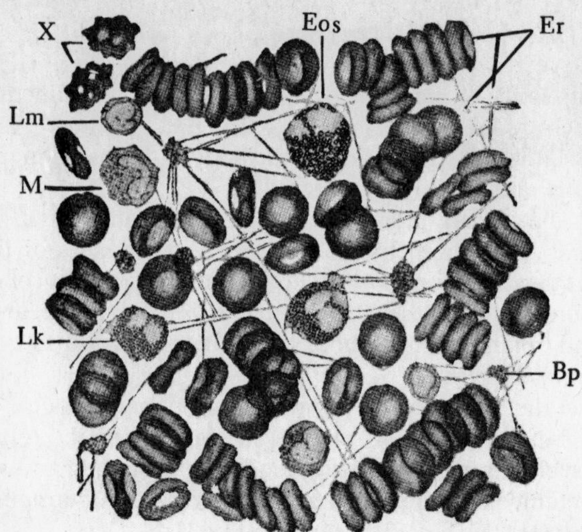

Fig. 9-1. Human blood. (*Bp*) Platelets. (*Er*) Erythrocytes. (*Eos*) Eosinophil. (*Lk*) Neutrophil. (*Lm*) Lymphocyte. (*M*) Monocyte. (*X*) Crenated erythrocytes. (Reprinted with permission of W. B. Saunders Co., Philadelphia, from A. A. Maximow and W. Bloom, *A Textbook of Histology,* 6th ed., 1952.)

ed, biconcave discs, circular in general form but possessing a dumbbell shape when seen from the side. In freshly drawn blood they tend to adhere to each other with their flat surfaces together, forming *rouleaux*, each of which resembles a stack of coins. *Size:* The average mean diameter is 7.7 μm; the average mean thickness is 1.9 μm. Red blood cells may show considerable variation in size within an individual. They are slightly larger in venous blood than in arterial blood, owing to osmotic changes. *Surface area:* A single red blood cell has a surface area of 120 μm^2. The total surface area of all the red blood cells in the body is about 3600 m^2. *Volume:* The volume of a single red blood cell is 0.85 μm^3.

NUMBER OF ERYTHROCYTES. The red blood cell *count,* or number of erythrocytes in the body, is expressed by *volume.* In males, there are generally about 5 million red blood cells per cubic millimeter; in females, about 4.5 million/mm^3. The total number of red blood cells in the average-sized person is roughly 35 trillion.

Variations in the number of red blood cells occur under certain conditions:

1. *With age.* At birth the erythrocyte count (averaging over 6 million per mm^3) is much higher than later in life.
2. With *time of day.* The count is lower during sleep, higher during activity.
3. *With altitude.* Persons living at altitudes 3,050 m (10,000 ft) or

more above sea level have red blood cell counts as much as 30 percent above normal. A temporary ascent to high altitudes causes a discharge of immature red blood cells from the bone marrow into the bloodstream.

4. With *muscular exercise or an increase in environmental temperature.* An increase in red blood cells occurs under such conditions. This is the result of release of concentrated blood from the spleen, which accompanies increased activity.

STRUCTURE OF ERYTHROCYTES. A nucleus is lacking in mature red blood cells. The *body* of the cell consists of a spongelike *stroma* which contains the respiratory pigment *hemoglobin.* Hemoglobin comprises 60 to 80 percent of the total solids of the cell; the remaining solids include other proteins (0.5 to 1.0 percent); phospholipids, such as lecithin and cephalin (0.4 percent); cholesterol (0.3 percent); and smaller quantities of inorganic salts, urea, amino acids, and creatine. The cell membrane consists of proteins in combination with lipid substances. It possesses a high degree of selective permeability, being relatively impermeable to sodium and calcium ions but readily permeable to potassium ions. Red blood cells are so flexible and elastic that they can pass through the extremely small capillaries.

HEMOGLOBIN. This is the most important constituent of the red blood cell. It is a conjugated protein consisting of a colored iron-containing portion (*hematin*) and a simple protein (*globin*). Hemoglobin has the ability to unite readily with oxygen, a process called *oxygenation,* in which an unstable compound, *oxyhemoglobin,* is formed. In so doing, it acquires a bright red color—the color characteristic of arterial blood. This occurs in the passage of the blood through the lungs, where the oxygen tension is high.

Human blood contains approximately 15 g of hemoglobin per 100 ml of blood (16 g in males, 14 g in females). The total amount of hemoglobin present in the body averages 1 kg. The oxygen-carrying capacity of 1 g of hemoglobin is 1.34 ml. The amount of hemoglobin present can be determined with a *hemoglobinometer,* an apparatus by which the color of a sample of blood is compared with known standards. There are several other methods of hemoglobin determination.

FUNCTION OF ERYTHROCYTES. The primary function of red blood cells is to carry oxygen to the tissues. This is accomplished through the ability of the iron-containing hemoglobin to combine readily with oxygen in the formation of oxyhemoglobin. Hemoglobin also plays an important role in the transportation of carbon dioxide from the tissues, a part of the carbon dioxide (8 to 10 percent) combining directly with the hemoglobin to form *carbohemoglobin.* The hemoglobin also serves in an indirect way in carbon dioxide carriage.

DEVELOPMENT OF ERYTHROCYTES. In the adult, red blood cell formation (*erythropoiesis*) takes place in the red bone marrow (*myoloid tissue*) located principally in the vertebrae, ribs, sternum, diploe of cranial bones, and proximal ends of the humerus and femur. The red blood cells arise from stem cells (*proerythroblasts*) that possess a nucleus and lack hemoglobin. By mitosis these give rise to *erythroblasts,* which gradually produce hemoglobin. In successive development, the hemoglobin content increases and the cells become smaller, reaching a stage called a *normoblast.* Normoblasts multiply but eventually reach a stage at which degenerative changes occur in their nuclei; the nuclei are extruded from the cell, sometimes entirely but usually in fragments. At this stage, they are *young erythrocytes,* the cytoplasm of which contains a delicate reticular structure giving them the name of *reticulocytes.* The reticulocytes lose their reticular pattern before the cells enter circulation as *mature erythrocytes.*

The number of immature red blood cells in bone marrow and mature red blood cells in circulation, collectively called *erythron,* remains fairly constant. However, marked changes may occur as a result of variations in oxygen tension, loss of blood, hemolysis, or the effects of disease. There is strong evidence for a humoral factor that circulates in the blood and controls the rate of red blood cell formation. This factor, called *erythropoietin,* is a glycoprotein hormone produced principally, but not exclusively, in the kidneys. Its formation depends upon another substance, a *renal erythropoietic factor* (REF) or *erythrogenin,* produced by kidney cells. It is thought that REF is an enzyme that acts on a serum substrate or the renal erythropoietin precursor, converting one or the other into active erythropoietin capable of accelerating erythropoiesis, especially in conditions that necessitate a rapid increase in red blood cells.

LIFE HISTORY OF ERYTHROCYTES. Lacking a nucleus, red blood cells can live for only a limited time. Estimates of the amount of bile pigment excreted daily indicated that in a normal adult approximately 20 million red blood cells are destroyed every minute. Bile pigments are formed from the disintegration products of hemoglobin. The daily loss of hemoglobin averages about 20 g. The total number of red blood cells and the hemoglobin content of the blood at any one time depends upon the balance between red cell formation in the bone marrow and red cell destruction.

The life span of a red blood cell is not definitely known, but various studies, including some making use of radioactive isotopes, indicate that it is approximately 120 days. Blood cells are lost by the processes of hemolysis and fragmentation, which occur throughout the circulatory system, and phagocytosis of whole cells and cell fragments, which takes place in the cells of the reticuloendothelial tissues, especially those in the spleen, liver, and bone marrow.

In the blood-destroying organs, the hemoglobin breaks down into an iron-free portion (*globin*) and an iron-bearing portion (*hematin*). The latter is decomposed into bilirubin and an iron compound. Both are carried to the liver, where the bilirubin is excreted in the bile as one of the bile pigments, while the iron, if not needed for the formation of new red blood cells, is stored.

Leukocytes. White blood cells (WBC), or *leukocytes,* are found in the blood and lymph and, to a limited extent, in the tissues and tissue fluids. They differ from red blood cells in the following respects: They possess a nucleus, lack hemoglobin, have the power of active ameboid movement, are usually larger in size, and are much less numerous. They average about 7000/mm³, with a normal range of 5,000 to 11,000/mm³. (Fig. 9-2).

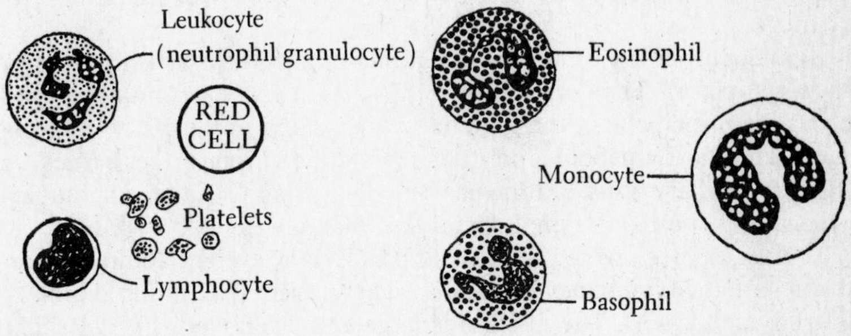

Fig. 9-2. The formed elements of the blood. (Reprinted with permission of the University of Chicago Press from A. J. Carlson and V. E. Johnson, *The Machinery of the Body,* 4th ed., 1953.)

CLASSES OF LEUKOCYTES. White blood cells are of two general classes, granulocytes and agranulocytes. Granulocytes are subdivided into neutrophils, eosinophils, and basophils, agranulocytes into lymphocytes and monocytes.

Granulocytes. These cells, which are formed in myeloid tissue (bone marrow), are characterized by the presence of specific types of granules in their cytoplasm that differ in size and in their reaction to various stains. On the basis of the staining reaction of the granules, there are three types of cells: neutrophils, eosinophils, and basophils.

1. *Neutrophils,* also called *polys* or *polymorphs,* average 10 to 12 μm in diameter. Their nucleus usually consists of three to five lobes, hence their name *polymorphonuclear leukocytes* (PMN). The nucleus of cells of females may also contain a drumstick-shaped piece of chromatin, the *Barr body,* by which chromosomal sex can be determined. The Barr body is an X chromosome. The cytoplasm of neutrophils is filled with fine granules (*azurophilic* and *specific*) that stain with neutral dyes. With Wright's stain, they are light purple. The *azurophilic granules* are the lysosomes of polymorphs and contain a number of hydrolytic enzymes and peroxidase; *specific granules* contain a bactericidal substance and

an enzyme, alkaline phosphatase. These substances are involved in the destruction of bacteria that are phagocytosed. Neutrophils are the most numerous of leukocytes, comprising 60 to 70 percent of the total. They are actively phagocytic.

2. *Eosinophils* average 10 to 15 μm in diameter and usually possess a bilobed nucleus. The cytoplasmic granules, large and coarse, stain with a bright red dye, eosin (an acid dye). They are relatively few in number, making up only 2 to 5 percent of the white cell count.

3. *Basophils* average about 10 μm in diameter and have a large and indistinctly lobed nucleus. The granules are irregular in shape and stain with basic stains, taking on a purplish blue color when stained with Wright's stain. Basophils comprise only 0.5 percent of the leukocyte count.

Agranulocytes. These relatively undifferentiated cells lack granules in their cytoplasm. They reproduce by mitosis, a mode of multiplication occurring principally in the lymphatic organs and in the connective tissues. They are ameboid and readily migrate through the tissues or through capillary walls. On the basis of cell size and structure, agranulocytes are of two types, lymphocytes and monocytes.

1. *Lymphocytes,* which are spherical cells, may average about 8 μm in diameter (small lymphocytes) or may reach a size of 12 μm in diameter (large lymphocytes). The nucleus is large and usually possesses a slight indentation; it occupies a major portion of the cell. The cytoplasm is homogeneous and stains a pale blue with the usual blood stains. Small lymphocytes constitute 20 to 25 percent of the white cell count. Large lymphocytes are similar to small lymphocytes excepting in size; they are numerous in children, but in adults they occur in insignificant numbers. Lymphocytes develop in bone marrow and lymphatic tissue, especially the lymph nodes.

2. *Monocytes,* also called *large mononuclear leukocytes,* average 12 to 15 μm in diameter. The nucleus is large and horseshoe-shaped, usually indented on one side. The cytoplasm stains grayish blue with Wright's stain. Monocytes constitute about 3 to 8 percent of the white cell count. Monocytes are produced principally in bone marrow.

Functions of Leukocytes. All leukocytes, but especially the neutrophils, exhibit the property of ameboid movement; that is, they are able to protrude cytoplasmic projections (pseudopodia) into which the protoplasm of the cell flows, thus bringing about changes in the shape and position of the cell. As a consequence, they are not confined within blood or lymph vessels but are found widely distributed in the tissues, particularly the connective tissues. Leukocytes are constantly migrating through the capillary walls to the tissues (*diapedesis*); from the tissues they may reenter the bloodstream.

Phagocytosis. Phagocytosis (from the Greek *phagein,* "to eat") is the process, exhibited by all classes of leukocytes, of engulfing substances of a particulate nature, especially bacteria. It is very pronounced in the neutrophils. The ability of leukocytes to ingest constitutes one of the body's primary defenses against invasion by infectious organisms. Leukocytes produce powerful proteolytic enzymes that may either act within the cell to digest phagocytosed substances or be discharged and act extracellularly on the surrounding tissue.

Functions of Granulocytes. Neutrophils, the most abundant of the granulocytes, function primarily as phagocytes in acute inflammation, details of which are given in the following paragraph. The specific functions of eosinophils and basophils are not definitely known. Both types are slightly ameboid and tend to leave the bloodstream and act in the tissues. *Eosinophils* seem to be involved in anaphylactic reactions, since they are abundant at the sites of allergic reactions and in the blood of individuals who suffer from allergies. Eosinophils increase in number in persons suffering from asthma and hay fever and in persons infected with certain animal parasites, as in trichinosis. They decrease in number in acute infections, after the administration of corticosteroids, and following stress, such as in burns or in postoperative states. Eosinophils are thought to phagocytose antibody-antigen complexes. *Basophils* are thought to function in the storage of heparin, histamine, and serotonin. They increase in number in Hodgkin's disease, in certain viral infections, and in chronic sinus infections. They decrease in number following steroid therapy, under certain stress conditions, and in hyperthyroidism.

Inflammation. When living tissue is injured in any way, the body reacts to neutralize or eliminate the deleterious agent, counteract any injurious effects, and repair the damage done. The various bodily responses that result constitute *inflammation.* Inflammation occurs when tissues are subjected to physical or chemical damage or when they are invaded by pathogenic agents (viruses, bacteria, fungi, protozoa). The characteristic signs and symptoms of inflammation are *redness, local heat, swelling, pain,* and usually *altered function.* The redness, heat, and swelling are due to increased blood flow through the vascular bed. Vasodilatation occurs, and blood vessels become congested. Increased capillary permeability leads to exudation of blood plasma into the surrounding tissues. This results from the liberation of *histamine* and other amines from mast cells. With the slowing of blood flow, leukocytes, especially neutrophils, begin to stick to the endothelium of the venules and migrate through the endothelial walls into the tissues. They accumulate in large numbers and become actively phagocytic, especially if bacteria are present. The directional movement of neutrophils toward bacteria, presumably resulting from chemical substances they produce, is called *chemotaxis.* Chemical substances produced in bacterial infections that are involved in the inflammatory response include a number of *amines* (vasoamines), of which histamine is of special importance;

serotonin, released from blood platelets; and various *kinins,* especially *bradykinin,* which is a powerful pain-producing agent. Bradykinin develops from the interaction between substances present in the plasma and in tissues.

In most infections, a cream-colored fluid called *pus* usually accumulates. Pus consists of dead polymorphs, living and dead bacteria, plasma exudate, and the products of tissue breakdown. If the area of injury is below the surface, an *abscess* (cavity filled with pus) may be formed; if it is on the skin or a mucosal surface, it constitutes an *ulcer.* In many infections, substances called *pyrogens* are formed; these, when absorbed into the bloodstream, are carried to the hypothalamus, where they affect the temperature-control center, resulting in elevation of body temperature (fever or pyrexia). Other substances affect the bone marrow, causing the release of mature leukocytes into the bloodstream. In many acute infections, a *leukocytosis-inducing factor* (LIF) develops; it stimulates the production and release of immature leukocytes into the bloodstream, and the leukocyte count rises. If the count exceeds 10,000, a condition of *leukocytosis* exists. Following healing, the count returns to normal. A reduction in the number of white blood cells, a condition called *leukopenia,* may result from pathologic conditions, the administration of marrow-depressant drugs, or radiation.

Function of Agranulocytes. Lymphocytes are primarily involved in immunologic reactions. There are two basic types, long-lived *T lymphocytes* (or T cells) and short-lived *B lymphocytes* (or B cells). The precursors of T lymphocytes develop in the thymus gland very early in life; B lymphocytes are so named because they are formed in the *bursa of Fabricus,* a lymphatic structure found only in birds. In humans their origin is uncertain, but it is thought that they originate in depots of lymphatic tissue comparable to the bursa of Fabricus. Evidence is also strong that B lymphocytes also originate in bone marrow.

Following their formation, both types of cells migrate by way of the bloodstream to lymphatic tissue, especially lymph nodes and the spleen, where they proliferate and acquire the ability to produce antibodies. Both types are able to develop into blast cells that are capable of dividing and forming more small lymphocytes. B cells, with T cells cooperating, have the ability to develop into *plasma cells,* also called *immunoblasts.* Plasma cells are the source of *immunoglobulins,* which function as antibodies, of primary importance in the immune responses of the body, especially those involved in resistance to bacterial infection. T lymphocytes have the ability to suppress or assist in the production of antibodies by B cells or plasma cells. T cells are responsible for immunity against viral, fungal, and certain parasitic infections. They are capable of destroying cancer cells and transplanted tissue cells and are of primary importance in the rejection of tissue grafts. Certain T cells, also called *killer cells,* confer *cell-mediated immunity;* B cells are involved in *hu-*

moral immunity. T cells are long-lived, existing for many days, years, or even a lifetime; B cells are short-lived, existing for only a few hours or days.

Monocytes are motile cells that can readily migrate through capillary walls into the tissues, where they are transformed into macrophages which, with polymorphs, actively engage in phagocytic activities. They are especially active in disposing of cellular debris left following inflammation. Monocytes originate in bone marrow. Their average life span is three days.

Fate of Leukocytes. Many white blood cells die and disintegrate in circulating blood or in the tissues and are phagocytized by macrophages in the connective tissue, liver, and spleen. In infections, many die and are destroyed by cytolytic enzymes in pus. Some are lost by migration through the epithelium into the lumen of the intestine.

Blood Platelets (Thrombocytes). These are small, colorless, disc-shaped bodies present in circulating blood. They originate as small, cytoplasmic protrusions that are pinched off of giant cells *(megakaryocytes)* present in the bone marrow. Platelets average about 3 μm in diameter; they range in number from 250,000 to 400,000/mm^3; their length of life is about 10 days. In freshly drawn blood, they quickly clump together, agglutinating in clusters. Platelets lack a nucleus or nuclear material.

Platelets play an important role in *hemostasis,* the prevention of bleeding. First, they induce contraction of injured vessels and, through their adhesiveness, form a plug that helps to seal and close the opening; second, they play a major role in coagulation through the liberation of a substance that is essential in the formation of thromboplastin, which initiates the clotting process.

BLOOD CLOTTING AND HEMORRHAGE

Within the blood vessels, circulating blood is of fluid consistency. If, however, it ceases to circulate or escapes from a blood vessel, it soon loses its fluidity and sets into a jellylike mass. It is then said to have *clotted* or *coagulated.* A *clot* consists of a mass of tangled strands of *fibrin* in which are enmeshed red and white blood cells. If the clot stands for a time, it undergoes shrinkage, in the course of which a clear, straw-colored fluid is exuded; this fluid is called *blood serum.* It is similar to plasma except that it is lacking in fibrogen.

Coagulation. The clotting of blood is an extremely complex process consisting of a series of enzymatic reactions involving a number of plasma proteins. The enzymes within circulating blood are inactive, but if blood is shed or placed in contact with a wettable surface such as glass, a series of reactions that result in coagulation is initiated. Some 13 factors, designated by Roman numerals I to XIII, may be involved. The exact role of each is not known, but the basic reactions in the process are believed to be as described here.

ESSENTIAL SUBSTANCES. Four substances seem to be essential for the clotting of blood:

1. *Prothrombin:* a substance present in circulating blood, capable of being converted to *thrombin.*
2. *Thromboplastin:* an enzyme present in platelets and tissue, formed from the interaction of several factors.
3. *Calcium:* as a cation.
4. *Fibrinogen:* a protein present in blood plasma.

STEPS IN COAGULATION.

1. *Thromboplastin* (III) is formed from the interaction of several substances derived from disintegrating platelets, injured tissues, and plasma proteins.
2. *Prothrombin* (II) is converted to *thrombin* through the action of thromboplastin, calcium ions (IV), and certain plasma factors.
3. *Soluble fibrinogen* (I) is converted to insoluble *fibrin* through the action of thrombin.

In brief, the reactions are:

$$\text{Prothrombin} + \text{calcium ions} + \text{thromboplastin} \rightarrow \text{Thrombin}$$
$$\text{Fibrinogen} + \text{thrombin} \rightarrow \text{Fibrin}$$

Several other factors involved in the process have been identified, but their exact roles have not been determined. Blood does not normally clot within a blood vessel because prothrombin is maintained in an inactive state by the presence of a plasma inhibitor, *antiprothrombin.*

A number of tests have been devised to detect defects in the clotting mechanism. These include determination of coagulation time, bleeding time, prothrombin time, clot retraction time, and the thromboplastin generation test. *Coagulation time* is the time required for the blood to clot when removed from the body. It averages from 8 to 12 min but may vary with the type of test used. *Bleeding time* is the time required for the blood to clot sufficiently to close a small opening made by pricking the skin. It averages about 2½ min. These and other tests determine the effectiveness of the clotting mechanism.

VITAMIN K AND COAGULATION. Vitamin K, the "coagulation vitamin," plays a vital role in the clotting of blood. It is found in green-leaf foods and is produced by certain bacteria. In the absence or a deficiency of vitamin K, the liver fails to produce the necessary prothrombin and the blood fails to clot or clots very slowly. Absorption of this vitamin depends on the presence of bile; consequently, obstructive jaundice is usually accompanied by delayed clotting of the blood. Deficiency of this vitamin in foods is of little significance in adults, since it is produced by bacteria in the intestine. In newborn infants, however, the prothrombin level may fall during the first few days, giving rise to hemorrhagic attacks.

METHODS FOR HASTENING COAGULATION.

1. *Application of heat.* Cloths moistened with a hot physiological salt solution are effective.

2. *Application of styptics or hemostatics.* These are substances such as alum, ferric chloride, zinc chloride, and silver nitrate that act *(a)* through their astringent action, constricting the walls of the blood vessels, and *(b)* by precipitating plasma proteins, thus interfering with the flow of blood.

3. *Provision of a rough surface,* such as sterile gauze or cotton.

4. *Injection of tissue extracts* that contain thromboplastic substances.

5. *Injection of epinephrine* into the bloodstream, which brings about the rapid clotting of blood. Epinephrine has the same effect when liberated naturally as a result of emotional excitement or muscular exercise.

METHODS OF DELAYING COAGULATION. Several substances, known as anticoagulants, act to delay blood clotting. They are *sodium* and *potassium oxalate,* which act by precipitating calcium; *fluorides* and *citrates,* which form soluble compounds with calcium, thus suppressing its ionization; *hirudin,* a substance secreted by the buccal glands of the medicinal leech *(Hirudo)*; *heparin,* a substance formed in the liver and present in the blood; and *dicoumarin,* a substance originally isolated from spoiled sweet clover but now produced synthetically that reduces the prothrombin content of the blood. In collecting blood for experimental, analytical, or transfusion purposes, one can retard coagulation by preventing contact of the blood with foreign substances or injured tissues, by cooling it, by collecting it in a receptacle in which the surfaces have been treated with a nonwetting agent such as paraffin or silicone, or by adding anticoagulants. Some anticoagulants, especially oxalates, should not be used in blood that is to be transfused.

INTRAVASCULAR COAGULATION. Clots *(thrombi)* may form within a blood vessel or the heart, a condition called *thrombosis.* This process is initiated by the adherence of platelets to the vessel wall. This can result from mechanical injury, as from cuts or blows; inflammation, as in thrombophlebitis; degenerative changes within a vessel, as in atherosclerosis, a slowing or stasis in blood flow with eddying movements, as in varicose veins, or increased coagulability of the blood, as may result from snake venom, foreign sera, or excess epinephrine.

A *thrombus* may occlude a vessel and deprive tissues of their blood supply. If a small vessel is involved, the tissues may obtain their blood supply through collateral circulation from anastomosing vessels; if the vessel is large, the blood supply may cease, leading to local death of tissues *(necrosis)* or general death if a vital organ such as the brain, heart, or lungs is involved. *Infarction* usually follows, with a mass of scar tissue (an *infarct*) developing in the area involved. This occurs frequently in the heart following coronary occlusion.

If a thrombus becomes detached and moves along with the blood, it is called an *embolus.* An embolus may travel to any part of the body and produce an *embolism,* which results in obstruction of the flow of blood. An embolus may consist of a transported clot, a mass of air bubbles or fat globules, a bacterial mass, or a collection of animal parasites. If the embolus is a clot, the condition is referred to as *thromboembolism.*

Clots, wherever formed, tend to be slowly reabsorbed and disappear. This is accomplished through the removal of fibrin *(fibrinolysis),* which results from the action of *plasmin,* a proteolytic enzyme present in plasma in an inactive form, *plasminogen.*

Hemorrhage. The normal milieu of whole blood is the circulatory system—the closed system of arteries, veins, and capillaries—and the heart. *Hemorrhage* or bleeding is the *abnormal* escape of blood from this system. An *external* hemorrhage occurs when a vessel is punctured and blood escapes from the body; an *internal* hemorrhage occurs when blood escapes from a vessel and accumulates in surrounding tissues, in the cavity of a visceral organ, or in a body cavity.

EFFECTS OF HEMORRHAGE. Some of the effects of severe hemorrhage are (1) oxygen lack, owing to reduction in the number of red blood cells; (2) fall in blood pressure, owing to decrease in blood volume; (3) increase in heart rate; (4) reduction of the force of the heartbeat; and (5) shock, with severe hypotension (low blood pressure). If the blood loss exceeds about 30 percent of blood volume, transfusion of blood is usually necessary.

REACTIONS OF THE BODY TO LOSS OF BLOOD. Loss of blood through hemorrhage gives rise to the following reactions by the body at large:

1. The *clotting mechanism* is brought into action, and the walls of the injured vessel contract to close the opening.
2. Reflex *vasoconstriction* on an extensive scale occurs, especially in the blood vessels leading to organs that are not essential to life (skin and muscles, for example). This reduces the capacity of the circulatory system and tends to restore normal blood pressure to facilitate functioning of the vital organs.
3. Restoration of *blood volume* is accomplished temporarily by (*a*) contraction of the spleen, which forces into circulation the blood stored by that organ and (*b*) withdrawal of fluid from the tissues. The latter accounts for the great thirst that usually follows severe hemorrhage.
4. Replacement of *plasma proteins* and *red blood cells.* The time required for this depends on the amount of blood lost, the diet, responsiveness of the liver and bone marrow, and other factors. In a normal, healthy subject who has lost a pint of blood, replacement occurs within three to four weeks.

REGULATION OF THE COMPOSITION OF THE BLOOD

One of the striking features of blood is its constancy of composition. It is continuously receiving and giving up materials to cells, yet fluctuations in its physical nature and its chemical composition are confined to extremely narrow limits. This tendency of the organ systems and tissues of the body to maintain a balanced state is called *homeostasis.*

Reaction of Blood. The acidity or alkalinity of blood depends on the concentration of hydrogen (H^+) ions and hydroxyl (OH^-) ions. Normally the blood is slightly alkaline, having a pH between 7.30 and 7.45. Acids (carbonic, lactic, hydrochloric) are being constantly produced in the body during metabolic processes, yet the blood remains relatively constant. This is accomplished by a number of mechanisms, the more important being (1) elimination of carbon dioxide by the lungs, (2) elimination of acid substances by the kidneys, and (3) action of buffers in the blood.

Buffers are chemical substances in blood that prevent a change in pH when acids or bases are added. A *buffer system* consists of a weak acid and a salt of that acid (for example, carbonic acid and sodium bicarbonate). Buffer action may be illustrated as follows: If hydrochloric acid is added to the blood, it reacts with sodium bicarbonate to form sodium chloride and carbonic acid, thus:

$$HCl + NaHCO_3 \rightarrow NaCl + H_2CO_3$$

In the foregoing reaction, the strong acid (hydrochloric acid) has been replaced by a weak acid (carbonic acid), which, being volatile, is easily disposed of through the loss of carbon dioxide from the lungs. The bicarbonate has thus served as an *alkali reserve,* and its presence has prevented a change in the pH of the blood when the strong acid has been added. The principal buffers of the blood are sodium bicarbonate, disodium phosphate, and the plasma proteins.

Effect of Respiration. Respiration is so regulated that the tension of carbon dioxide in the blood is maintained at a constant level. With increased production of carbon dioxide, breathing is increased. When the amount of carbon dioxide is reduced, as in forced ventilation or by the addition of bases to the blood, the rate of respiration decreases. The ratio of carbon dioxide to plasma bicarbonate remains fairly constant, at 1:20.

Hydrogen Ion Concentration (pH). The pH range compatible with life lies between 6.8 and 7.8. Fixed acids are not normally present in the blood, but in certain pathologic states, such as diabetes and nephritis, they may appear. The presence of fixed acids depletes the alkali reserve, and *acidosis* develops. However, since the pH rarely falls below 7.0, the term *acidosis* does not imply a true acidity of the blood, a condition that occurs only under extreme pathologic conditions. *Alkalosis* means an increased accumulation of alkali reserve, a condition that may occur as a result of the ingestion of excessive amounts of alkalies or the loss of gastric hydrochloric acid in prolonged vomiting. The development of acidosis or alkalosis in which the pH of the blood remains unchanged is referred to as being *compensated;* if the pH increases or decreases, the state is referred to as being *uncompensated.*

TISSUE FLUID

Tissue fluid lies outside the blood circulatory system. It occupies the spaces in the various tissues (such as the spaces between cells and between the fibers of connective tissues). It is the "middleman" of the body, serving as the intermediary by which substances in the bloodstream are transmitted to the cells.

Tissue fluid is formed from the blood as the result of *hydrostatic blood pressure,* which tends to force the plasma through the thin endothelial cells of the capillary walls or through the spaces between these cells. The chief process in its formation is *filtration.* Tissue fluid resembles blood plasma but is lacking in red blood cells. It contains all the blood proteins, but in slightly different concentrations. Another factor in the formation of tissue fluid is *osmotic pressure.* Salts, glucose, and other crystalloids in blood plasma exert little osmotic pressure within capillaries, since they diffuse readily through their walls. However, the osmotic pressure exerted by proteins acts to draw water from the tissues into the capillaries. At the arterial end of a capillary, the force of filtration exceeds that of osmosis, and the flow is outward into the tissues. As the blood passes along the capillary, the concentration of plasma proteins rises, increasing the osmotic pressure. At the same time, blood pressure decreases; consequently, at the venous end of the capillary, osmotic pressure exceeds hydrostatic pressure, and fluid is drawn from tissues into capillaries, with resorption of tissue fluid into the bloodstream.

LYMPH

Outside the capillaries, the tissue fluid bathes the cells of the tissues. It is collected into a network of extremely fine vessels (the *lymph capillaries*) or into small spaces (*lacteals*) in the villi of the intestines. These converge into larger *lymph vessels,* which conduct the fluid away from the periphery. The fluid within these lymph vessels (or lymphatics) is called *lymph.* Lymphatics are more permeable than are capillaries; in fact, their distal ends are regarded as being open, thus permitting ready access to the lymph of colloidal substances and particulate matter that may enter the tissue spaces. All lymph vessels lead eventually to the *right lymphatic* and *thoracic ducts,* which empty into the right and left subclavian veins, respectively, where the lymph reenters the bloodstream.

Lymph is a clear, viscid fluid resembling blood plasma in its concentration of salts and some other substances. The same blood proteins are present, but in variable concentrations. Red cells are lacking, but white cells, especially lymphocytes, are numerous. Lymph functions in the return to circulation of "leaked" fluid from the capillaries. It serves to transport fats that have been absorbed during digestion.

The *flow* of lymph is brought about by *differences of pressure* at the two ends of the lymphatic system. Pressure in the main lymph ducts is low; accordingly, lymph tends to flow from the peripheral vessels centrally to the main lymph ducts. Supplementary factors that aid in movement of the lymph are (1) contraction of skeletal muscles, (2) respiratory movements, and (3) contraction of smooth muscles in the villi of intestines. *Valves* in lymphatic vessels prevent backflow of the lymph.

PRACTICAL CONSIDERATIONS

Hematology. The study of blood constitutes the science of *hematology.* A knowledge of the constituents of blood and their relationship to various pathologic conditions is of primary importance to physicians in the diagnosis of disease. As a consequence, the laboratory is assuming a position of increased importance in medical science. Laboratory methods are being refined and perfected, and with the increase in the use of automated machines—for example, the Coulter counter and the Autoanalyzer—more complete tests can be performed on more samples with a greater degree of accuracy than was previously possible. However, the cost of these instruments is high, limiting their use to well-financed institutions whose laboratories handle a large volume of samples.

Enumeration of Blood Cells (Blood Count). The apparatus used to determine the number of red and white blood cells is called a *hemocytometer.* This comprises (1) a *counting chamber,* a glass slide marked with cross-rulings; (2) a *cover glass;* and (3) two *pipettes,* one containing in its bulb a red bead, the other a white bead. The former is used for counting red cells, the latter for counting white cells.

For the *red cell count,* blood is drawn into the pipette with the red bead up to the 0.5 mark; then it is diluted by drawing in physiologic saline solution or Hayem's solution until the mixture reaches the 101 mark, which gives a dilution of 1:200. The pipette is shaken vigorously so as to mix the blood and the saline or Hayem's solution thoroughly. Then a drop of this diluted blood is placed on the counting chamber and allowed to run under the cover glass. The cross-rulings on the slide and the space between the cover glass and the slide are so constructed that each small space contains 1/4,000 mm^3. The red cells within the area of 80 small spaces are counted. The resultant number, multiplied by 10,000, gives the total number of red cells per cubic millimeter ($4{,}000 \div 80 \times 200 = 10{,}000$).

For the *white cell count,* the pipette with the white bead is used. Again, blood is drawn up to the 0.5 mark. Tuerk's solution (1 percent glacial acetic acid, to which a tinge of gentian violet or methyl green has been added) is used as the diluting agent. The acid hemolyzes the red cells while the white cells are stained slightly. Diluting fluid is drawn up to the 11 mark, giving a dilution of 1:20. Cells are counted in the 16 large spaces seen in the four corners of the ruled area on the slide, and the average number in a corner is determined. The cubic content of a large space is 25/4,000 mm^3. Therefore the white cell count (per cubic millimeter) is the average number in a corner area multiplied by 200 $(4{,}000 \div [16 \times 25] \times 20 = 200)$.

In some instances, a *differential* count is required in order to determine the numbers and percentages of the various kinds of leukocytes. The conditions that alter the numbers of various types of leukocytes are (1) *leukocytosis* (increase in the number of white blood cells; principally neutrophils; it occurs in infections caused by pyogenic organisms and in pneumonia, whooping cough, and other infectious diseases); (2) *eosinophilia* (increase in the number of eosinophils, associated with allergic conditions, such as asthma, anaphylactic shock, and certain skin diseases and with infestations by animal parasites, such as *Ascaris, Trichinella,* or hookworm); (3) *lymphocytosis* (increase in the number of lymphocytes, occurring in chronic, inflammatory states following acute infections); (4) *leukemia* (see discussion on page 298); and (5) *leucopenia* (reduction in the number of leukocytes, occurring in typhoid fever, tuberculosis, and influenza; it also results from the effects of certain poisons, such as benzol, aminopyrine and its compounds, and other toxic substances).

Blood Transfusion. Transfusion is the introduction of whole blood or some of its constituents into the bloodstream. Conditions that may necessitate transfusion are (1) loss of blood after hemorrhage, (2) reduction of blood volume in circulation, which occurs in shock, (3) decrease in red blood cells or plasma proteins, (4) deficiency in blood coagulating factors, and (5) need to replace blood in certain hemorrhagic diseases.

Fresh blood less than 12 hr old may be used in some cases of hemophilia or erythroblastosis fetalis. However, most transfusions utilize blood from a *blood bank,* blood that has been collected and stored either in a liquid or frozen state. Since whole blood may contain substances that may cause adverse reactions, in most cases it is desirable to transfuse only the blood fraction needed. Whole blood can now be fractionated into its various constituents, some of which are concentrated red cells, packed white cells, whole plasma, platelet-rich plasma, concentrated albumin, fibrinogen, or various gamma globulin fractions. Red cells in whole blood stored at 4°C (39°F) survive for only a period of three to four weeks; in frozen preparations stored at either −80°C or −150°C, the life of red cells is extended for months.

TRANSFUSION OF WHOLE BLOOD. There are several dangers involved in the transfusion of whole blood. These include (1) the possible transmission of infectious diseases such as malaria, homologous serum jaundice (viral hepatitis), and mononucleosis; (2) the possibility that the blood of the donor may contain foreign proteins to which the recipient is allergic; and (3) the possible incompatibility of the donor's blood with that of the recipient.

Compatibility. The normal blood plasma of one subject may contain substances that would cause *agglutination* (clumping together) of the blood cells of the other subject. Such bloods are said to be *incompatible,* since an individual receiving incompatible blood may go into a state of shock and die. Human blood is classified in four groups, namely O, A, B, and AB (formerly designated I, II, III, and IV). The reactions of these groups are shown in the following table:

Cells of Donor Groups	*Reaction to Serum of Recipient Groups*			
	O	A	B	AB
O	−	−	−	−
A	+	−	+	−
B	+	+	−	−
AB	+	+	+	−

\+ agglutination
− nonagglutination

A person whose blood falls into group O is said to be a *universal donor;* the cells of group O blood are not usually agglutinated by the serum of any other group. A person whose blood falls into group AB is called a *universal recipient,* since the serum of this blood type does not agglutinate the blood cells of any other group. Among individuals of the white race, the groups are distributed roughly as follows: group O, 43 percent; group A, 40 percent; group B, 13 percent; group AB, 4 percent. The proportions in the various groups vary in different races. In races of the Far East, the proportion in group A decreases and that in group B increases; among American Indians, group O is the predominant group.

Although the ABO system of blood groups is of greatest clinical importance, other blood groups, including M, N, P, and others, are known to exist and, in rare cases, are clinically significant, especially for a person who has received many transfusions.

THE RH FACTOR. The blood of most persons contains an antigen designated the *Rh factor.* Rh is short for *rhesus monkey,* on which relevant experiments have been conducted. This antigen acts as an agglutinogen. There are several Rh antigens, of which the D antigen is the most important. When the blood of a person possessing this factor (an Rh-positive) is transfused into the blood of an individual without this factor (an Rh-negative), the recipient develops antibodies (anti-Rh agglutinins). If a subsequent transfusion of Rh-positive blood is performed, a hemolytic reaction occurs, due to the presence of this agglutinin.

In pregnant women such a reaction may take place after a single transfusion. This is believed to be due to the development of Rh agglutinin in the mother's blood as a result of a previous transfusion or the presence in the fetus's blood of an Rh-positive factor inherited from the father. In infants, a blood disorder, *erythroblastosis fetalis* (hemolytic disease of the newborn), may occur. This is due to the development in Rh-negative mothers of the anti-Rh agglutinin, which passes through the placenta to the fetus, causing the destruction of the red cells of the fetus (usually Rh-positive). The blood of the mother can be monitored; if titers of the anti-Rh agglutinin are high, treatment can be instituted. The life of an affected baby at term can often be saved by completely replacing its blood.

TRANSFUSION OF PLASMA. This method has the advantage over other techniques in that the blood products of many donors can be pooled and transfused indiscriminately. The liquid plasma can be kept under refrigeration indefinitely, or the plasma can be frozen or dried; transportation and storage difficulties are thus minimized. A new procedure of obtaining plasma called *plasmapheresis* is now utilized: Blood is removed from the donor, and red cells are separated from the plasma by centrifugation. The red cells, suspended in a saline substrate or other suitable medium, are reinjected into the donor. This method avoids wasting red blood cells and enables the donor to be bled more frequently, even at weekly intervals.

TRANSFUSION OF CRYSTALLOIDS. Solutions of glucose or of physiologic saline solution are sometimes injected for conditions that involve dehydration of tissues. Their effects are usually transitory, since the fluid passes quickly through capillary walls into the tissues.

Edema. Ordinarily, the rate of formation of tissue fluid and lymph and the rate at which lymph reenters the bloodstream either through resorption in the capillaries or by way of the lymph ducts is approximately equal. But if there is a disturbance in the balance between these two processes whereby the formation of tissue fluid increases, it will tend to accumulate in the tissue spaces, giving rise to swelling or *edema.* When this condition is generalized over the body, it is called *anasarca* or *dropsy.* When it is localized in the abdominal cavity, it is called *ascites.*

CAUSES OF EDEMA. The principal factors in the causation of edema are obstruction in lymph flow, increased capillary permeability, increased capillary pressure, decrease in plasma proteins, or an excess in salt and water content of the body. (1) *Obstruction in lymph flow* may result from blockage of the lymph channels or lymph nodes by neoplasms, the presence of parasites (as in filariasis or elephantiasis), inflammatory conditions, or the removal of lymph nodes. (2) *Increased capillary permeability* may result from inflammation, especially that resulting from bacterial infections, the effect of various metabolic conditions such as acidosis; anaphylactic reactions, as in hay fever; anoxia, especially that resulting from carbon monoxide poisoning; or poisoning from various drugs, such as barbiturates. (3) *Increased capillary pressure* may result from abnormally high venous pressure, as in congestive heart failure; obstruction in venous return, as by a thrombus or tumor; or from the effects of gravity, as in *postural edema,* which results from standing in one position too long. (4) *Decrease in plasma proteins* may result from malnutrition (inadequate intake of proteins and vitamins, especially thiamine); malabsorption of proteins, as in the malabsorption syndrome; defective synthesis of proteins by the liver; or loss of proteins in the urine in kidney disease. (5) *Excessive salt and water content of the body* may result from excess salt intake, especially sodium, impaired renal excretion, as in kidney disease, or excessive secretion of salt-active adrenocortical steroids.

DISORDERS AND DISEASES INVOLVING THE BODY FLUIDS

Common diseases involving the formation and composition of the body fluids are anemia, hemophilia, leukemia, purpura, and thrombocytopenia.

Anemia. In this condition there is reduction in the number of red blood cells, a reduction in the amount of hemoglobin, or both. Symptoms of severe anemia are weakness, pallor of skin, loss of weight, shortness of breath, tachycardia, gastrointestinal disorders, enlargement of the spleen, degenerative changes in the nervous system, and many others. Most of these symptoms result from an inadequate supply of oxygen to the tissues.

Anemia may result from external or internal hemorrhage, excessive destruction of erythrocytes, defective red cell formation in the bone marrow, and other causes.

Anemia Resulting from Hemorrhage. Acute blood loss may occur from accident or from intentional bleeding, as in a blood donation. A temporary loss of a liter or more leads to temporary anemia. *Chronic blood loss* may result from gastrointestinal ulcer or carcinoma, bleeding hemorrhoids, hookworm infestation, or other intestinal disorders.

Anemias Resulting from Excessive Destruction of Erythrocytes. Destruction of erythrocytes may be *intravascular* (within blood vessels) or *extravascular* (within the spleen, lymph nodes, or bone marrow). Hemolysis may result from hereditary disorders, such as sickle-cell anemia and thalassemia, or it may be acquired as a result of blood incompatibility or the effects of infectious agents, drugs, or chemicals.

Anemias Due to Defective Red Blood Cell Production. These may result from a deficiency in an essential nutrient, such as iron, copper, cobalt, folic acid, or vitamin B_{12}, or a deficiency in erythropoietin, a hormone produced by the kidney.

Pernicious anemia is a megaloblastic anemia resulting from deficient absorption of vitamin B_{12}, an *extrinsic factor,* resulting from the absence of an *intrinsic factor* produced by mucosal cells in the fundus and body of the stomach and normally present in gastric juice. Atrophy of the gastric mucosa and achlorhydria are characteristic of this condition.

Other Causes of Anemia. Anemia may result from (1) inhibition of the bone marrow (*aplastic anemia*), which may result from ionizing radiations or the effects of certain drugs or chemicals; (2) replacement of hemopoietic tissue in bone marrow by fatty or fibrous tissue or neoplastic tissue, as in leukemia; (3) defective heme synthesis; (4) excessive destruction of blood cells by the spleen; or (5) depression of bone marrow function by a humoral agent produced by the spleen, as may occur in hypersplenism.

Hematoma. A hematoma is a localized mass of blood, usually clotted, in an organ, space, or tissue. It may be the result of an injury, such as a blow, or other factors that may cause the rupture of a blood vessel.

Hemophilia. This is the condition in which the blood fails to clot within the normal time. Persons so afflicted are known as "bleeders." They may bleed excessively from minor injuries, such as tooth extraction or even a small cut. Large black-and-blue spots appear under the skin from even light blows. Internal bleeding into joints or the gastrointestinal tract may occur. Hemophilia is caused by a deficiency in antihemophilic globulin (factor VIII).

Hemophilia is hereditary and is due to a sex-linked factor; that is, the gene (hereditary determiner) is carried in the X chromosome. If a male "bleeder" mates with a normal female, none of the offspring will manifest the condition. However, the normal daughters, called "carriers," may transmit the predisposing gene to some of their children, but the condition will develop only in the sons, never in the daughters. As a consequence, hemophilia is almost entirely restricted to males; at the present time, there are records of only three authentic cases of hemophilia in females.

Leukemia. *Leukemia* is a neoplasmic disorder of blood-forming tissue involving primarily the leukocyte series. In everyday language, it is cancer of the blood. There is an excessive production of leukocytes, the white cell count often reaching into the hundreds of thousands. Immature cells may appear in the blood in large numbers. There are many types of leukemia, named after the specific type of white cell involved (e.g., *eosinophilic, monocytic,* or *lymphocytic leukemia*). Proliferation of neoplastic cells results in hyperplasia of the bone marrow and concomitant enlargement of the spleen, liver, and lymphatic tissue. Normal blood-forming cells in the bone marrow are crowded out, with a reduction in the formation of erythrocytes and consequent anemia. Impaired functioning of the leukocytes results in increased susceptibilty to infection and reduced resistance to disease.

Leukemia may be acute or chronic. *Acute leukemia* occurs principally in young persons and until recently was usually fatal within a year; however, many childhood leukemias are now being arrested by the use of anticancer drugs. Chronic leukemia usually occurs in later adulthood and is more amenable to treatment, which may prolong life for a number of years.

The causes of leukemia, as with other types of cancer, are varied. In experimental animals, leukemia may be induced by viruses, irradiation, and carcinogenic chemicals. In humans, it may result from excessive doses of radiation or prolonged exposure to certain chemicals such as benzene. In most cases, its cause is unknown.

Lymphomas. These comprise a heterogeneous group of neoplasms involving tissues of the reticuloendothelial and lymphatic systems. The two principal types are Hodgkin's disease and lymphosarcoma. *Hodgkin's disease* involves primarily the lymph nodes but also the spleen, liver, and bone marrow. Giant multinucleated reticulum cells are characteristic, along with proliferation of granulocytes and plasma cells. *Lymphosarcomas* involve primarily the lymph nodes, and lymphocytes are the principal cells involved. Both Hodgkin's disease and lymphosarcomas present a picture much like leukemia.

Purpura. This is a group of disorders characterized by the appearance of purplish or brownish red discolorations resulting from small hemorrhages that occur in the skin and brain and in the mucosa of the respiratory, gastrointestinal, and genitourinary tracts. If the hemorrhages are small and punctate, they are called *petechiae;* if they are large, forming bruises or black-and-blue marks, they are called *ecchymoses.*

Purpura is a platelet disorder. There may be a deficiency in numbers (thrombocytopenia) resulting from inadequate production or increased platelet destruction as occurs in hypersplenism. Purpura often follows viral diseases.

10: THE CIRCULATORY SYSTEM

The circulatory system comprises all structures concerned with transportation of body fluids from one region of the body to another. It has two main divisions, a cardio-vascular system and a lymphatic system. Structures comprising the *cardio-vascular system* are the *heart,* which by contraction forces blood through the blood vessels; *arteries* which conduct blood from the heart to tissues, with their smaller branches called *arterioles; veins,* which conduct blood from tissues toward the heart, with their smaller branches called *venules;* and *capillaries,* extremely small vessels that connect arteries and veins. Structures comprising the *lymphatic system* are *vessels* that collect and conduct lymph and *lymphatic organs* consisting of lymphatic tissue. The *conducting vessels* include *lymph capillaries,* minute vessels within tissues (in intestinal villi, they are called *lacteals*); *lymph vessels,* an elaborate system of vessels that carry lymph from the tissues; and *main lymph ducts,* the *right lymphatic* and *left thoracic ducts,* that carry lymph from lymph vessels to the bloodstream. The two main lymph ducts empty into the right and left subclavian veins, respectively, near their junctions with the internal jugular veins, to form the innominate (brachiocephalic) veins. The *lymphatic organs* include nonencapsulated lymph nodules, lymph nodes, the tonsils, the spleen, and the thymus.

THE CARDIOVASCULAR SYSTEM

Heart (Fig. 10-1). The heart, a hollow muscular organ, lies between the lungs in the lower median portion of the thoracic cavity, or middle mediastinum. It is about the size of a man's fist, averaging 12 cm in length, 9 cm in width, and 6 cm in thickness. It weighs about 300 g in the male and 250 g in the female. Its shape is roughly that of an inverted cone. From the superior portion, the "base," emerge the vessels that enter and leave this organ. The inferior, rounded portion, or "apex," points down slightly anteriorly and to the left.

PERICARDIUM. The heart is enclosed in a double-walled fibroserous sac, the *pericardium.* The inner layer (*visceral pericardium*) closely invests the heart and constitutes the outermost layer of the heart wall. The outer layer (*parietal pericardium*) consists of fibrous connective tissue lined with mesothelium. The space between these two layers contains a serous fluid (*pericardial fluid*) that provides smooth surfaces for easy movement of the heart during its expansion and contraction.

WALL OF THE HEART. The wall is composed of three layers: (1) the

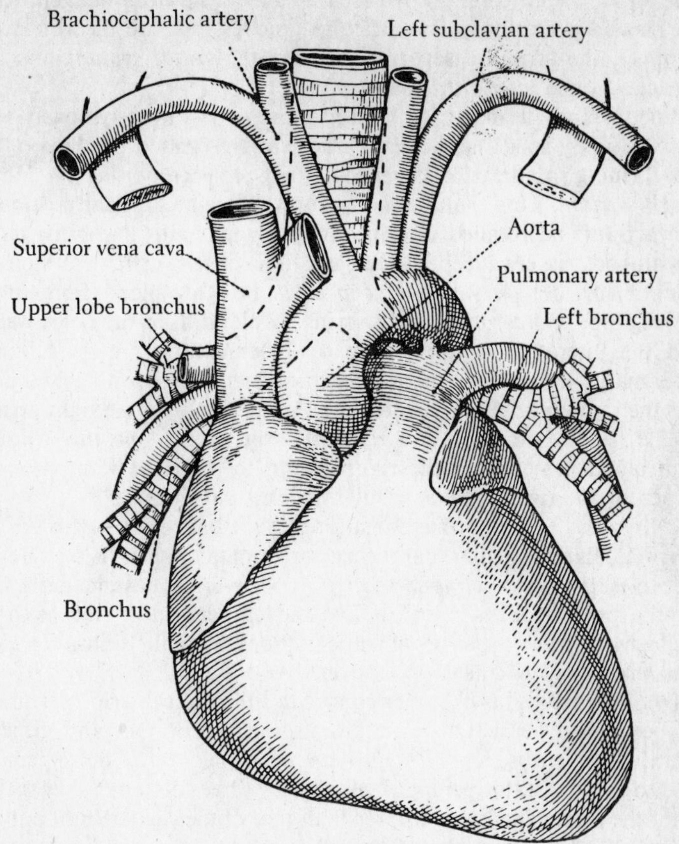

Fig. 10-1. The heart, anterior view. (Reprinted with permission of Macmillan, London and Basingstoke, from W. J. Hamilton, *Textbook of Human Anatomy,* 1976.)

outermost *epicardium* (the visceral pericardium already mentioned), composed of mesothelium and elastic fibers; (2) *myocardium,* a thick middle layer of cardiac muscle fibers grouped together into bands that tend to run obliquely, this layer being much thicker in the ventricles than in the atria and thickest in the left ventricle; and (3) *endocardium,* the innermost layer, made up of endothelium and a thin layer of underlying connective tissue. The endocardium lines the heart cavities and covers the valves; it is continuous with the endothelium of blood vessels entering and leaving the heart.

CHAMBERS OF THE HEART. The heart has four compartments or *chambers:* the left and right *atria* and the left and right *ventricles. Exter-*

nally, an atrioventricular groove separates the atria from the ventricles; anterior and posterior interventricular grooves separate the ventricles. *Internally,* the atria are separated by the *interatrial septum,* the two ventricles by the *interventricular septum.*

Sometimes the term *auricle* is used more or less synonymously with *atrium.* Strictly speaking, however, *auricles* are limited portions of the atria, forming small earlike diverticula that project externally.

Right Atrium. This chamber constitutes the right superior portion of the heart. It is thin-walled, and into it empty the veins that bring blood from all body tissues except the lungs. These vessels, which enter on the posterior surface, are *superior vena cava,* bringing blood from upper body regions; and *coronary sinus,* bringing blood from the heart wall.

On the interatrial septum is an oval depression, the *fossa ovalis,* which marks the position of the *foramen ovale,* an opening that connects the right and left atria in the fetal heart. Between the right atrium and the right ventricle is the *right atrioventricular opening,* through which blood passes from the right atrium to the right ventricle. It is guarded by the right atrioventricular or *tricuspid valve.*

Left Atrium. This chamber constitutes the left superior portion of the heart. It is slightly smaller that the right atrium, and its walls are thicker. It receives the four *pulmonary veins,* which bring oxygenated blood from the lungs. It opens inferiorly into the left ventricle by means of the *left atrioventricular opening,* which is guarded by the left atrioventricular or *bicuspid valve* (the *mitral valve*).

Right Ventricle. This chamber constitutes the right inferior portion of the apex of the heart. Leading from its superior surface and directed upward to the left is the *pulmonary artery,* which carries deoxygenated blood to the lungs. Its opening is guarded by three *semilunar valves.* The wall of the right ventricle is thicker than that of the right atrium but has only one-third the thickness of the left ventricle.

Left Ventricle. This chamber constitutes the left inferior portion of the apex of the heart. Its walls are very thick, for by its contractions the blood is forced through the aorta to all parts of the body except the lungs. From its superior surface the aorta arises, its opening being guarded by three *semilunar* (*aortic*) *valves.*

The inner layers of the walls of both ventricles present irregular surfaces. Muscular ridges (*trabeculae carnae*) and the *papillary muscles,* to which the *chordae tendineae* are attached, project inwardly. The chordae tendineae are fibrous cords attached to the free edges of the bicuspid and tricuspid valves.

VALVES OF THE HEART (Figs. 10-2 and 10–9). Two types of valves are found in the heart, atrioventricular and semilunar.

The *atrioventricular valves* are thin, leaflike structures or *cusps* located at the atrioventricular openings. The bases of these leaflets form a ring about the opening, so that blood readily passes between them. Their free

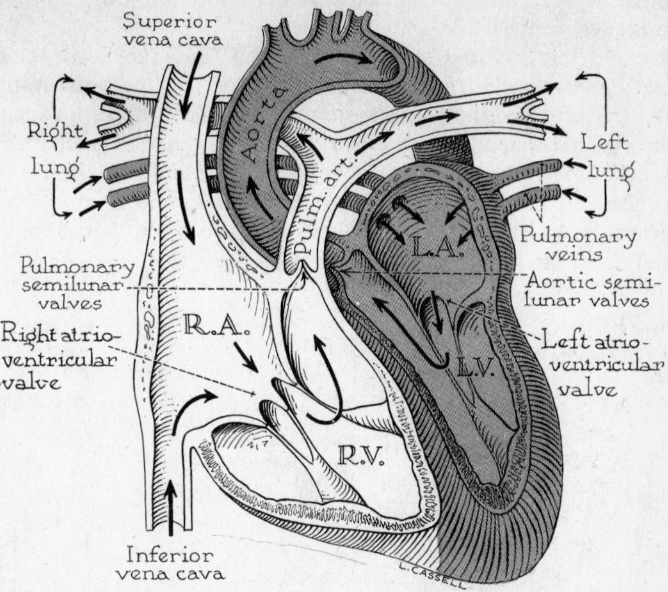

Fig. 10-2. Diagram of the heart, showing chambers, valves, and direction of blood flow. (Reprinted with permission of W. B. Saunders Co., Philadelphia, from B. G. King and M. J. Showers, *Human Anatomy and Physiology,* 6th ed., 1969.)

edges project inferiorly into the ventricular cavities, and from these edges of the valves the *chordae tendineae* extend to the papillary muscles.

The *right atrioventricular* or *tricuspid valve* consists of three *cusps;* the *left atrioventricular valve* (*mitral valve*), two.

Upon contraction of the atria, blood readily flows through the opening between the leaflets, spreading them apart. When the ventricles contract, blood is forced upward against the outer surfaces of the valves, causing them to approximate each other, thus closing the opening and preventing return of the blood to the atria.

The *semilunar valves* are pocketlike structures that surround the openings to the aorta and to the pulmonary artery. They are arranged in sets of three. On contraction of the ventricles, the blood is forced through the opening between the valves, forcing them against the walls of the vessels and permitting blood to enter the arteries. On cessation of ventricular contraction, back pressure (resulting from peripheral resistance and the elasticity of the stretched arterial walls) tends to force the blood back into the heart, but this is prevented by the valvular action that closes the openings. The *pulmonary semilunar valves* are located at the point where the pulmonary artery exits from the right ventricle; the *aortic semilunar valves* are located at the point where the aorta exits from the left ventricle.

Conductile Tissue of the Heart (Fig. 10-3). In the heart there is a special *impulse-conducting system* that is concerned with the initiation of the heartbeat and the coordination of the beat of the various parts of the heart. This system includes the *sinoatrial (SA) node,* the *atrioventricular (AV) node,* the *atrioventricular bundle* (*bundle of His*) and its branches, and *Purkinje fibers,* which continue to and terminate in the myocardium of the ventricles. This system and the role it plays in the regulation of the heartbeat is described on page 349.

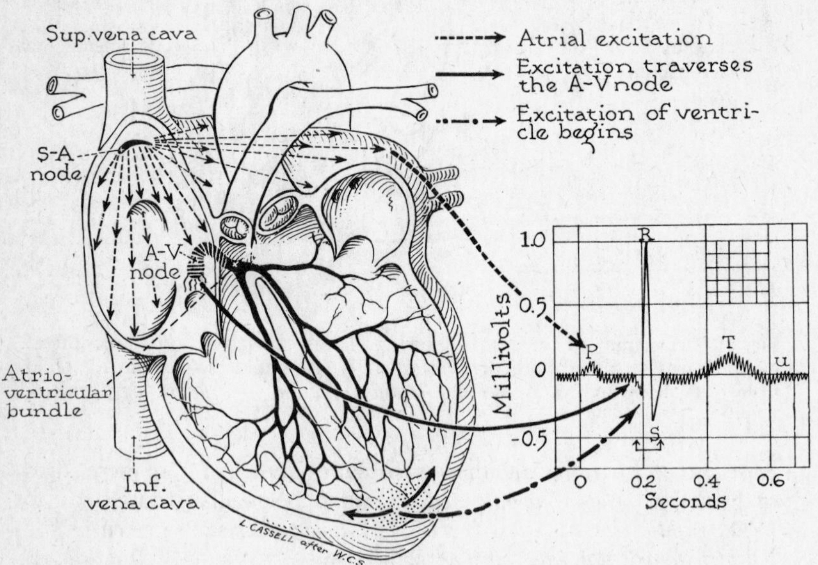

Fig. 10-3. Diagram of the heart, showing the impulse-conducting system. (Reprinted with permission of W. B. Saunders Co., Philadelphia, from B. G. King and M. J. Showers, *Human Anatomy and Physiology,* 6th ed., 1969.)

Arteries. An artery is an elastic muscular tube (Figs. 10-4 and 10-5) that conducts blood from the heart.

Layers of the Artery Wall. The wall of an artery has three layers:

1. *Tunica intima* or *interna,* consisting of an inner layer of endothelium resting on the *internal elastic membrane* of fibroelastic tissue. This layer forms the lining of the artery.

2. *Tunica media,* the middle and thickest layer, consisting of smooth muscle cells, elastic and collagenous fibers, and thin reticular fibers.

3. *Tunica adventitia,* the outermost layer, consisting of elastic and collagenous fibers that pass into the adjoining connective tissues without a sharp line of demarcation.

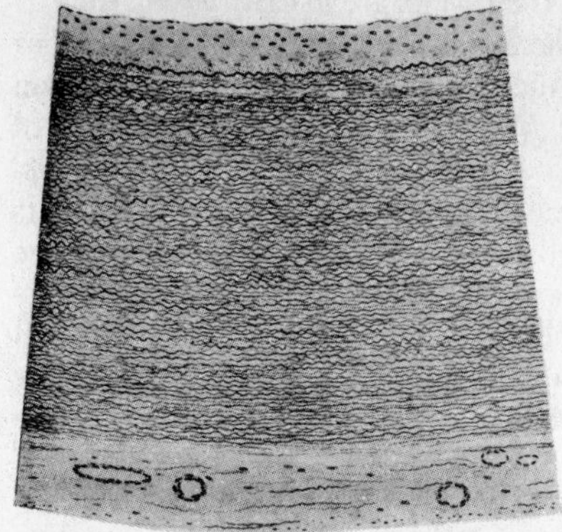

Fig. 10–4. Section through the wall of a large artery. Note vasa vasorum in the tunica adventitia. (Redrawn from Sobotta.)

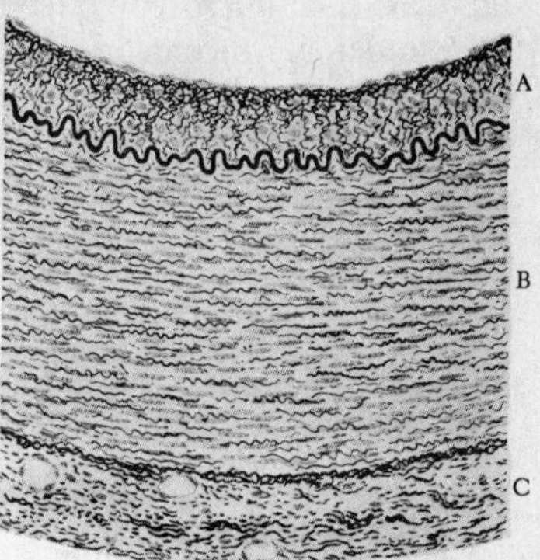

Fig. 10–5. Section through the wall of a medium-sized artery. (*A*) Tunica intima. (*B*) Tunica media. (*C*) Tunica externa.

(Reprinted with permission of Macmillan, London and Basingstoke, from W. J. Hamilton, *Textbook of Human Anatomy,* 1976.)

TYPES OF ARTERIES. There are three types of arteries:

1. *Elastic arteries* are large in caliber and include the aorta, innominate, subclavian, part of the common carotid, and the pulmonary. Elastic fibers predominate in their tunica media. Arteries of this type are called *conducting* arteries.

2. *Muscular arteries* are *continuations* of the elastic arteries and similar to them in structure, except that muscular fibers predominate in their tunica media. These are the principal *distributing* arteries.

3. *Arterioles* are very small branches of the muscular arteries that lead to the capillaries. They average 0.3 mm or less in diameter and are barely visible to the unaided eye. The smallest arterioles, called *precapillary* or *terminal arterioles,* continue as *metarterioles,* in which the muscle coat is replaced by a layer of discontinuous, noncontractile cells called *perivascular cells* or *pericytes.*

Arteries and their accompanying veins are generally located on the flexor sides of limbs, thus being protected against injury or excessive stretch resulting from movement of the limbs. Veins and lymph vessels are usually placed less deeply than the arteries.

Veins. A vein is a vessel that conducts blood away from the tissues and toward the heart. Veins increase progressively in size from the tissues to the heart, their course in general paralleling that of arteries. In structure, veins resemble arteries, except that their walls are thinner and less elastic. In microscopic sections their walls are usually collapsed and their lumen is irregular in shape, whereas arteries are usually circular or ovoid in form and their lumen is much smaller, with the tunica intima presenting a corrugated appearance. Veins possess the same three layers that are found in arteries, but their boundaries are usually indistinct. The tunica media contains much less muscular and elastic tissue.

Many veins, especially those of the extremities, contain *valves,* thin semilunar-shaped flaps of connective tissue that form pocketlike structures. They are attached to the inner surface of the walls of the vessel with their free edges directed toward the direction of blood flow. They are usually arranged in pairs. Such valves prevent a backflow of the blood.

Capillaries. These minute vessels form the terminations of arterioles. Leading to the venules, they form complicated anastomosing networks of vessels in the tissues. The wall of a capillary consists of a single layer of squamous epithelial cells; this endothelium is a continuation of that lining the arteries and veins. A basement membrane completely encloses each capillary. Capillaries average about 8 μm in diameter, and each possesses a lumen barely large enough for red blood cells to pass through in single file. Pericytes surround some capillaries.

There are two types of capillaries, continuous and fenestrated. In *continuous capillaries,* the borders of endothelial cells are contiguous, but the junctions are of a type that permits the interchange of water and other small molecules between the blood and tissue fluid. Leukocytes can also migrate between the endothelial cells. In *fenestrated capillaries,* there are small openings (*fenestrae* or *fenestrations*) between adjoining endothelial cells. The fenestrae, although covered by an extremely thin diaphragm and basement membrane, permit the relatively free interchange of small molecular substances between the blood and tissue fluid. In most capillaries, minute pinocytic vesicles are extremely numerous on both inner and outer surfaces of the endothelial cells. By pinocytosis, macromolecules (proteins, for example) can be transported through the endothelial cell membranes.

Blood flows slowly through the capillaries. This is due to their very narrow diameter and their greater cross-sectional area as compared with the artery that supplies them.

Sinusoids. These are minute blood vessels found in certain organs, such as the liver, spleen, adrenal glands, and bone marrow. They differ from capillaries in that their diameter is generally larger and more variable, and their walls, instead of being formed by a continuous layer of endothelium as in capillaries, are formed by irregularly placed phagocytic and nonphagocytic cells. The phagocytic cells possess the ability to store vital dyes and to phagocytose bacteria. Furthermore, sinusoids lack a layer of connective tissue, which usually surrounds capillaries;

consequently, they lie in closer proximity to the cells of the organs in which they are found. In certain of the blood-forming organs of the body, such as the spleen and bone marrow, the walls of the sinusoids have openings through which cells can enter or leave the bloodstream.

Connections Between Arteries and Veins. In general, blood passes from an artery to a vein through a single set of capillaries in the tissues. However, there are a number of exceptions to this general plan, some of which are listed here.

1. From an artery directly to a vein (as in the sole of the foot or the palm of the hand). Such a connection is called an *arteriovenous anastomosis.*
2. From an artery through capillaries to an arteriole and then through a second capillary net to a vein (as in the kidneys).
3. From an artery through capillaries to a vein and then through a network of sinusoids to a vein (as in the portal circulation through the liver).
4. From an artery through capillaries to sinusoids and then to a vein (as in the adrenal gland and spleen).
5. From an artery through spaces to a sinus and then to veins (as in the placenta of a gravid uterus).

GENERAL PLAN OF CIRCULATION

Blood coming from the body tissues enters the right atrium through the venae cavae. Contraction of the right atrium forces blood past the tricuspid valve into the right ventricle. From this point begin the two subdivisions of the the cardiovascular system: the pulmonary and systemic circulations. As the terms imply, the former serves the lungs, the latter the various systems of the body.

Pulmonary Circulation. Contraction of the right ventricle forces blood past the semilunar valves into the pulmonary artery, which carries it to the lungs. In passing through the capillaries of the lungs, the blood is oxygenated and carbon dioxide is given off. Pulmonary veins carry the blood back to the left atrium of the heart.

Systemic Circulation. Contraction of the left atrium forces blood past the bicuspid valve into the left ventricle. Contraction of the left ventricle forces blood past the semilunar valves into the aorta. From the aorta, blood is carried to capillaries of all tissues by arteries and arterioles. From the tissues, capillaries lead to venules, which, in turn, lead to veins, which empty into the superior and inferior venae cavae. The venae cavae empty into the right atrium.

It can thus be seen that the blood makes a double circuit in its travels, since following pulmonary circulation it reenters the heart at the left atrium, where it began its course throughout the body as the systemic circulation.

VESSELS OF THE CARDIOVASCULAR SYSTEM

The principal arteries and veins comprising the pulmonary and systemic circuits are described in the following sections (Fig. 10-6). The *pulmonary circuit* includes vessels to and from the lungs; the *systemic circuit* includes vessels to and from all the other organs of the body.

Vessels of the Pulmonary Circuit. The *pulmonary artery* carries deoxygenated blood to the lungs. It arises as a short, thick trunk from the superior surface of the right ventricle, passes diagonally upward to the left, and crosses the root of the aorta. It divides into *right* and *left pulmonary arteries,* branches of which enter the right and left lungs, respec-

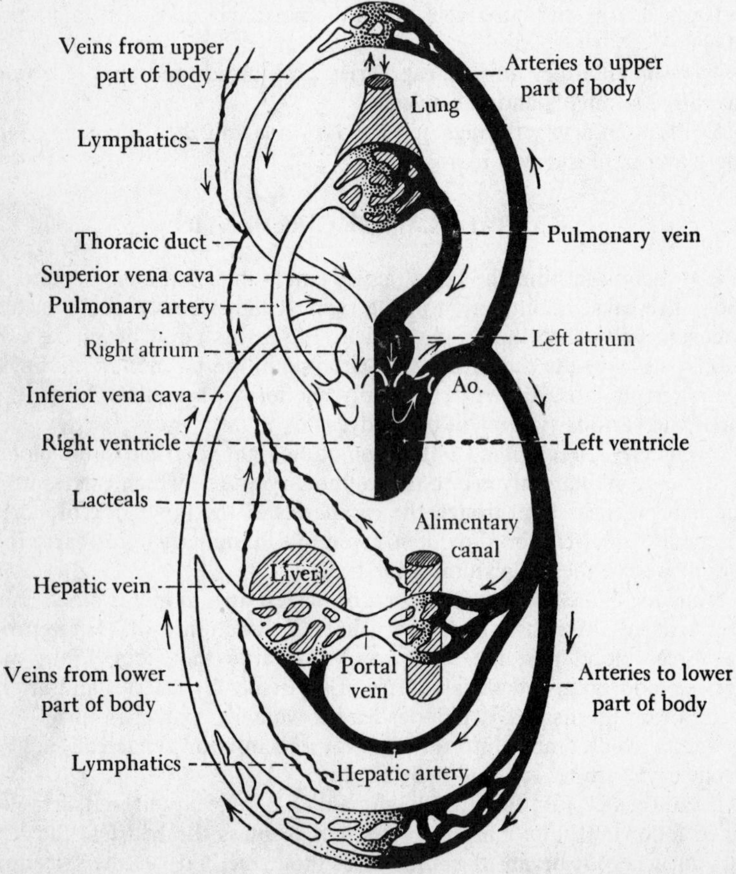

Fig. 10-6. Diagram of blood circulation. Arrows show direction of flow. The two arrows above the lung show interchange of air. (Reprinted with permission of C. V. Mosby Company from J. F. McClendon, *Physiological Chemistry,* 1946.)

tively. At its bifurcation, a short connective-tissue structure, the *ligamentum arteriosum,* connects with the aortic arch. This cord represents the vestige of an artery, the *ductus arteriosus,* that was functional in the fetus.

Pulmonary veins are four in number, two from each lung. They carry oxygenated blood from the lungs to the left atrium.

Vessels of the Systemic Circuit. The systemic division of the cardiovascular system consists of the aorta (with its numerous branches) and four principal veins (with their branches).

SYSTEMIC ARTERIES. The main arterial trunk carrying blood from the heart to the tissues is the *aorta.* It emerges from the superior surface of the left ventricle, passes upward under the pulmonary artery as the *ascending aorta,* turns to the left as the *aortic arch,* then turns downward as the *descending aorta.* The descending aorta, lying close to the bodies of the vertebrae, passes downward through the diaphragm and continues to the level of the 4th lumbar vertebra, where it terminates by dividing into the two common iliac arteries. The portion of the aorta above the diaphragm is the *thoracic aorta;* that below is the *abdominal aorta.* The principal branches of the various portions of the aorta are as follows (Figs. 10-7 and 10-8).

Ascending aorta	Coronary	
Arch of the aorta	Brachiocephalic (Innominate)	Right common carotid Right subclavian
	Left common carotid	
	Left subclavian	
	Visceral branches	*Parietal branches*
Thoracic aorta	Bronchial	Intercostal
	Esophageal	Subcostal
	Pericardial	Superior phrenic
	Mediastinal	
	Visceral branches	*Parietal branches*
Abdominal aorta	Celiac	Inferior phrenic
	Superior mesenteric	Lumbar
	Testicular or ovarian	
	Middle suprarenal	
	Renal	
	Inferior mesenteric	
	Terminal branches	
	Middle sacral	
	Common iliac	Ext. iliac Hypogastric (Int. iliac)

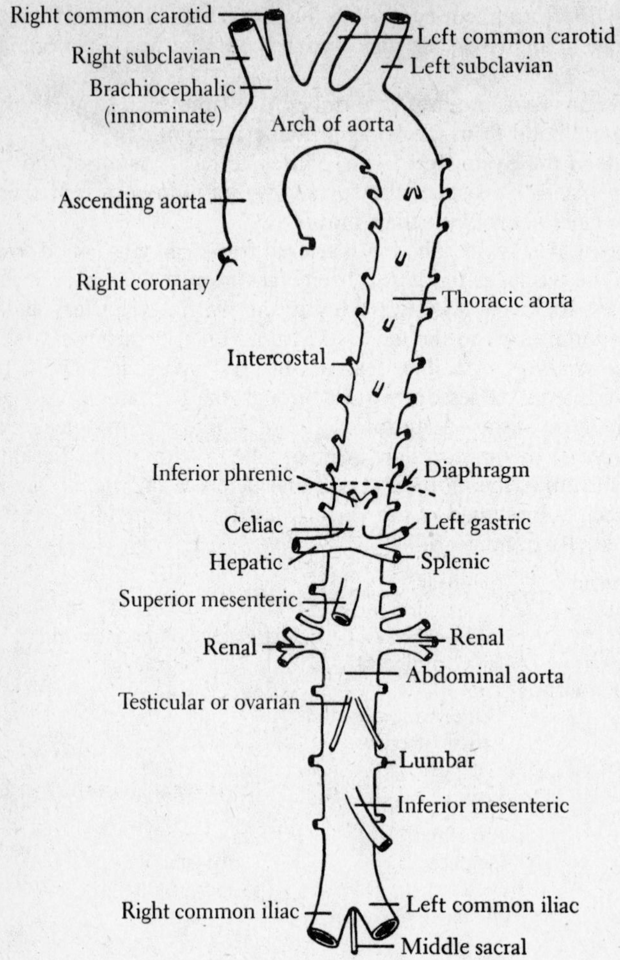

Fig. 10-7. Aorta and branches. (Reprinted with permission of C. V. Mosby Company and the author from C. C. Francis, *Introduction to Human Anatomy,* 2nd ed., 1954.)

Ascending Aorta. The branches of the ascending aorta are the *left* and *right coronary arteries,* which arise close to the origin of the aorta, their openings being in the pockets of the aortic valves. They supply the heart (Fig. 10-9).

Arch of the Aorta. The branches of the aortic arch are the innominate, left common carotid, and left subclavian arteries (Figs. 10–8 and 10-10).

Brachiocephalic (Innominate) Artery. This is the first branch of the

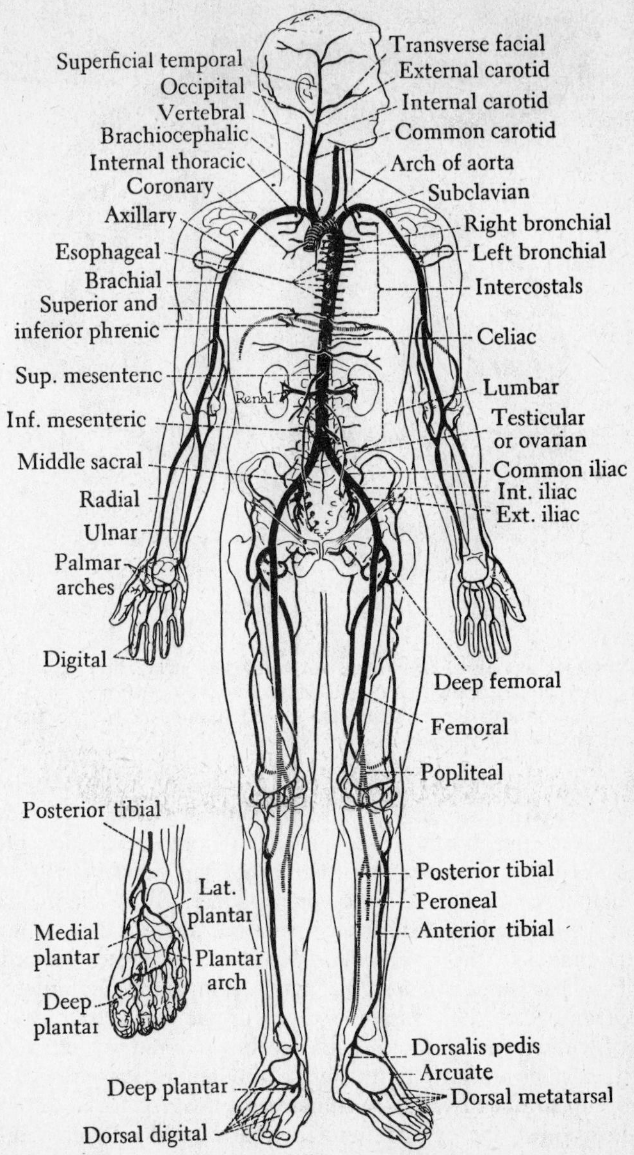

Fig. 10-8. The arterial system. (Reprinted with permission of W. B. Saunders Co., Philadelphia, from B. G. King and M. J. Showers, *Human Anatomy and Physiology*, 6th ed., 1969.)

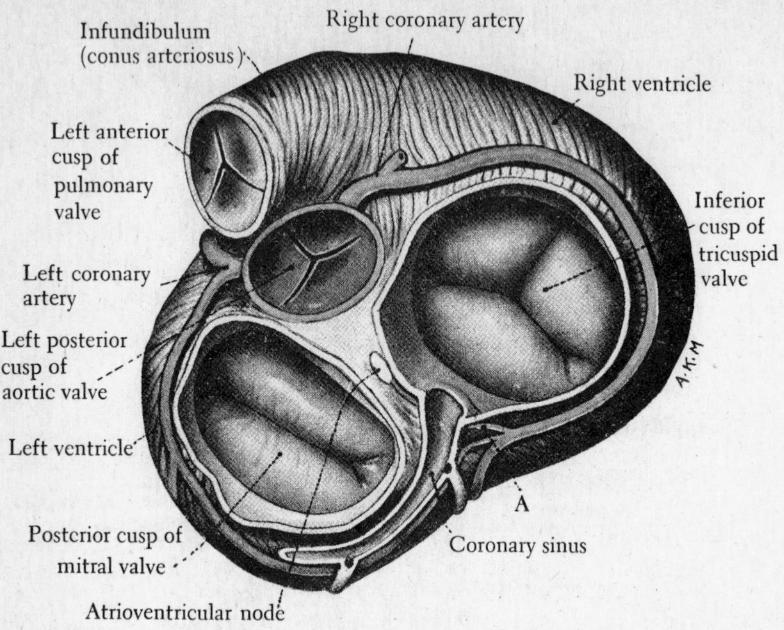

Fig. 10-9. Cross section of the heart, just below the atria, showing valves and coronary arteries. (*A*) branch of right coronary artery to atrioventricular node. (Reprinted with permission of Macmillan, London and Basingstoke, from W. J. Hamilton, *Textbook of Human Anatomy,* 1976.)

aortic arch. It passes superiorly and diagonally to the right, crossing the trachea in its course. As it passes out of the thorax, it divides into two branches: the right common carotid and the right subclavian arteries.

The right *common carotid artery* passes superiorly beside the trachea to about the level of the 3rd cervical vertebra, where it divides into the following branches: the *external carotid,* which continues upward, giving off branches supplying muscles, glands, skin, and other structures of the neck and face, then passes in front of the ear as the *superficial temporal artery,* which supplies the scalp; and the *internal carotid,* which continues upward in the neck and enters the cranial cavity through the carotid foramen, terminating at the *circulus arteriosus.* At the bifurcation of the external and internal carotid arteries, there is a slight dilatation, the *carotid sinus.* Closely associated with the sinus is the *carotid body.* These last two structures, about 5 mm in length, contain the endings of sensory nerves (presso- and chemoreceptors), which play an important role in the regulation of circulation and respiration.

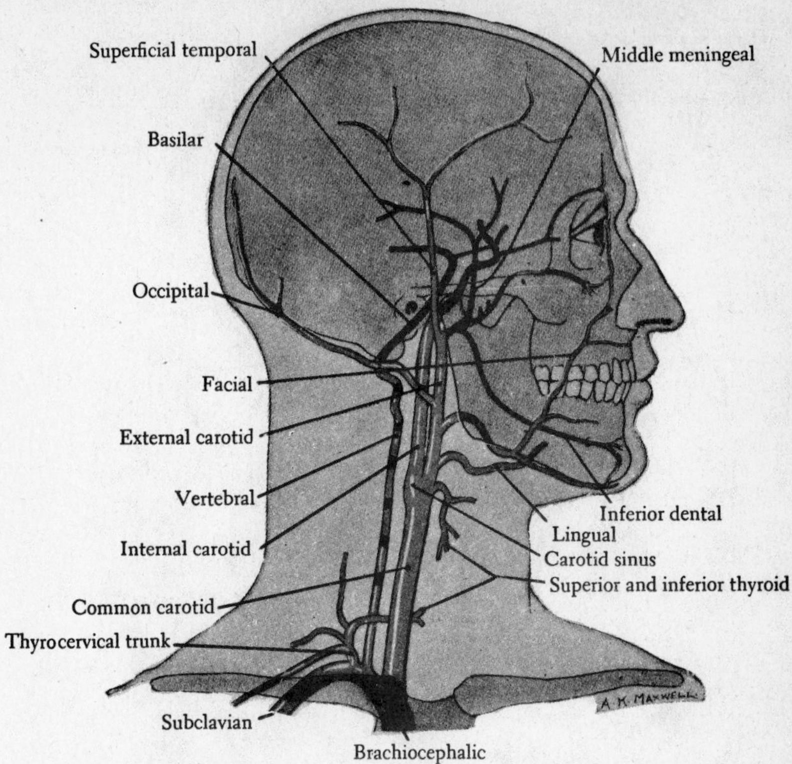

Fig. 10-10. Main arteries of the head and neck. (Reprinted with permission of Macmillan, London and Basingstoke, from W. J. Hamilton, *Textbook of Human Anatomy,* 1976.)

The *circulus arteriosus* (Fig. 10-11) consists of an anastomosing group of blood vessels located at the base of the brain. It surrounds the optic chiasm and the hypophysis and is formed of the following vessels: posteriorly, the proximal portions of the *posterior cerebral arteries;* laterally, the *posterior communicating, internal carotid,* and the proximal portions of the *anterior cerebral arteries;* anteriorly, the *anterior communicating arteries.* From this circle arteries extend into the meninges and the brain tissue. The *circulus arteriosus* receives its blood from the internal carotid and basilar arteries.

The *right subclavian artery* passes laterally to the arm. Two branches of it are the *vertebral* and *internal mammary* (*thoracica interna*) *arteries.* In the region of the axilla, it becomes the *axillary artery,* which continues into the arm. Other branches supply the region of the shoul-

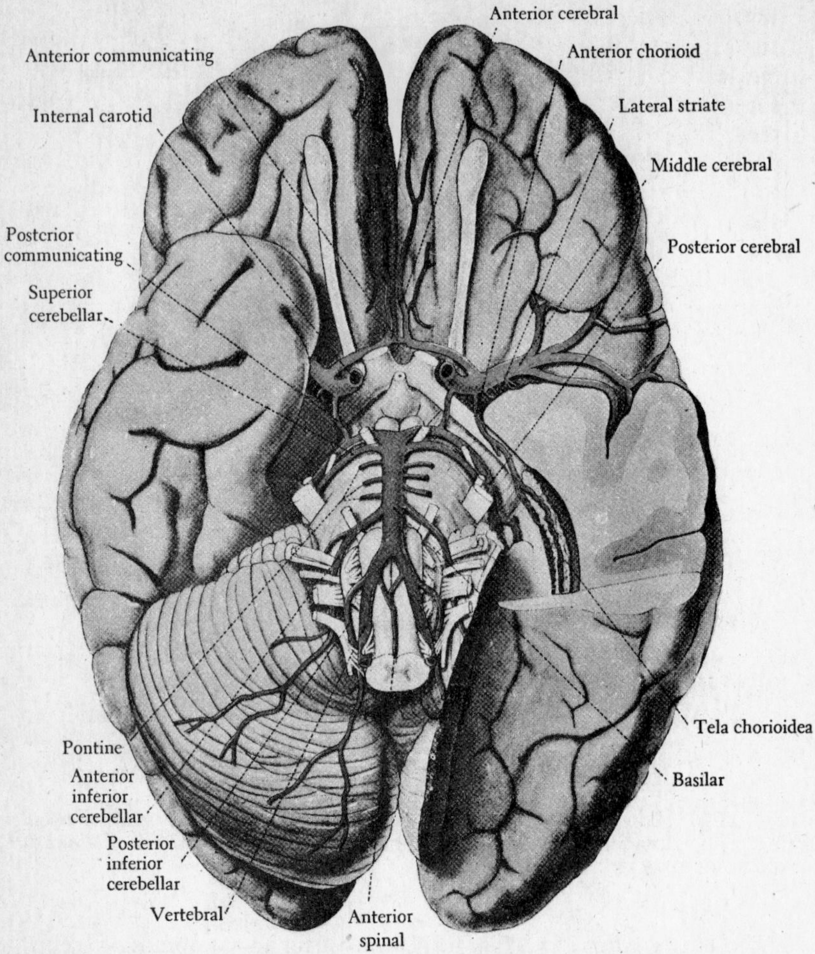

Fig. 10-11. Arteries at the base of the brain. Part of the left cerebellar hemisphere has been removed. (Reprinted with permission of Macmillan, London and Basingstoke, from W. J. Hamilton, *Textbook of Human Anatomy,* 1956.)

der, axilla, and upper part of the thorax. In the arm, the right axillary continues as the *brachial artery,* which, in the region of the elbow, divides into the *ulnar* and *radial arteries.* These continue into the distal portion of the arm, giving off branches to the wrist and hand. In the hand, the ulnar and radial arteries are connected by the *deep* and *superficial palmar arches,* branches of which supply the digits.

The *vertebral artery* arises from the subclavian artery near its origin. It passes upward through the neck, through the foramina in the transverse processes of the cervical vertebrae. At the 1st cervical vertebra it passes around the atlas and enters the cranial cavity through the foramen magnum. Within the skull, it unites with its mate from the opposite side to form the *basilar artery,* which joins the *circulus arteriosus* at the base of the brain.

Left Common Carotid Artery. The second branch of the arch of the aorta, the left common carotid, arises close to the innominate. It passes up the left side of the neck, following a course and possessing branches similar to those of the right common carotid.

Left Subclavian Artery. This artery arises from the aortic arch close to the left common carotid. It passes upward and out of the thorax, where it turns abruptly to the left to supply the left arm. It possesses the same branches as does the right subclavian.

Thoracic Aorta. The branches of the thoracic portion of the descending aorta supply either the internal organs or the body wall.

Visceral Branches. These branches of the thoracic aorta include the *bronchial* (one right and two left), *esophageal* (four or five), *pericardial* (two or three), and *mediastinal arteries,* which supply the bronchi and lungs, esophagus, pericardium, and mediastinum, respectively.

Parietal Branches. *Intercostal arteries* (nine pairs) pass laterally and supply the thoracic wall, carrying blood to the chest muscles, the vertebrae, the pleurae, and the meninges of the spinal cord. *Subcostal arteries* (one pair) pass to the body wall. *Superior phrenic arteries* (one pair) supply the upper surface of the diaphragm.

Abdominal Aorta. This portion of the aorta has many branches; it serves the entire lower region of the body. Its branches consist of three groups: visceral, parietal, and terminal. The branches of each group, listed in the order in which they arise, are as follows:

Visceral Branches of the Abdominal Aorta. These arteries supply the internal organs of the abdominal cavity. They include the following arteries:

1. The *celiac artery,* a very short artery, emerges from the aorta a short distance below the diaphragm. It divides into three branches: the *left gastric,* or *coronary,* which supplies a portion of the stomach and the lower end of the esophagus; the *hepatic,* which supplies the liver, gallbladder, and portions of the stomach and duodenum; and the *splenic,* which supplies the spleen and portions of the stomach and pancreas.
2. The *superior mesenteric artery* arises immediately below the celiac. It supplies the major portion of the small intestine and the first portion of the large intestine. It has many branches: inferior pancreaticoduodenal, 12 or more intestinal arteries, the right and middle colic arteries, and the ileocecal artery.
3. The *middle suprarenal arteries* supply the suprarenal glands.
4. The *renal arteries* supply the kidneys.
5. The *internal testicular arteries* in the male and the *ovarian arteries*

in the female arise just below the renal arteries. They lead to the testes and ovaries, respectively.

6. The *inferior mesenteric artery* supplies the colon and the rectum. Branches are the left colic, sigmoid, and superior rectal.

Parietal Branches of the Abdominal Aorta. These arteries supply the structures of the body wall in the lower portion of the trunk. They include the following arteries:

1. The *inferior phrenic arteries* supply the lower surface of the diaphragm. Branches also pass to the adrenal gland.

2. The *lumbar arteries* (four pairs) pass to the body wall, where they supply the muscles, skin, vertebrae, and the spinal cord and its meninges.

Terminal Branches of the Abdominal Aorta. These branches supply portions of the pelvis and the entire lower extremities. They include the following arteries:

1. The *middle sacral artery* arises near the point of bifurcation of the abdominal aorta. It supplies the sacrum and the coccyx.

2. *Common iliac arteries* arise from the bifurcation of the abdominal aorta. Each continues a short distance and then divides into two branches: the *hypogastric* (*internal iliac*) and the *external iliac arteries.* The hypogastric passes into the pelvic cavity, where it divides into an *anterior* and a *posterior* branch. These supply the muscles of the pelvic wall, urinary bladder, reproductive organs (including the external genitalia and uterus), and structures on the medial side of the thigh. The *external iliac* passes from its origin (the common iliac) under the inguinal ligament, where it leaves the pelvic cavity as the femoral artery.

3. The *femoral artery,* a continuation of the external iliac, passes through the thigh, lying medial and posterior to the femur. It gives off several branches, the largest being the *profunda femoris.* Above the popliteal fossa (the space directly posterior to the knee joint), the femoral artery becomes the popliteal artery. The *popliteal artery* passes through the popliteal fossa and into the lower leg. Just below the knee, the *anterior tibial artery* branches off from it and passes downward to the ankle joint, where it continues as the *dorsalis pedis artery.* The main portion of the popliteal artery continues to the foot as the *posterior tibial artery;* in the foot, this divides into the *medial* and *lateral plantar arteries.*

4. The *peroneal artery,* a large branch of the posterior tibial, arises just below the border of the popliteus muscle and passes along posterior and lateral sides of the leg to the foot, where it divides into small branches.

Systemic Veins (Fig. 10-12). These are the veins that conduct blood from the tissues to the heart. The principal systemic veins are (1) the *coronary sinus,* (2) the *superior vena cava,* (3) the *inferior vena cava,* and (4) the *portal vein,* which drains the abdominal viscera. The last-named vein differs from the others in that the blood it carries from the tissues passes through a second capillary network (sinusoids) in the liver before it enters the inferior vena cava.

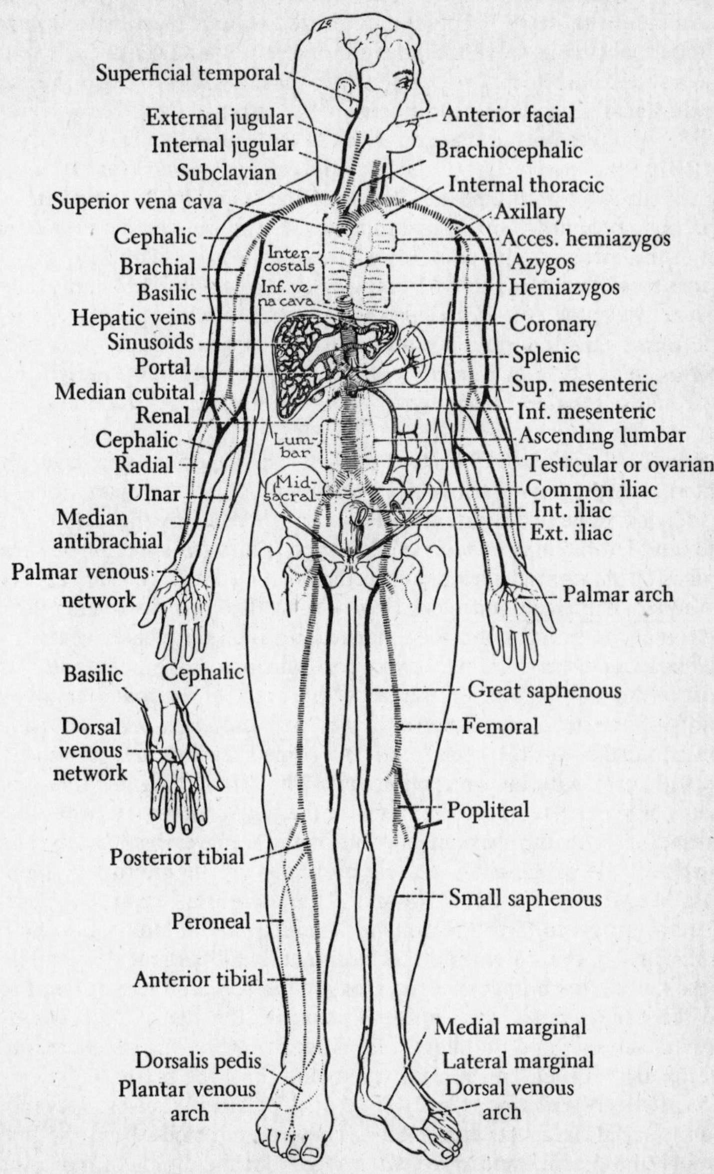

Fig. 10-12. General view of the venous system. Deep veins are cut across. (Reprinted with permission of W. B. Saunders Co., Philadelphia, from B. G. King and M. J. Showers, *Human Anatomy and Physiology,* 6th ed., 1969.)

Venous channels may be classified as (1) *superficial veins,* which lie in the superficial fascia directly beneath the skin; (2) *deep veins,* which accompany the more deeply located arteries; and (3) *venous sinuses,* channels lacking in valves and devoid of a muscular layer in their walls. The last-named, which are found only in the cranial cavity, are spaces in the dura mater lined with endothelium; the larger of these venous sinuses lie in grooves on the inner surfaces of the cranial bones.

Coronary Sinus and Its Branches. The *coronary sinus* (Fig. 10-6) is a short vein lying on the posterior side of the heart. Its opening into the right atrium, guarded by an incomplete valve, is situated near the opening of the inferior vena cava. It receives most of the blood from heart tissues through one of the following vessels: the *great, small,* and *middle cardiac veins;* the *posterior vein;* and the *oblique vein.*

Cardiac veins that do not empty into the coronary sinus are *anterior cardiac veins,* three or four in number, which empty into the right atrium; and the *smallest cardiac veins,* most of which empty into the atria, a few into the ventricles.

Superior Vena Cava and Its Branches. The superior vena cava (Fig. 10-13), a large venous trunk that returns blood to the heart from the head, neck, upper extremities, and thorax, is formed by the union of the right and left brachiocephalic veins. It opens into the superior posterior surface of the right atrium at about the level of the 3rd costal cartilage.

Veins of the Head and Neck (Fig. 10-14). The principal veins draining the head and neck are the external and internal jugular veins. The *external jugular vein* drains the external portions of the head, including most of the scalp and face. It arises at the angle of the mandible by the union of the *posterior facial* and *posterior auricular veins.* It passes downward through the neck, receiving some branches and terminating at its junction with the subclavian vein. The *internal jugular vein* is the principal vein of the neck. It arises at the jugular foramen, where it is continuous with the sigmoid portion of the transverse sinus. It passes down the side of the neck in close proximity to the internal and common carotid arteries. At the level of the sternum, it unites with the subclavian vein to form the *brachiocephalic* (*innominate*) *vein.* The internal jugular vein receives blood from nearly all parts of the head and neck. Among the branches it receives are the *common facial vein,* from the face; the *lingual vein,* from the tongue; the *pharyngeal;* and the *thyroid* (superior and middle) *veins.* From the *transverse sinus* the internal jugular vein receives most of the blood from the brain.

Sinuses of the Cranial Cavity (Fig. 10-15). These sinuses are venous channels that lie in the dura mater. They receive blood from the brain through the *cerebral* and *cerebellar veins,* from the diploe of the cranial bones through the *diploic veins,* and from the sinuses and external portions of the skull through the *emissary veins.*

Superior Sagittal Sinus. This sinus lies in the midsagittal plane, com-

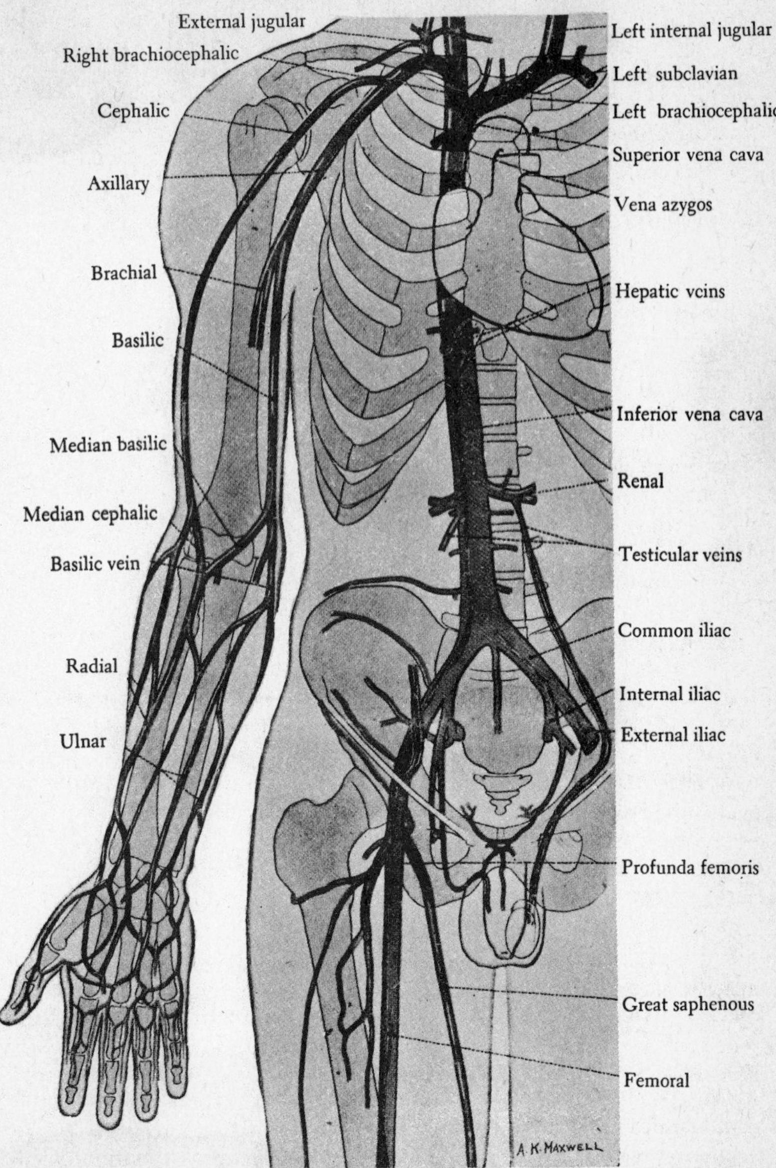

Fig. 10-13. The venae cavae and their branches. (Reprinted with permission of Macmillan, London and Basingstoke, from W. J. Hamilton, *Textbook of Human Anatomy,* 1976.)

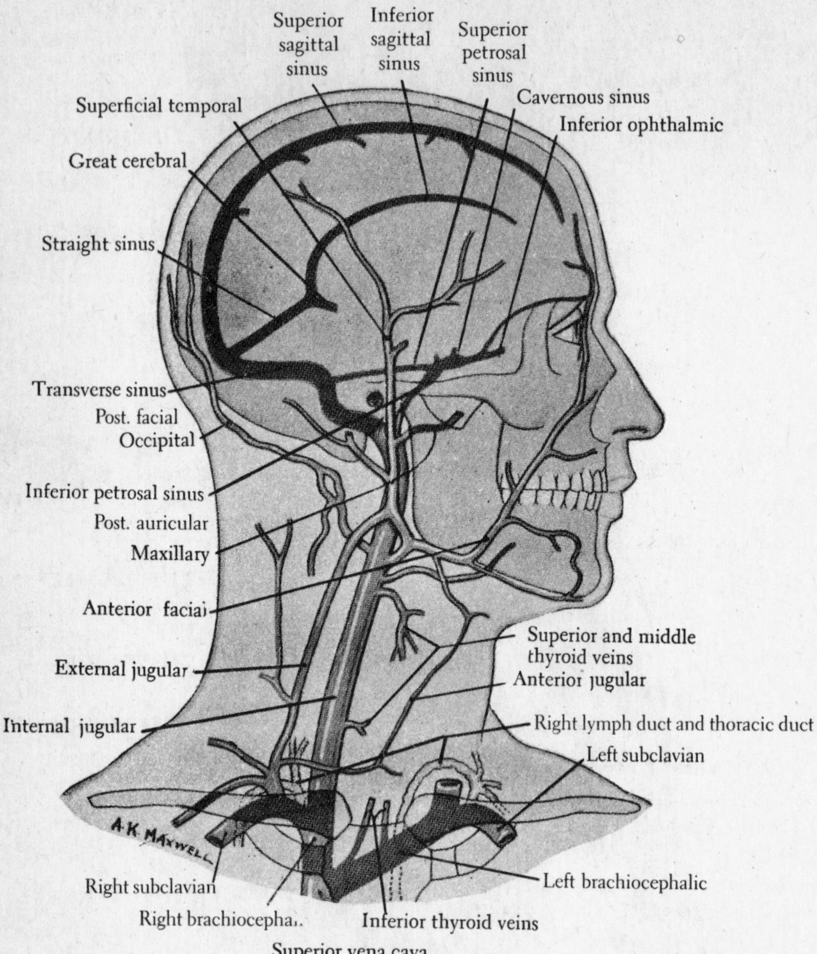

Fig. 10-14. Veins of the head and neck, showing the superior vena cava and its branches. (Reprinted with permission of Macmillan, London and Basingstoke, from W. J. Hamilton, *Textbook of Human Anatomy,* 1976.)

mencing at the foramen caecum and lying in the convex portion of the falx cerebri. It terminates posteriorly, internal to the posterior occipital protuberance at its junction with the transverse sinuses.

Inferior Sagittal Sinus. This sinus lies along the inferior margin of the falx cerebri. It unites with the great cerebral vein to form the *straight sinus.*

Straight Sinus. This sinus lies along the line of junction of the falx cerebri and the tentorium cerebelli. It empties into the left transverse sinus.

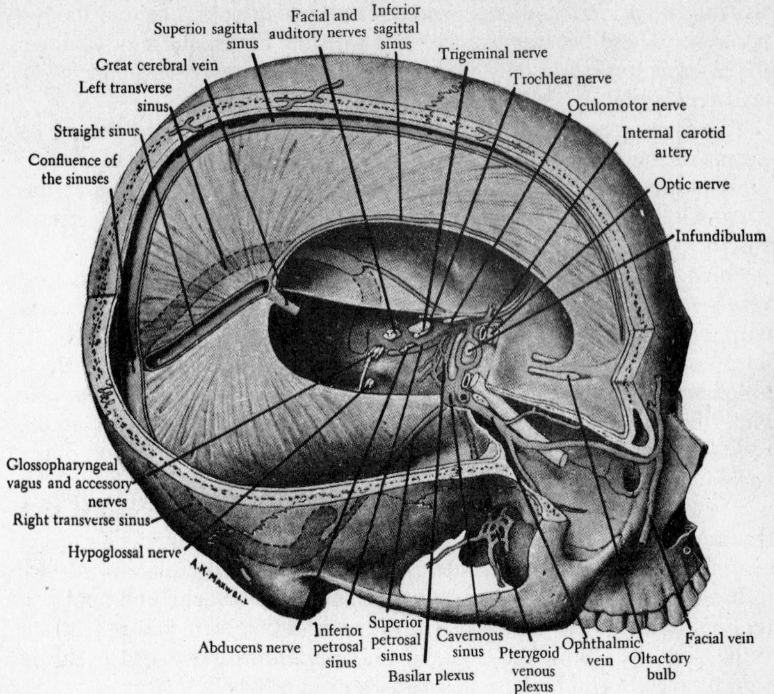

Fig. 10-15. Venous sinuses of the cranial cavity. (Reprinted with permission of Macmillan, London and Basingstoke, from W. J. Hamilton, *Textbook of Human Anatomy,* 1976.)

Occipital Sinus. This sinus lies in the midline, inferior to the confluence of the sinuses into which it empties. It arises near the foramen magnum. The occipital is the smallest of the sinuses.

Transverse Sinuses. These are large sinuses arising at the internal occipital protuberance. Each passes laterally, then curves anteriorly and downward as the sigmoid portion to pass through the jugular foramen, where each is continuous with the internal jugular vein. Each transverse sinus receives blood from the straight sinus and several small veins at the base of the brain and skull.

Confluence of the Sinuses. This is a dilated region located at a point where the superior sagittal, occipital, and transverse sinuses converge.

Several smaller sinuses of the cranial cavity, constituting the anterior-inferior group, include the following:

Cavernous Sinuses. These lie lateral to the body of the sphenoid bone, posterior to the supraorbital fissure, through which each receives blood from the orbit via the *superior ophthalmic vein.* The two cavernous sinuses are connected with each other by the *intercavernous sinuses:* the *anterior intercavernous sinus,* which passes anterior to the hypophysis, and the *posterior intercavernous sinus,* which passes behind the hypophysis. These two vessels, together with the cavernous sinuses, form the *circular sinus.* Laterally each cavernous sinus receives a small *sphenoparietal sinus.*

The cavernous sinuses communicate with the transverse sinuses by means of the *superior petrosal sinus* and with the *internal jugular vein* through the *inferior petrosal sinus.* A basilar plexus lies over the basilar portion of the occipital bone and connects the two inferior petrosal sinuses.

Veins of the Upper Extremity (Fig. 10–13). The *right* and *left subclavian veins* drain the upper extremities, each terminating at its junction with the internal jugular vein to form the brachiocephalic vein. Near its termination each subclavian receives the external jugular. At its angle of union with the internal jugular, the left subclavian receives the *thoracic duct* (the principal lymph duct) and the right subclavian receives the *right lymphatic duct.* Each subclavian vein is formed by the union of the cephalic and axillary veins.

The *cephalic vein* arises from the dorsal arch on the radial side (thumb side) of the hand. It passes upward along the radial side of the arm, turning anteriorly and then passing along the anterolateral surface of the elbow and continuing laterally to the biceps brachii and finally to its junction with the axillary vein, where the two form the subclavian vein. Near the elbow, the cephalic vein communicates with the basilic vein by means of a branch, the *median cubital* (basilic) vein.

The *axillary vein* is a continuation of the subclavian vein. It is formed by the union of the brachial (deep) vein, and the basilic (superficial) vein. The *brachial vein* lies alongside the brachial artery. It originates at the elbow from the union of the *radial* and *ulnar veins.* The *basilic vein* originates near the wrist from the union of the dorsal veins of the hands. It passes up the ulnar side of the forearm on the posterior surface, then swings anteriorly in the region of the elbow and continues along the medial side of the biceps brachii to its termination at the point where it joins with the brachial vein.

The deep veins and the superficial veins are listed here:

Deep Veins	*Superficial Veins*
subclavian	cephalic
axillary	basilic
brachial	
ulnar	
radial	

Veins of the Thorax. Two large veins, the *left* and *right brachiocephalic (innominate)* are formed by union of the internal jugular and subclavian veins, which return blood from the head, neck, and upper extremities. The left is the larger of the two.

Each brachiocephalic vein receives the *deep cervical, vertebral, internal mammary,* and *inferior thyroid veins.* In addition, the left brachiocephalic receives the *left superior intercostal vein* and veins from the thymus, trachea, esophagus, and pericardium.

The *azygos vein* is a large vein originating in the abdomen from a vessel, the *ascending lumbar vein.* It passes through the diaphragm and continues upward along the *right* side of the vertebral column to the level of the 4th thoracic vertebra, where it turns forward to enter the superior vena cava. It receives the *right subcostal,* several *right intercostals,* the *right superior intercostal, hemiazygos,* and *accessory hemiazygos veins,* and *bronchial veins* from the lungs.

The *hemiazygos vein* has a course comparable to that of the azygos, excepting that it lies to the *left* of the vertebral column. It receives the *left subcostal vein* and veins from the esophagus, mediastinum, and lungs. It empties into the azygos vein.

Inferior Vena Cava and Its Branches (Fig. 10–13). This, the largest vein in the body, receives most of the blood from the regions of the body below the level of the diaphragm. The inferior vena cava is formed by the union of the two *common iliac veins* at about the level of the 4th lumbar vertebra. It passes upward to the right of and alongside the aorta, through the diaphragm, and enters the right atrium. The inferior vena cava receives blood from the following veins: *inferior phrenic* (from the diaphragm), *hepatic* (from the liver), *right suprarenal* (from the adrenal gland), *renal* (from the kidneys), *right testicular* or *ovarian* (from the right testis or right ovary), *lumbar* (from the body wall), and *common iliac* (from the pelvic region and the lower extremities). There are no valves in the inferior vena cava.

Inferior Phrenic Veins. These two small veins, *left* and *right,* drain the diaphragm. The left often has two branches, one connecting with the left renal or suprarenal, from which it may receive blood.

Hepatic Veins. These consist of two groups (upper and lower) that enter the vena cava just below the diaphragm. They receive blood from the *sublobular veins* of the liver. Blood received by the liver from the hepatic artery and the portal vein is returned by the hepatic veins to the inferior cava.

Suprarenal Veins. These drain the adrenal glands. The *right* empties into the inferior vena cava, the *left* into the left renal vein.

Renal Veins. These thick but short veins lie anterior to the renal arteries. They receive blood from the kidneys. The *left* also receives blood from the *left testicular* or *ovarian, left inferior phrenic,* and *left suprarenal veins.* It opens into the vena cava at a higher level than does the right renal vein.

Testicular and Ovarian Veins. In the male, testicular (spermatic) veins originate from branches in the testes and epididymides that unite to form a convoluted structure, the *plexus pampiniformis,* which constitutes the main bulk of the spermatic cord. Three or four vessels from this plexus pass through the inguinal canal and unite to form the *testicular vein,* which on the right side ascends and enters the inferior vena cava near the renal vein and on the left side enters the left renal vein. In the female, ovarian veins, similar in their terminations to the testicular veins, receive blood from the ovaries, uterine tubes, and uterus. During pregnancy they become much enlarged.

Lumbar Veins. Four pairs of lumbar veins lead from the abdominal wall to the inferior vena cava. They collect blood from the skin and muscles, the vertebrae, and the meninges of the spinal cord. They are connected by a longitudinal vein, the *ascending lumbar,* which passes upward and connects with the azygos and hemiazygos veins, on the right and left sides, respectively.

Common Iliac Veins. These are the two large veins that unite to form the inferior vena cava. Each is formed by a union of the *external iliac* and the *hypogastric (internal iliac) veins.* Each also receives the *iliolumbar* and *lateral sacral veins.*

The *external iliac vein,* which continues into the leg as the *femoral vein,* receives the following branches: the *inferior epigastric* (from the anterior body wall) and the *deep circumflex iliac* (from muscles and skin of the abdomen).

The *hypogastric (internal iliac) vein* receives the following branches:

1. Veins having their origin outside the pelvis, among them the *superior gluteal veins* (from the buttocks), *inferior gluteal veins* (from the upper part of the posterior portion of the thigh), *internal pudendal veins* (from the genitalia), and *obturator* (from the upper median region of the thigh).
2. Veins having their origin in venous plexuses of the pelvic viscera, among them, in the male, the *middle hemorrhoidal (rectal)* (from the urinary bladder, prostate gland, and seminal vesicles) and *vesical* (from the lower part of the bladder and the prostate gland), and in the female, the *uterine* and *vaginal veins.*
3. Veins that originate in front of the sacrum, including the *lateral sacral veins.*

Veins of the Lower Extremity. Both superficial and deep veins are found in the lower extremity.

The principal *superficial* veins of the leg are:

1. The *great saphenous vein,* lying on the anterior and medial surfaces of the leg. It originates at the dorsum of the foot and extends the entire length of the leg, terminating just inside the *hiatus saphenous* at the femoral vein. It is the longest single vein in the body. The great saphenous vein drains the foot and superficial portions of the leg.
2. The *small saphenous vein* lies on the posterior side of the lower leg. It originates posterior to the lateral malleolus and passes upward to the popliteal space (fossa) behind the knee, where it divides into two

branches, one opening into the popliteal vein, the other into the deep femoral vein. It drains the lateral and dorsal portions of the foot and the posterior portion of the lower leg.

The principal *deep* veins of the leg are:

1. The *posterior tibial vein,* which arises behind the medial malleolus by a junction of the *medial* and *lateral plantar veins* of the foot.

2. The *anterior tibial vein,* arising as a continuation of the *vena comitantes* and the *dorsalis pedis* of the foot.

3. The *popliteal vein,* arising at the lower border of the popliteus muscle by a union of the posterior and anterior tibial veins, passes upward through the popliteal fossa for a short distance; then, together with other femoral vessels, it passes through an aponeurotic opening in the adductor magnus muscle. At this point it becomes the femoral vein. It receives several small tributaries, including the *small saphenous vein.*

4. The *femoral vein* (a direct continuation of the popliteal vein), which continues through the proximal two-thirds of the thigh, lying close to the femoral artery. It receives the *great saphenous, deep femoral,* and *profunda femoris veins.* It passes under the inguinal ligament through the femoral ring, where it continues in the pelvic cavity as the *external iliac vein.*

PORTAL SYSTEM. The portal system (Fig. 10-16) includes the veins that receive blood from the stomach, small and large intestines, gallbladder, pancreas, and spleen. Blood from these organs passes through the *portal vein* to the liver. In the liver the portal vein divides into two trunks, which in turn terminate in small capillary-like vessels called *sinusoids,* where the blood comes into intimate contact with the liver cells. From the sinusoids, blood enters the *central veins* of the lobules of the liver, which in turn lead to *sublobular veins.* These unite into larger vessels and finally empty into *hepatic veins,* which conduct blood to the inferior vena cava.

In the portal system, blood passes through two *capillary nets:* (1) the capillaries of the stomach, intestines, spleen, pancreas, or gallbladder, and (2) the sinusoids of the liver.

The main trunk of the *portal vein* is formed by a union of its two principal tributaries, the superior mesenteric and splenic veins. The *superior mesenteric* drains the small intestine, cecum, and ascending and transverse colon; the *splenic* or *lineal* drains the spleen, pancreas, and part of the stomach. Just before entering the portal vein, the splenic vein receives a branch, the *inferior mesenteric vein,* which drains the descending and sigmoid colon and rectum. Other veins that contribute to the portal vein are the *coronary* or *gastric,* from the stomach and lower esophagus; the *pyloric,* from the pyloric portion of the stomach; the *cystic,* from the gallbladder; and the *paraumbilical,* from the body wall in the region of the umbilicus.

THE VERTEBRAL VENOUS SYSTEM. This system comprises all the

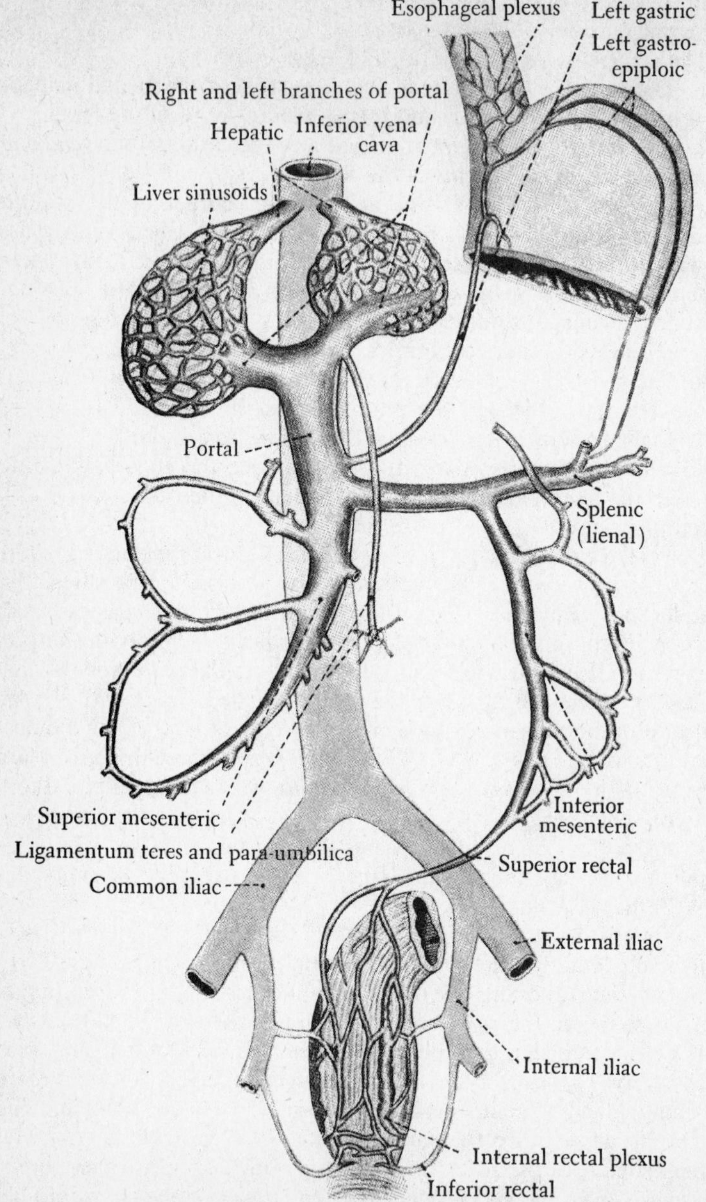

Fig. 10-16. Portal circulation. (Reprinted with permission of Macmillan, London and Basingstoke, from W. J. Hamilton, *Textbook of Human Anatomy,* 1976.)

veins of the head and neck, the trunk wall, the major small veins that return blood from the veins themselves (*venae vasorum*) of the extremities, and the veins of the vertebral column. Although it communicates with the caval veins, the system also bypasses them. The system is an enormous, valveless venous lake in which the pressure is low and varies with pressure changes in the thoracic and abdominal cavities so that the direction of flow in the system also varies. Plexuses of the system are formed at various parts of the body. Knowledge of this system is particularly important in the understanding of tumor cells. Formerly inexplicable metastases no longer remain mysterious. How would a carcinoma originating in the neck or the breast, for example, reach the bladder or the pelvic bones or the lower extremities? Directly through the vertebral venous system and/or through the adjacent plexus, as for example, through the especially capacious pelvic plexus.

SYSTEMIC CIRCUITS

In the systemic circulation, blood leaves the heart through the aorta, passes through arteries and arterioles to the capillaries of the tissues, then through venules and the veins to the venae cavae or the coronary sinus to reenter the heart. Within this general pattern, however, the blood may follow any of several courses in completing the circuit. The principal possibilities are as follows:

1. Circuit through the heart (coronary circuit).
2. Circuit through the upper extremity.
3. Circuit through the neck and head.
4. Circuit through the thorax.
5. Circuit through the digestive organs (including the liver).
6. Circuit through the kidneys (renal circuit).
7. Circuit through the pelvis and lower extremity.

1. *Circuit Through the Heart (Coronary Circuit)*

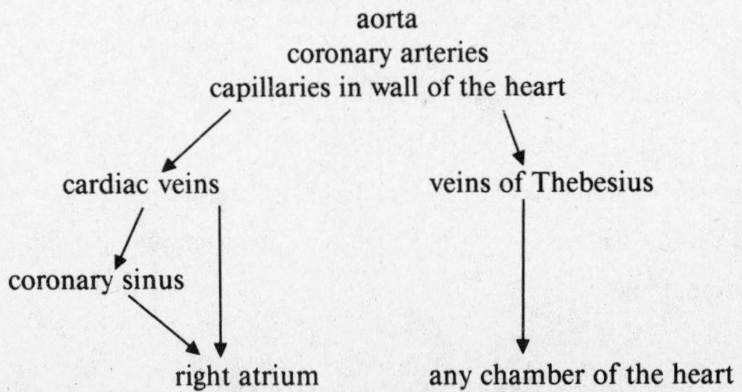

SYSTEMIC CIRCUITS (Cont.)

2. *Circuit Through the Upper Extremity*

ascending aorta
arch of aorta
brachiocephalic (innominate) artery
subclavian artery
axillary artery
brachial artery
smaller arteries and arterioles
capillaries of the tissues
venules and smaller veins
basilic or cephalic vein
axillary vein
subclavian vein
brachiocephalic (innominate) vein
superior vena cava
right atrium

3. *Circuit Through the Neck and Head*

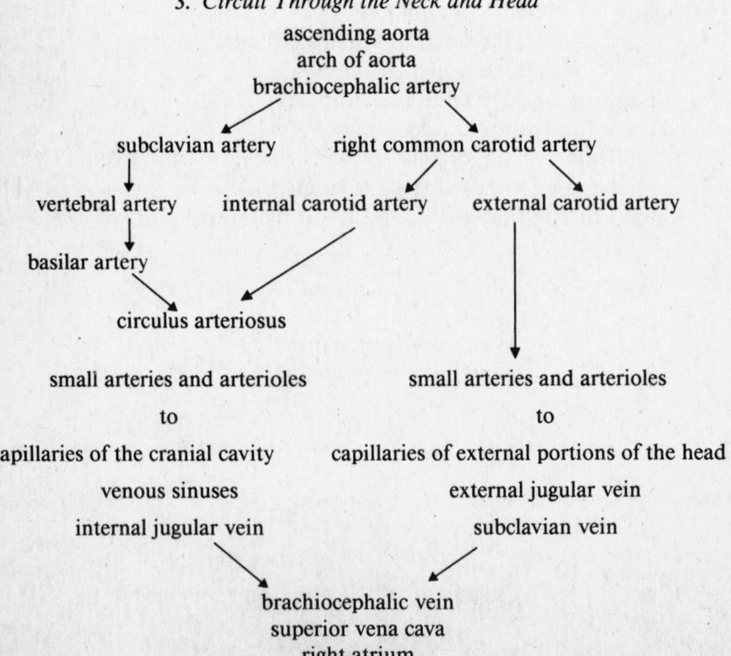

SYSTEMIC CIRCUITS (Cont.)

4. *Circuit Through the Thorax*

ascending aorta
arch of the aorta
thoracic aorta

bronchial artery
↓
small arteries and arterioles within lung
↓
capillaries of bronchi
↓
bronchial veins
azygos or brachiocephalic vein
superior vena cava
right atrium

intercostal artery
↓
small arteries and arterioles of thoracic wall
↓
capillaries of muscles, vertebrae, and ribs
↓
intercostal veins
azygos or sup. intercostal veins
superior vena cava
right atrium

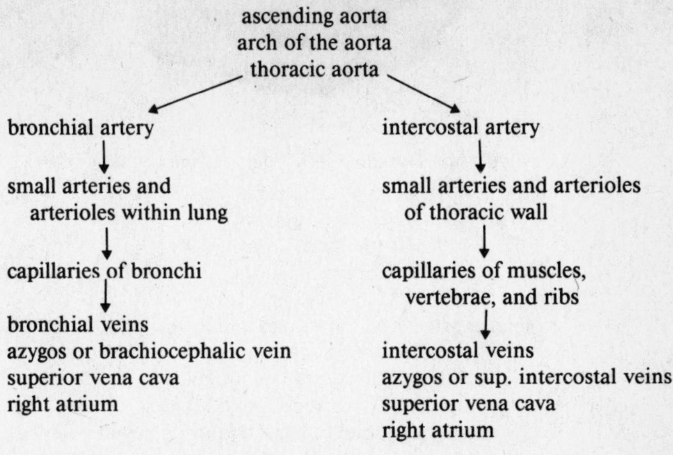

5. *Circuit Through the Digestive Organs (Including the Liver)*

ascending aorta
aortic arch
thoracic aorta
abdominal aorta

→ celiac artery
→ hepatic artery → liver → sinusoids of liver
→ splenic artery → through smaller arteries and arterioles to capillaries of → spleen, pancreas → through smaller venules and veins to → splenic vein → portal vein
→ left gastric artery → stomach → coronary vein → portal vein

→ superior mesenteric artery → upper intestine → superior mesenteric vein → portal vein

→ inferior mesenteric artery → lower intestine → inferior mesenteric vein → portal vein

portal vein
sinusoids of liver

SYSTEMIC CIRCUITS (Cont.)

5. *Circuit Through the Digestive Organs* (*Cont.*)

small veins within liver
hepatic veins
inferior vena cava
right atrium

6. *Circuit Through the Kidney* (*Renal Circuit*)

ascending aorta
aortic arch
thoracic aorta
abdominal aorta
renal artery
smaller arteries and arterioles within the kidney
afferent artery to glomerulus
efferent artery from glomerulus
capillary plexus about renal tubules
venules and smaller veins within kidney
renal vein
inferior vena cava
right atrium

7. *Circuit Through the Lower Extremity*

ascending aorta
aortic arch
thoracic aorta
abdominal aorta
common iliac artery
↙ ↘
external iliac artery → femoral artery → popliteal artery / posterior tibial artery → smaller arteries and arterioles → capillaries of the tissues → great and small saphenous veins / popliteal vein → femoral vein ↓

hypogastric (internal iliac) artery → smaller arteries and arterioles → capillaries of the tissues ↓

SYSTEMIC CIRCUITS (Cont.)

7. *Circuit Through the Lower Extremity* (*Cont.*)

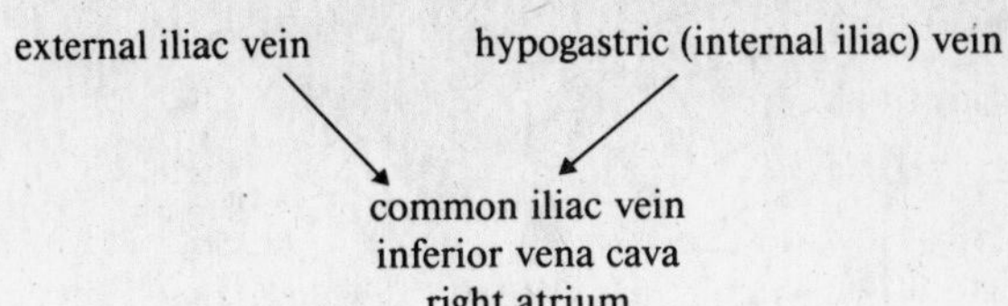

inferior vena cava
right atrium

These seven circuits describe the course normally followed by the blood. If, however, blood vessels become obstructed or are ligated, a *collateral circulation* may be established by mean of anastomoses or interconnections between the vessels.

FETAL CIRCULATION

The human fetus lives under conditions radically different from those of postnatal life. Developing in the fluid-filled amniotic sac within the uterus of the mother, it secures its oxygen and food through an organ called the *placenta.* In the placenta the blood of the mother is brought into close contact with the fetal blood vessels, so that an interchange of substances can occur between fetal blood and maternal blood. Oxygen and nutritive substances are received by diffusion into the fetal blood, and waste products are given up. The fetus is connected to the placenta by the *umbilical cord,* which contains two *umbilical arteries* and one *umbilical vein.* The placenta is continuous with the uterine endometrium.

In the fetus, the lungs and the digestive tract are nonfunctional; consequently, their blood supply is limited. The lungs are collapsed, and little blood goes to them because of an arrangement whereby most of the blood bypasses the lungs. This consists of (1) an opening, the *foramen ovale,* in the interatrial septum of the heart, by which the blood in the right atrium can pass directly to the left atrium, and (2) a vessel, the *ductus arteriosus,* which connects the pulmonary artery directly with the aorta.

The fetal heart begins to beat early in fetal life (about the fourth week). Blood circulates through the fetus in much the same manner as in the adult, with the exception of structures mentioned in the preceding paragraph and the placental circulation. Blood reaches the placenta through the *umbilical arteries,* which are branches of the *hypogastric* or *internal iliac arteries.* In the placenta, blood passes through the capillaries of the villi, then returns to the fetus by way of the *umbilical vein,* which enters the fetus at the umbilicus. Most of the blood passes directly to the inferior vena cava by way of the *ductus venosus;* the remainder passes indirectly by way of the liver through the sinusoids and the hepatic veins. (See Fig. 10-17.)

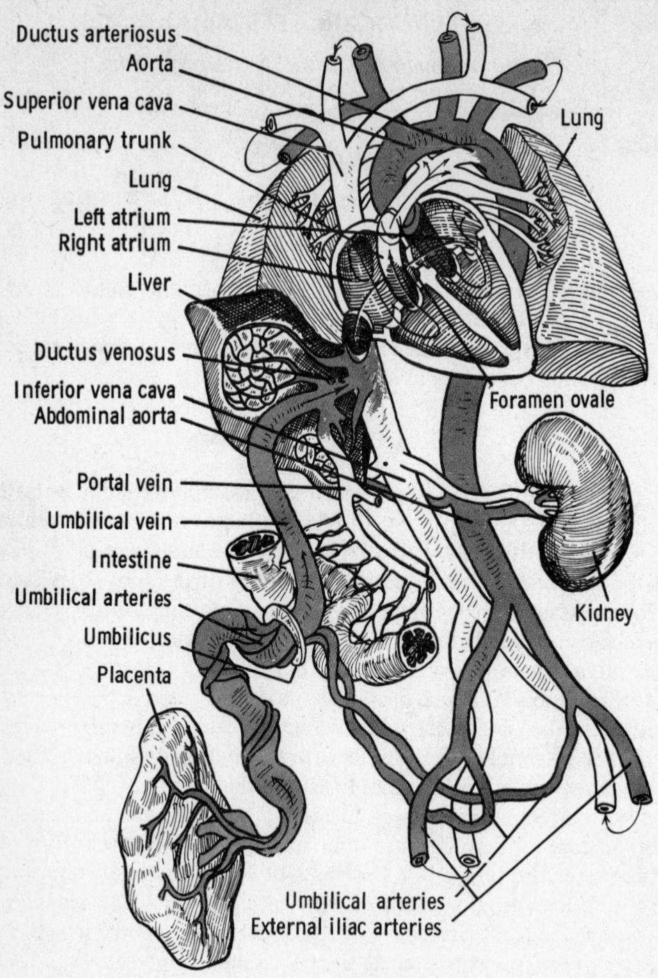

Fig. 10-17. Diagram of the fetal circulation shortly before birth. Course of the blood is indicated by arrows. (Reprinted by permission of J. B. Lippincott Co., from E. M. Greisheimer and M. P. Wiedeman, *Physiology and Anatomy,* 9th ed., 1972.)

Changes in Circulation at Birth. At birth, with the first inspiration, the lungs become expanded and begin functioning. Placental circulation ceases, the placenta is discharged from the uterus, and the connection of the newborn with the placenta is severed by the cutting of the umbilical cord. Correlated with these major changes, the following circulatory changes take place:

1. The ductus arteriosus closes, its closure being brought about by the

contraction of the smooth muscles in its wall and the gradual growth of tissue of the intima. Its complete occlusion occurs normally about 8 to 10 weeks after birth, after which it persists as a fibrous cord, the *ligamentum arteriosum.*

2. The foramen ovale closes, its position being indicated later by a depression, the *fossa ovalis,* located on the interatrial septum.

3. The umbilical vein atrophies and becomes a fibrous cord that persists as the round ligament of the liver, the *ligamentum teres.*

4. The ductus venosus atrophies and is transformed into the fibrous *ligamentum venosum,* which is superficially embedded in the wall of the liver.

5. The umbilical arteries atrophy and become the *lateral umbilical ligaments,* two fibrous cords from the bladder to the umbilicus.

These changes usually take place gradually, sometimes requiring several days (5 to 10) or even a few months for completion. Occasionally, the obliteration of an embryonic structure may not be completed for years, or it may not occur at all. For example, the foramen ovale may fail to close. This is not of physiologic significance, since the existence of a valvelike structure prevents backflow of blood into the right atrium. However, if the ductus arteriosus fails to close, a patent ductus arteriosus results, and a circulatory imbalance occurs. This causes unoxygenated blood to pass through the foramen ovale from the right to the left atrium. In such cases, the major portion of the blood fails to pass through the lungs, proper oxygenation does not occur, and the infant becomes cyanotic, the so-called *blue baby.* A similar result may occur from incomplete development of the interventricular septum.

THE LYMPHATIC SYSTEM

The lymphatic system is the part of the circulatory system that conveys lymph from its source, the body tissues, to the point where it reenters the bloodstream. The structures of the lymphatic system are of two kinds: *conducting vessels,* which are concerned with the conduction of lymph, and *lymphatic organs,* which are composed principally of lymphatic tissue. The former comprise lymph capillaries, lymph vessels, and lymph ducts; the latter, the lymphatic nodules, lymph nodes, spleen, tonsils, and thymus.

Circulation of Lymph. As already stated, the chief function of the lymphatic system is to return to the bloodstream the fluid filtered out of the blood capillaries. This is the *tissue fluid,* which carries to the tissues the materials that the cells utilize, and to which the cells give up waste products. When this fluid enters the lymph capillaries, it is referred to as *lymph.* The lymph passes through the lymph vessels and the main lymph ducts to reenter the bloodstream. In its course, it passes through one or more lymph nodes.

The lymphatic system has no pumping mechanism, as exists in the cardiovascular system. The flow of lymph is brought about by the following factors:

1. Difference in pressure between the lymph vessels originating in the tissues and points where the main lymph ducts empty into the subclavian veins. The pressure at the capillary end of lymph vessels is higher than that at the venous end. Valves at the openings of the main lymph ducts prevent blood from entering the lymphatic system.

2. Pressure resulting from the continuous formation of lymph, which is the result of *effective filtration pressure* (the difference between hydrostatic pressure and osmotic pressure in the capillaries). This forces tissue fluid into the lymphatic capillaries. Since lymph originates from tissue fluid, which comes from blood, pressure within the capillaries is partly responsible for lymph pressure. Accordingly, the heartbeat is the prime, though indirect, cause of the movement of lymph.

3. Intestinal movements (movements of the villi and peristaltic contractions). Each villus contains smooth muscle fibers, which, by contraction, force chyle from the lacteals into the lymphatic vessels. *Chyle* is the material filling the lymphatic vessels of the intestines; it consists of lymph and absorbed food products, especially emulsified fats.

4. Rhythmic, peristaltic contraction of smooth muscles fibers in segments of lymph vessels between valves.

5. Muscular activity. Contraction or passive massage of skeletal muscles tends to propel the lymph onward in the lymphatic vessels. Lymph vessels between muscles are compressed, and the lymph is forced onward. Backflow is prevented by the presence of valves in the larger vessels.

6. Respiratory movements. Expansion of the lungs upon inspiration exerts a pressure on lymph in the thoracic ducts, forcing lymph onward in these vessels. It also increases negative intrathoracic pressure, which produces an aspiratory effect that acts to suck lymph from the smaller vessels into the larger vessels. Descent of the diaphragm also compresses the abdominal contents, exerting a pressure on all abdominal vessels; this causes lymph to flow into the main thoracic duct from the abdominal lymphatics.

Lymphatic Vessels. The vessels conducting lymph include lymph capillaries, lymph vessels, and the main lymph ducts (Fig. 10-18).

LYMPH CAPILLARIES. The lymph capillaries lie in the connective tissue spaces of most organs and tissues. They are thin-walled vessels, usually slightly larger in diameter than the blood capillaries. Lymph capillaries originate as blind tubes, are extensively branched, and anastomose freely. Their diameter is not uniform, as is that of blood capillaries; they tend to be constricted at some points and dilated at others. Lymph capillaries are located among blood capillaries but are independent of them. They are extremely numerous in tissues underlying sur-

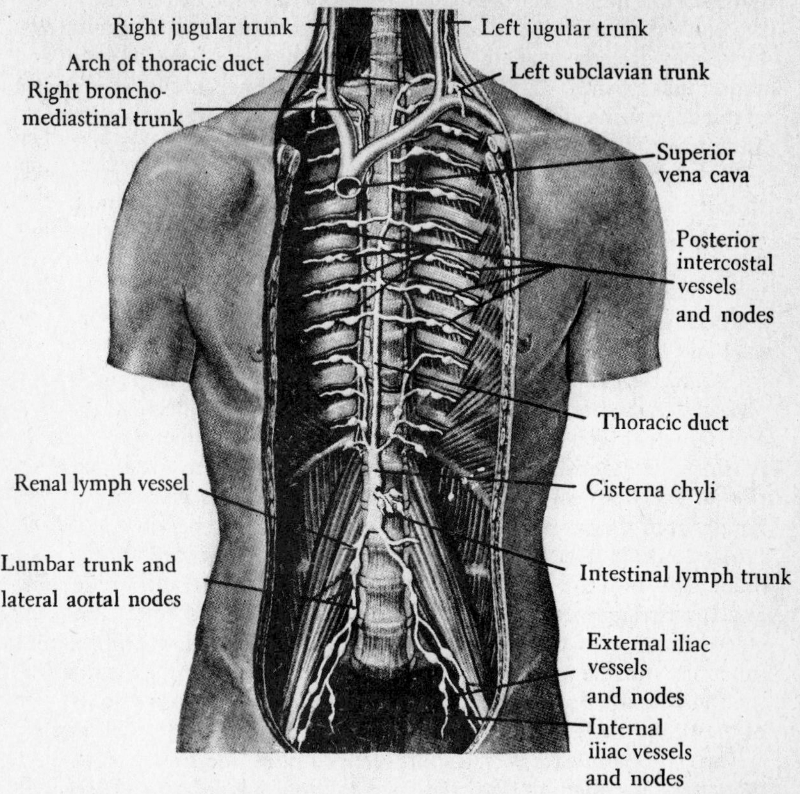

Fig. 10-18. The principal lymphatic vessels. (Reprinted with permission of Macmillan, London and Basingstoke, from W. J. Hamilton, *Textbook of Human Anatomy,* 1976.)

face membranes, such as the skin and the mucous membranes of the respiratory and digestive tracts, less numerous in organs such as muscles. They are entirely absent in deeply located tissues, such as those of the central nervous system and the bone marrow. The wall of a lymph capillary consists of a single layer of endothelial cells. Lymph capillaries in the villi of the intestines are called *lacteals.*

Lymph Vessels. Lymph vessels are larger vessels that collect lymph from capillaries. Their walls are thin and transparent; lymph can ordinarily be seen within them. They tend to lie in the subcutaneous or the connective tissue. The principal lymph vessels follow the course of arteries and veins. They form two sets of vessels, a superficial set and a deep set. The *superficial* lymph vessels lie near the surface and in general follow the course of the superficial veins; the *deep vessels* are more remote from the surface, being found deep within the extremities, under the mucous membranes of the digestive, respiratory, and urogenital systems, and directly beneath the serous membranes of the thoracic and abdominal cavities. The deep lymph vessels in general follow the course of the deep veins. They are more numerous than the veins, they anastomose freely, and their trunks tend to remain uniform in diameter.

Most lymph vessels have *valves.* These occur in pairs and are placed with their free edges toward the direction of the flow of the lymph. The vessels tend to become distended just above each valve, so that they present a "beaded" appearance. By contraction, smooth muscle fibers in the distended portions aid in the flow of lymph.

MAIN LYMPH DUCTS. The main lymph ducts include the *thoracic duct* and the *right lymphatic duct.*

The *thoracic duct* is the principal lymph duct, conveying lymph from all parts of the body except the right side of the head, neck, and thorax and the right upper extremity. It arises in the abdomen at a dilated portion (the *cisterna chyli*) which lies to the right of the aorta, just anterior to the body of the 2nd lumbar vertebra. This cisterna receives lymph from the vessels draining the lower limbs, the pelvic region, and the abdominal organs. The thoracic duct passes upward through the diaphragm into the thorax, continuing along the aorta and the esophagus. It passes posterior to the brachiocephalic vein and enters the neck, where it turns to the left and enters the *left subclavian vein* near its junction with the left internal jugular vein. Just before its termination, the thoracic duct receives the principal lymph vessels from the left side of the head, the left upper extremity, and the left side of the thorax.

The *right lymphatic duct,* a short duct, empties into the *right subclavian vein* at its junction with the right internal jugular vein. It receives lymph from the right side of the head, the right upper extremity, and the right side of the thorax.

LYMPHATIC ORGANS

Lymphatic organs are structures composed principally of lymphatic tissue. They include the lymph nodes, spleen, tonsils, and thymus gland.

Lymphatic Tissue. *Lymphoid* or *lymphatic tissue* is not one of the primary tissues of the body; it is rather a specialized form of connective tissue consisting of a supporting framework or *stroma* made up of a network of *reticular fibers* and *reticular cells.* The fibers, when treated by silver-impregnation methods, stain selectively and hence are called *argyrophil fibers.* The cells of the stroma include *primitive reticular cells* and *phagocytic reticular cells* (macrophages). The former comprise a syncytial network in which the reticular fibers are embedded; the latter have the ability to engulf foreign particles and store vital dyes, thus forming a part of the *reticuloendothelial system.*

In the meshes of the reticulum forming the stroma are large numbers of *free cells,* which consist largely of lymphocytes of varying sizes. Plasma cells are also present.

Lymphatic Nodules. These minute masses of lymphatic tissue form the structural unit of lymphatic tissue. Many are found in loose connective tissue, especially that lying beneath the mucous membranes that line the digestive, respiratory, and urogenital tracts. Each nodule is a mass of densely packed lymphocytes. In some nodules, a *germinal center,* a lightly staining area containing cells undergoing mitosis, may be present. The nodules produce lymphocytes, and plasma cells are present in some indicating the presence of B and T lymphocytes (see page 286) essential for the immune reactions of plasma cells. The nodules may exist as individual or solitary nodules, or they may be concentrated in groups that are (1) nonencapsulated, as in Peyer's patches of the ileum; (2) partially encapsulated, as in a tonsil; or (3) encapsulated, as in lymph nodes, the spleen, and the thymus gland.

Lymph Nodes (Fig. 10-19). Lymph nodes are small, ovoid or bean-shaped bodies found at intervals in the course of the lymphatic vessels. Each node consists of *lymphatic tissue* embedded in a *stroma* consisting of reticular fibers and reticular cells, and the whole is enclosed in a fibrous connective tissue *capsule.* On one side of the node is a depression, the *hilus,* where blood vessels enter and leave and efferent lymph

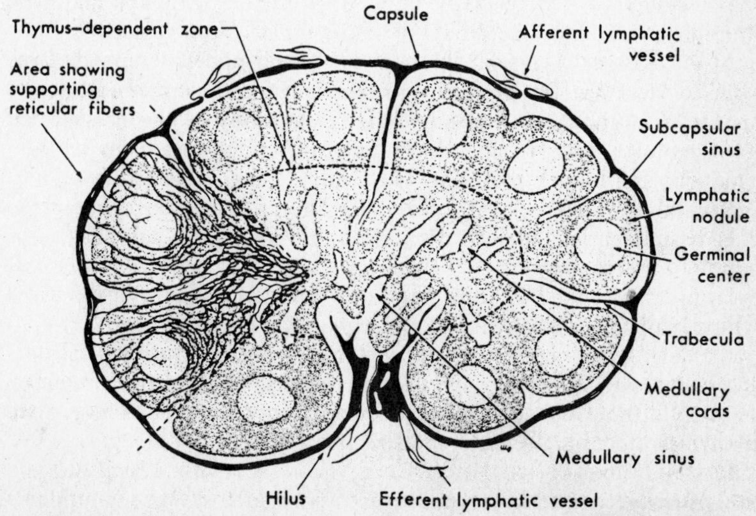

Fig. 10-19. Diagram of a lymph node. (Reprinted with permission of J. B. Lippincott Co. from A. W. Ham and D. H. Cormack, *Histology,* 8th ed., 1979.)

vessels make their exit. Extending inward from the capsule are septa or *trabeculae,* which divide the node into compartments. The outer, compact region of the node constitutes the *cortex;* the inner, more diffuse portion, the *medulla.* In the cortical region can be seen masses of densely packed cells consisting mostly of lymphocytes. These are the *cortical* or *primary lymph nodules.* In the center of each nodule is a lighter-staining area, the *germinal center,* where lymphocytes originate. The tissue of the *medulla* is similar to the tissues in the cortex, except that the lymphocytes are not arranged in the form of nodules but form *medullary cords.* These cords anastomose, forming a network enclosing spaces, the *medullary sinuses.*

In each node is a region designated the *thymus-dependent zone,* which is seeded with T lymphocytes from the thymus. This zone includes the inner portion of the cortex and the outer portion of the medullary cords. Its development and maintenance depend upon the thymus. This region produces lymphocytes that serve as *killer cells,* cells that destroy foreign antigenic cells. In other portions of the node, lymphocytes develop from B cells, which originate in the bone marrow. These lymphocytes are capable of developing into plasma cells, which are the source of humoral antibodies.

Vessels entering a lymph node are called *afferent vessels.* These penetrate the capsule and empty into the *marginal sinus,* a space between the capsule and the cortical nodules. From this sinus the lymph flows centrally past the nodules into the *sinuses* of the *medulla* and from here to the *efferent vessels,* which leave the node at the hilus. Lymph may pass through several nodes before it enters the main lymph vessels.

In the passage of lymph through the lymph nodes, lymph receives lymphocytes that are primarily responsible for the immunologic responses (cell-mediated and humoral) of the body to the presence of antigens (e.g., foreign cells, cancer cells, infectious organisms or their products, and foreign proteins). The lymph is also subjected to a filtering action in which such particulate substances as bacteria, carbon particles, tissue debris, and vital dyes are removed through the phagocytic action of macrophages.

Lymph nodes tend to occur in groups in various regions of the body, where lymphatic vessels from the peripheral portions of the body converge. The lymph enters the nodes through afferent vessels, traverses the nodes, and leaves through efferent vessels, which carry the lymph onward to other nodes or to larger lymphatic vessels. For this reason, the lymph nodes are referred to as *regional nodes.*

LYMPH NODES OF THE HEAD, FACE, AND NECK. In the *head and face* are several groups of lymph nodes, each consisting of varying numbers of nodes. Their names indicate the general position of the groups: occipital, posterior auricular, anterior auricular, parotid, facial, deep facial, lingual, and retropharyngeal. In the *neck* are the submaxillary, submental, superficial cervical, anterior cervical, and deep cervical groups. The *deep cervical group* contains large nodes that form a chain along the carotid artery beside the pharynx, trachea, and esophagus. Efferent lymph vessels from this group unite to form the *jugular trunks,* which empty into the main lymph ducts.

LYMPH NODES OF THE UPPER EXTREMITY. These are concentrated mainly in the *axillary* group, which is located in the region of the armpit. This group includes 20 to 30 large nodes through which passes lymph from the afferent vessels of the arm, the thoracic wall, and the mammary gland. Efferent vessels lead to the *subclavian trunks,* which empty into the main lymph ducts.

LYMPH NODES OF THE LOWER EXTREMITY. These are concentrated mainly in the *inguinal group.* This group includes 12 to 20 large nodes that receive afferent vessels from the lower limbs, external genitalia, buttocks, and lower portion of the abdominal wall. Efferent vessels lead to the external iliac nodes.

LYMPH NODES OF THE ABDOMEN AND PELVIS. There are two groups of lymph nodes in the abdominal and pelvic regions, the parietal and the visceral.

Parietal Nodes. The *parietal groups,* which lie beneath the peritoneum close to the aorta, include:

1. *External iliac nodes,* numbering 8 to 10, lying along the external iliac artery. They receive vessels from the inguinal nodes, the lower portion of the abdominal wall, the inner portion of the thigh, and the reproductive and excretory organs.
2. *Hypogastric* (*internal*) *iliac nodes,* lying along the hypogastric vessels. They receive vessels from the pelvic organs, the perineum, the external genitalia, the buttocks, and the posterior portion of the thigh.
3. *Common iliac nodes,* 4 to 6 nodes that lie along the common iliac artery. They receive vessels from the hypogastric and external iliac nodes.
4. *Epigastric nodes,* lying along the inferior epigastric vessels.
5. *Iliac circumflex nodes,* a small group lying along the deep iliac circumflex vessels.
6. *Lumbar nodes,* a large group of nodes chainlike along the lower portions of the aorta and the inferior vena cava. They include three paired groups: the *lateral aortic, preaortic,* and *retroaortic nodes.* These receive afferent vessels from the abdominal and pelvic viscera, from the lower lymph nodes, and from the abdominal wall. Their efferent vessels empty into the *intestinal* and *lumber trunks,* which empty into the *cisterna chyli.*

Visceral Nodes. The *visceral groups* lie along the arteries supplying the visceral organs. They include:

1. *Inferior gastric nodes,* which lie along the greater curvature of the stomach in the pyloric region.

2. *Hepatic nodes,* which lie along the hepatic and gastric arteries. They drain the liver, gallbladder, and portions of the stomach, duodenum, and pancreas. The hepatic nodes empty into the preaortic nodes.

3. *Pancreaticolienal nodes,* a group lying along the splenic artery on the superior surface of the pancreas. They drain the stomach, spleen, and pancreas and empty into the preaortic nodes.

4. *Superior mesenteric nodes,* from 100 to 150 in number, consisting of three groups: *mesenteric, ileocolic,* and *mesocolic nodes.* The afferents drain the jejunum, ileum, cecum, vermiform appendix, and the ascending and transverse colon. The efferents lead to the preaortic nodes.

5. *Inferior mesenteric nodes,* groups of nodes which lie along the branches of the inferior mesenteric artery. Their afferents drain the lower portions of the alimentary canal; their efferents lead to the *preaortic nodes.*

The afferent lymph vessels of the mesenteric nodes have their origin in the alimentary canal and play an important role in the absorption of food. Large quantities of fats are carried by the lymph; consequently, the contents of the lymph vessels, called *chyle,* have a milklike appearance, and the course of these vessels can be followed readily, especially after a meal that is rich in fats. The mesenteric nodes constitute the largest group of lymph nodes in the body. Their efferents lead to the preaortic nodes and through them to the *intestinal trunks,* which empty into the *cisterna chyli* at the lower end of the thoracic duct.

LYMPH NODES OF THE THORAX. The lymph nodes of the thorax, like those of the abdomen and pelvis, consist of parietal and visceral groups.

Parietal Nodes. The *parietal groups,* lying along the inner surface of the thoracic wall, include:

1. *Sternal nodes,* lying in the intercostal spaces near the sternum. They receive afferent vessels from the mammary glands and the anterior abdominal wall. Efferent vessels empty into the *bronchomediastinal trunk,* which may open either directly into the bloodstream at the junction of the internal jugular and subclavian veins or into the main lymphatic ducts.

2. *Intercostal nodes,* lying in the intercostal spaces just lateral to the vertebral column. They receive afferent vessels from the posterior and lateral portions of the thorax. The efferents from the lower nodes lead into a vessel that empties into the *cisterna chyli;* those from the upper nodes empty into the *main lymphatic ducts.*

3. *Diaphragmatic nodes,* lying on the superior surface of the diaphragm. They receive afferents from the diaphragm and the upper abdominal organs. Efferents pass to the *sternal* or *mediastinal nodes.*

Visceral Nodes. The *visceral groups,* lying close to the thoracic viscera, include:

1. *Mediastinal nodes,* consisting of two groups, the anterior and the posterior. They drain the thymus, pericardium, esophagus, diaphragm,

and the upper surface of the liver. Some efferent vessels lead to the *bronchiomediastinal trunks,* others to the *thoracic duct* and the *tracheobronchial nodes.*

2. *Tracheobronchial nodes,* consisting of four groups: tracheal, bronchial, bronchiopulmonary, and pulmonary. They lie along the trachea and the bronchi and receive afferents from these structures as well as from the lungs and the heart. Their efferent vessels may lead to the *bronchiomediastinal trunks,* which join the main lymphatic vessels, or they may open directly into the junction of the internal jugular and subclavian veins.

The Spleen (Fig. 10-20). The spleen is the largest lymphatic organ of the body. It is an elongated, dark red, ovoid body located beneath the diaphragm on the left side of the body, inferior and posterior to the

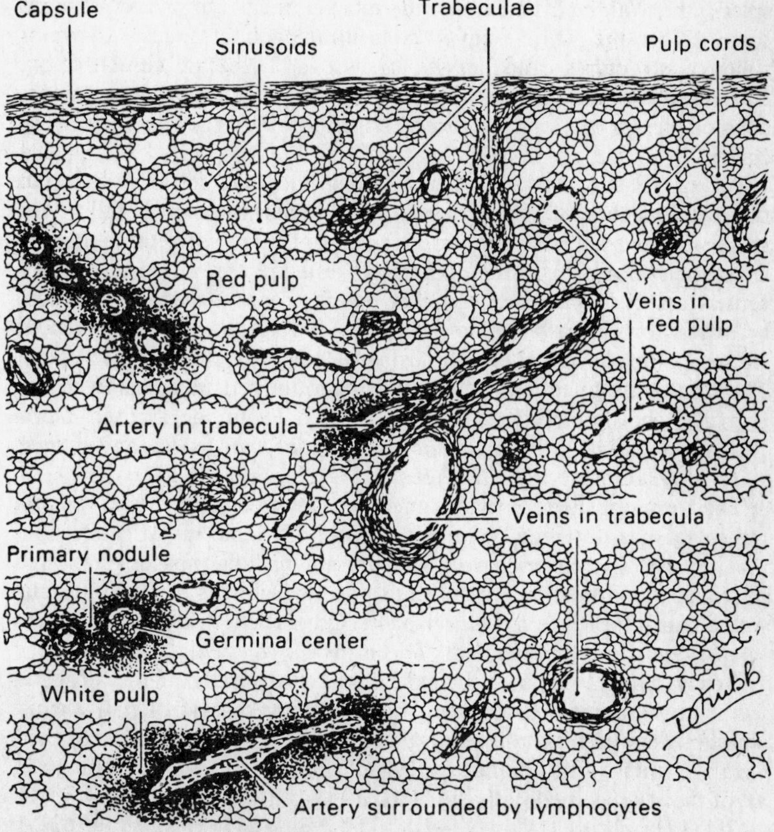

Fig. 10-20. Diagram of the spleen. (Reprinted with permission of J. B. Lippincott Co. from A. W. Ham and D. H. Cormack, *Histology,* 8th ed., 1979.)

stomach. It averages about 12 cm in length, 8 cm in width, 4 cm in thickness, and about 200 g in weight. Both the size and the weight are extremely variable.

Surfaces and Attachment of the Spleen. The spleen exhibits four surfaces: basal, phrenic, renal, and gastric. The gastric surface, which adjoins the stomach, contains the *hilus,* through which blood vessels enter and leave. The spleen is invested by a layer of peritoneum that is continuous with the abdominal peritoneum at the hilus, its only point of attachment. The spleen moves freely with movements of the surrounding viscera.

Splenic Pulp. The spaces between the trabeculae and connective tissue septa are filled with splenic pulp, a soft spongelike substance consisting of lymphatic tissue. Strands of this tissue form the *pulp cords.* Pulp infiltrated with red blood cells constitutes the *red pulp* of the spleen; the *white pulp* consists of the more compact lymphatic tissue. At some points the white pulp is concentrated about arteries, forming spherical structures called *splenic corpuscles.* These are similar to the nodules of lymph nodes, except that they contain within their substance a blood vessel, the *central artery.* In adults they may lack germinal centers.

Blood Vessels in the Spleen. Blood enters the spleen at the hilus through branches of the *splenic artery,* branches of which follow the trabeculae as *trabecular arteries.* Branches of these enter the pulp, and along their course are found nodules of white pulp, in which is located a central artery. Finally, the arteries divide into a number of straight branches, forming a brushlike *penicillus.* Blood then enters the *venous sinuses,* which form an anastomosing plexus throughout the red pulp. The sinuses are lined with cells belonging to the reticuloendothelial system that show marked phagocytic activity. From the sinuses, blood passes successively through *pulp veins, trabecular veins,* and *splenic veins,* which emerge from the hilus.

The nature of the terminal connections between arteries and veins in the spleen has not been definitely established. Some investigators hold that the terminal arteries open into the intercellular spaces of the red pulp and blood then enters the sinusoids through slits in the sinusoidal walls. This is the *open circulation theory.* Others believe that the arterioles open directly into the sinusoids. This is the *closed circulation theory.*

Nerve Supply of the Spleen. The nerve supply of the spleen is from the celiac plexus, through the secondary splenic (lienal) plexus, which is derived from the celiac ganglion and the right vagus nerve.

Functions of the Spleen. The spleen performs several functions, all of them concerned with the blood.

Blood Formation (*Hemopoiesis*). In fetal life and for a short time after birth, the spleen is a hemopoietic organ producing all types of blood cells. In the adult, however, only lymphocytes and plasma cells are produced. In certain pathologic conditions in which the bone marrow fails to function, the spleen may resume its hemopoietic function and produce red blood cells, a process called *extramedullary hemopoiesis.*

Blood Storage. The presence of an elastic tissue framework and smooth muscle cells enables the spleen to undergo marked and rapid changes in volume. Splenic contractions that occur under nervous or hormonal stimulation expel red blood cells and thus increase the number in circulation. The size of the spleen is much reduced during exercise and following hemorrhage or oxygen lack.

Filtration. Foreign substances such as bacteria and worn-out red blood cells, both whole and fragmented, are destroyed and removed from the circulation by the action of the phagocytic cells of the spleen.

Iron Storage. Iron is recovered from broken-down hemoglobin and temporarily stored to be used in the formation of new hemoglobin.

Bilirubin Production. Bilirubin, one of the bile pigments, is formed from hemoglobin freed from red blood cells broken down by the reticuloendothelial cells of the spleen.

Immunologic Action. The spleen serves as a site where lymphocytes are produced and where antigens in the blood can activate suitably programmed lymphocytes to develop into immunologically functioning cells capable of producing antibodies or becoming involved in cell-mediated immunologic reactions. The spleen plays an important role in the resistance of the body to bacterial and parasitic infections.

The spleen is not essential to life, for its functions, upon its removal, are taken over by other organs of the reticuloendothelial system, especially the bone marrow. One effect of removal is greater susceptibility to infectious diseases. Enlargement of the spleen occurs in a number of pathologic conditions, such as malaria, certain forms of anemia, and leukemia.

The Tonsils. The tonsils are masses of lymphatic tissue located in the walls of the pharynx. There are three pairs: *palatine tonsils,* in the oropharynx between the glossapalatine and pharyngopalatine arches; *pharyngeal tonsils,* in the median dorsal wall of the nasopharynx between the openings of the auditory tubes; and *lingual tonsils,* on the dorsum and sides of the posterior portion of the tongue.

STRUCTURE OF THE TONSILS. Each tonsil is a dense mass of lymphatic tissue lying in the connective tissue under the mucosa. Its free surface is covered by stratified squamous epithelium that is continuous with that of the pharynx. At many points on the surface there are deep indentations, or *crypts,* lined with stratified squamous epithelium. Bordering each crypt is a layer of lymphatic tissue in which are numerous *nodules* similar in structure to those of the lymph nodes.

Surrounding the tonsil on the basal side is a *capsule* of connective tissue from which fibers extend as *septa* separating the crypts, with their lymphatic tissue, from each other. The epithelium of the crypts is frequently infiltrated by masses of lymphocytes. Mucous glands of the pharynx are numerous about tonsils, and occasionally their ducts open into the crypts. Tonsils lack afferent lymphatic vessels but have lymph capillaries that constitute the beginnings of efferent lymph vessels.

FUNCTION OF THE TONSILS. The only known function of the tonsils is the formation of lymphocytes. This occurs in the germinal centers of the nodules. It has been thought that the tonsillar tissue, being strategically placed, serves to protect the body against invading microorganisms, but this has not been adequately demonstrated. Indeed, the reverse seems to be the case. The tonsillar crypts often provide lodging places for invading organisms, and their imperfect epithelium constitutes a portal of entry inviting local or general infections. This has given rise to the "focal infection" theory that many believe has been overstressed, resulting in unnecessary tonsillectomies. Hypertrophy of the tonsils, however, is common in childhood and may necessitate their removal. Hypertrophy of the pharyngeal tonsils is commonly referred to as "adenoids." Adenoid tissue may obstruct the nasal passageways, giving rise to mouth breathing, or the auditory tubes, causing ear disorders.

The Thymus (Fig. 10-21). The thymus or thymus gland is a lymphatic organ and an endocrine gland located in the mediastinal cavity anterior to and above the heart. Following birth, it grows rapidly until the second year; thereafter, growth is slow until the age of puberty, when it begins to undergo involution. In this process it decreases in size, the thymic tissue being replaced by adipose and connective tissue. It weighs 10 to 15 g at birth and 30 to 40 g at puberty. In old age the thymus enlarges.

STRUCTURE OF THE THYMUS. The thymus consists of two flattened symmetrical lobes closely applied to each other, each of which is enclosed in a capsule. From the connective tissue of the capsule, trabeculae divide each lobe into many lobules, each consisting of a cortex and a medulla.

The thymus is composed principally of lymphocytes and reticular cells. The latter originate from pharyngeal pouches in the embryo; hence, they are endodermal in origin. They are designated *epithelial reticular cells.* The *cortex* consists of densely packed, small lymphocytes called *thymocytes.* They arise by the division of stem cells that migrated to the thymus from the bone marrow. Reticular cells are few in number. In the *medulla,* the thymocytes are less numerous; hence, it stains lightly, and the reticular cells are more conspicuous. The medulla has, in addition, ovoid bodies called *thymic corpuscles (Hassall's corpuscles),* each of which consists of concentrically arranged cells enclosing a clear, hyalinized area of degenerated cells. Their significance is unknown. These bodies are characteristic of the thymus. The thymus lacks germinal centers. There are no lymph sinuses or afferent lymph vessels.

The blood supply to thymic tissue is unusual in that small arterioles

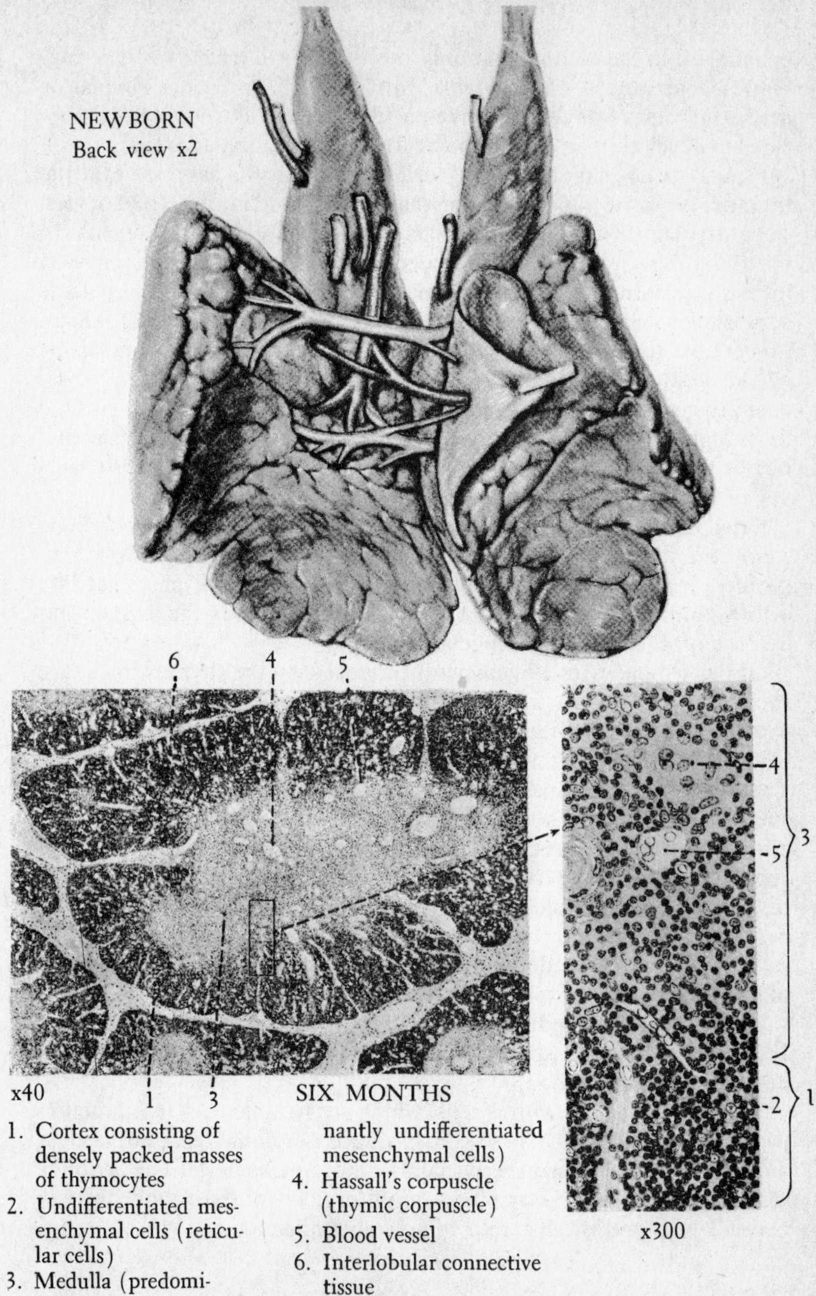

1. Cortex consisting of densely packed masses of thymocytes
2. Undifferentiated mesenchymal cells (reticular cells)
3. Medulla (predominantly undifferentiated mesenchymal cells)
4. Hassall's corpuscle (thymic corpuscle)
5. Blood vessel
6. Interlobular connective tissue

Fig. 10-21. Thymus. (From *Atlas of Human Anatomy,* Barnes & Noble, Inc., 1961.)

penetrate into the medulla but only capillaries into the cortex. The capillaries constitute a *blood-thymic barrier,* which prevents circulating macromolecules from entering the cortex. This enables cortical lymphocytes to develop in an environment free of circulating antigens.

FUNCTION OF THE THYMUS. Until recently, it was assumed that the thymus served no function other than the production of lymphocytes, since its removal in mature animals produced no untoward results. In the 1960s, however, it was discovered that if the thymus were removed from a newborn animal, this animal would not develop immunologic competence and would soon die. It would fail to reject grafts of foreign tissues, and its ability to develop antibodies against invading organisms would be impaired. As a consequence of these discoveries and subsequent investigations, the thymus has come to be regarded as one of the most important organs of the body, serving as the basic organ responsible for the development of the body's defense mechanisms. A brief survey of the way in which the thymus functions follows.

Early in embryonic development, the thymus receives hemopoietic stem cells from the yolk sac; these develop into lymphoblasts, which proliferate and differentiate into small lymphocytes. In postnatal life, cells from the bone marrow also establish themselves in the thymus and become progenitors of lymphocytes.

Of the lymphocytes (*thymocytes*) produced by the thymus, most are short-lived and die. A limited number, however (about 5 percent), in some manner acquire the ability to differentiate into mature cells capable of performing the functions associated with *cell-mediated immunity.* These long-lived cells pass by way of the bloodstream to lymphatic tissue located in lymph nodes, Peyer's patches, the spleen, and the bone marrow. In these organs, in thymus-dependent areas, they become precursor stem cells that give rise to T cells (thymus-derived cells). These T cells circulate in the blood and lymph, where they comprise about 80 percent of the lymphocytes. They are antibody-producing cells that are responsible for cell-mediated immunity, bringing about the destruction of cells of foreign transplants and of cancer cells.

The thymus also produces a number of important hormones (see Vol. 2), each of which has specific effects upon cellular and immunological development.

T cells also interact with B cells, which are lymphocytes that function in *humoral immunity.* These cells produce antibodies that act against most infectious organisms, especially pathogenic bacteria. The action is sometimes synergistic, sometimes inhibitory. The interrelationships between T cells and B cells are incompletely understood.

11: THE PHYSIOLOGY OF CIRCULATION

Transportation of the principal body fluids from one part of the body to another is the main function of the circulatory system. These fluids are the media through which gases, hormones, and a vast number of chemical substances (including nutrient material and metabolites) are carried through the body.

BLOOD PRESSURE

Blood, which is contained within a closed series of tubes, flows (*circulates*) because of inequalities of pressure in different parts of the circulatory system. The pressure that blood exerts within the blood vessels is called the *blood pressure.* This pressure depends mainly upon (1) the work of the heart, (2) peripheral resistance, (3) elasticity of the arterial walls, and (4) blood volume and viscosity.

Blood pressure is highest in the elastic arteries that are nearest to the heart; it is lower in the muscular arteries that are nearer to the periphery. As the blood passes successively through the smaller arteries, arterioles, capillaries, venules, and larger veins, its pressure becomes lower and lower until, in the vena cava, it is zero or slightly less (a "negative" pressure).

THE WORK OF THE HEART

The heart is a muscular pump that by rhythmic contractions, keeps the blood circulating throughout the pulmonary and systemic circulatory vessels.

Properties of Cardiac Muscle. Cardiac muscle (Fig. 6-4), like skeletal muscle, is striated, but the striations are less pronounced. It was formerly thought that cardiac fibers constituted a syncytium, but it is now known that the fibers are composed of individual cells joined together end to end. Some cells are bifurcated, giving the fibers a branched appearance. The fibers are bound together by tight junctions; this allows the action potential to spread from one fiber to another and throughout the heart so that the fibers of each chamber contract as a coordinated unit. The muscle fibers of the ventricles include four spiral bands so arranged that, during contraction, the transverse diameter and the base-apex diameter of the ventricles are both shortened. This results in the forceful expulsion of blood into the systemic and pulmonary circuits.

Cardiac muscle responds to the usual types of stimuli: thermal, electrical, chemical, and mechanical. It possesses properties characteristic of all muscle tissue: *irritability, conductivity, extensibility,* and *elasticity.* In addition, it possesses *automaticity* and *rhythmicity.*

AUTOMATICITY. This inherent property of cardiac muscle is not only characteristic of the heart as a whole; it can also be exhibited in a small portion cut from the organ. The heart of some animals, such as a frog or turtle, can be removed from the body and, if kept moist by physiologic salt solution, will continue to contract spontaneously (and rhythmically) for a considerable time. In the embryo the heart begins to contract even before nerve connections are established. Heart cells grown in tissue culture exhibit automatic rhythmic contractions.

RHYTHMICITY. Heartbeats occur regularly and rhythmically. If a beating heart is stimulated artificially by a series of electric shocks, the rhythm may be altered (slowed down, speeded up, or made irregular), but the muscle will not go into a state of sustained contraction, as does skeletal muscle. Neither summation nor tetanus occur in cardiac muscle.

Systole and Diastole. The ability of the heart to function as a pump depends on alternate contraction and relaxation. The contraction phase is known as *systole,* the relaxation phase as *diastole.* When the terms *diastole* and *systole* are used alone, they apply to the ventricle. They may, however, be specified, as *atrial systole, ventricular systole, atrial diastole,* and *ventricular diastole.*

Electrolyte Balance. The rhythmic contraction and relaxation of the heart are dependent upon maintenance of a proper balance of calcium, sodium, and potassium ions in the blood and tissue fluids. The effects of electrolyte imbalance are as follows:

1. *Calcium excess.* The heart stops in systole (a condition known as *calcium rigor*).
2. *Sodium excess.* The heart becomes progressively weaker and finally stops in diastole.
3. *Potassium excess.* The heart becomes weaker and finally stops in diastole (*potassium inhibition*).

REFRACTORY PERIOD. Cardiac muscle, if stimulated while contracting, will not respond to the stimulus. This period during which cardiac muscle cannot be stimulated is called the *absolute refractory period.* During relaxation, stimulation produces a response, but the reaction is below normal; consequently, this is called the *relative refractory period.* Stimulation of heart muscle can produce its full effects only when the period of relaxation is complete. Contraction results in loss of excitability (irritability), which can be restored only during the period of relaxation.

The refractory period for cardiac muscle is about 66 times that of skeletal muscle. In a heart that is beating at a normal rate (70 times/min) the refractory period is 0.3 sec.

EXTRASYSTOLE (PREMATURE BEAT). If, when the heart is beating normally, a stimulus is applied during or at the end of a diastole, a response (contraction) will occur before the next regular contraction is due. This is called *extrasystole* or *premature beat.* Such a premature beat is followed by a much longer diastole than is usual, an interval known as *compensatory pause.* Because the systole is usually weak and unnoticed, it seems as though the heart has "missed a beat." Extrasystoles may occur in the normal heart as the result of some abnormal stimulus and are observed in certain pathologic conditions.

THE ALL-OR-NONE LAW. This principle holds true for a single striated muscle fiber or a single nerve fiber, but not for muscles consisting of numerous individual fibers. In the heart, it applies to the organ *as a whole,* for cardiac muscle consists of fibers that are all interconnected. Accordingly, if the heart contracts at all, it contracts to its fullest extent. But this does not necessarily mean that all its contractions are of equal strength. Several factors, among them fatigue, effects of drugs, and degree of stretch, may alter the extent to which the heart muscle is capable of responding. It does mean that under a given set of conditions, an increase in the strength of a stimulus will fail to increase the force of contraction.

Impulse-Conducting System. The impulse-conducting system of the heart consists of atypical muscle fibers and *Purkinje fibers.* It includes the following:

SINOATRIAL NODE. The SA node is located in the wall of the right atrium near the entrance of the superior vena cava. From this node, impulses pass throughout the walls of the atria.

ATRIOVENTRICULAR NODE. The AV node is located in the lower part of the interatrial septum. From this node arises the *atrioventricular bundle (bundle of His),* which passes downward in the interventricular septum. After descending a short distance, this bundle divides into two trunks that branch repeatedly, sending fibers (*Purkinje fibers*) throughout the myocardium of the two ventricles.

The impulse that initiates the heartbeat arises spontaneously in the *sinoatrial node,* which as a consequence, is called the *pacemaker* of the heart. Heating of this node speeds up the heartbeat; cooling slows it down. From the SA node an *excitation wave* spreads throughout the musculature of the atria and to the atrioventricular node. The atria contract; at the same instant, the impulse is conducted through the AV node and the *bundle of His* and its branches to the ventricles, and slightly later the ventricles contract. The delay in passage of the impulse through the AV node gives the atria time to complete their systole before the ventricular systole begins. As a result, the two atria contract slightly ahead of the ventricles.

An interruption in the passage of the excitation wave through the conductile tissue of the AV node is called *heart block.* As a consequence, the chambers of the heart fail to beat in a coordinated manner. The atria beat normally, while the ventricles fail to beat in their normal sequence. When the interruption is in one of the branches leading from the AV node, the condition is called *partial block* or *intraventricular block;* this is seen in infections (e.g., rheumatic fever) and other pathologic conditions.

The Cardiac Cycle. The cardiac cycle includes the complete sequence of events occurring during a single heartbeat. The cycle in the human being, when the heart is beating at the normal rate of 70 times/min, has a duration of 0.86 sec. It is shorter when the heartbeat rate is faster, longer when it is slower.

A number of methods have been employed to establish the events of the cardiac cycle:

1. Gross observations of changes occurring in the heart.
2. Successive photographs or motion pictures of an exposed heart.
3. Graphic records obtained by (a) measuring and recording pressure changes within the cardiac chambers, (b) recording volume changes by means of a *cardiometer,* (c) recording the movements of threads that have been attached to the walls of the atria and the ventricles, (d) recording electrical variations by means of an *electrocardiograph,* (e) recording sound vibrations transmitted from the heart to the chest wall with a *phonocardiograph,* and (f) by use of ultrasound in *echocardiography.*

The time consumed in each of the phases of a single heartbeat when the rate is approximately 70 beats/min has been found to be as follows:

Atrial systole	0.1 sec	0.862 sec
Atrial diastole	0.762 sec	
Ventricular systole	0.379 sec	0.862 sec
Ventricular diastole	0.483 sec	

It is seen, then, that both the atrial and the ventricular diastolic times are longer than the systolic. This allows a short interval for rest, *the only rest period available to the heart,* which never stops beating (except for these very short intervals) from the time contractions begin in the embryo (at about the fourth week of development) to the time of death. In the human being who has passed his or her seventieth year, the heart has beaten *more than 2.5 billion times.* The basic factors involved in death are, of course, the cessation of the heartbeat and the consequent stoppage of circulation.

The Action of Cardiac Valves. The cardiac valves and their locations are as follows:

1. *Right atrioventricular* or *tricuspid valve,* between the right atrium and right ventricle.
2. *Left atrioventricular, bicuspid,* or *mitral valve,* between the left atrium and ventricle.
3. *Pulmonary semilunar valves,* at the orifice of the pulmonary artery from the right ventricle.

4. *Aortic semilunar valves,* at the orifice of the aorta from the left ventricle.

The *events of the cardiac cycle* (and the action of the valves that accompany them) are as follows:

While blood is entering the two atria, both atrioventricular valves are closed. The pressure in the atria rises until it exceeds the pressure in the relaxed ventricles, whereupon the atrioventricular valves open and blood flows to the ventricles; the valve leaflets float freely in the blood. Then atrial systole occurs, forcing *all* blood out of the atria into the ventricles. The ventricles begin their systolic contraction. As pressure rises in the ventricles, the force of the blood against the atrioventricular valves causes them to close. Ventricular systole continues until intraventricular pressure forces the blood past the semilunar valves, causing them to open. The ventricles are emptied of blood, which enters the pulmonary artery and the aorta. Ventricular diastole begins, and the intraventricular pressure drops rapidly. As soon as it is below the pressure in the pulmonary artery and the aorta, whose elastic walls have been distended, the blood tends to flow back into the ventricles. Backflow, however, is prevented by closure of the semilunar valves. Meanwhile, the atria are filling with blood, and ventricular diastole is being concluded. The cardiac cycle is now complete, and a new cycle is about to begin.

The primary function of the valves is to keep the blood flowing *in one direction.*

Heart Sounds. Through a stethoscope applied to the chest, two sounds produced by a beating heart can be heard distinctly. These, occurring close together, are described as "lub-dup"—the "lub" being softer, longer, and lower in pitch. The first sound is caused by closure of the atrioventricular valves and contraction of the ventricular muscle; the second, by closure of the pulmonary and aortic semilunar valves. In some individuals, a third sound can be heard immediately after the second. It is thought to be due to vibrations of the ventricular wall resulting from the sudden entrance of blood from the atria.

Diseased or Malformed Valves. Sometimes, as a result of abnormal embryonic development or inflammatory conditions (e.g., rheumatic fever, septicemia), the valve leaflets become deformed or partially destroyed. In bacterial infections, the endocardium covering the valves becomes inflamed, and excrescences called *vegetations* develop on the edges of the valves. These may be small, wartlike bodies or large, friable masses of fibrin. The leaflets may become thick and rigid, and the chordae tendineae may shorten. When these changes occur, the cusps do not close properly, and blood leaks past them. This backflow is referred to as *regurgitation,* and a valve so affected is said to be *incompetent* or *insufficient.* This condition gives rise to alterations in the heart sounds; instead of being clear and distinct, abnormal sounds called *murmurs* are heard. The occurrence of a faint murmur, however, does not necessarily mean that a valvular lesion exists, since a murmur may occur in the absence of any organic disease. As a murmur is principally the result of the turbulence of blood flow within the heart, conditions that increase the velocity of blood flow (e.g., hyperthyroidism) or a condition in which the viscosity of the blood is reduced (e.g., anemia) may give rise to a murmur.

Another condition is the abnormal narrowing of an orifice, called *stenosis.* It may occur at any of the openings of the heart and is the result of abnormal development or of disease, which causes the leaflets of the valves to lose their flexibility, become stiff, and fuse with each other. The flow of blood through the narrowed opening is impeded, resulting in greatly increased velocity. Stenosed valves are usually also incompetent.

Murmurs may also be caused by such septal defects as the presence of a persistent opening in the interatrial or interventricular septum. The mitral valve is the valve most often involved in valvular disorders.

Treatment of Cardiac Disorders. By the use of specialized diagnostic techniques, such as *angiography, cardiac catherization,* and *echocardiography,* cardiac disorders can be accurately diagnosed and often corrected by surgery. An *artificial pacemaker* may be installed when the heart beat is abnormally slow. A pacemaker is a small, battery-powered unit implanted under the skin and wired to the heart to regulate the rate and rhythm of contractions. Heart surgery, closed or open, may be resorted to. *Closed heart surgery* is performed without halting the action of the heart or exposing its interior. In *open heart surgery,* when the interior of the heart is exposed, the pumping action is halted during which time oxygenated blood is supplied by use of a *heart-lung machine* (a cardiopulmonary bypass). By heart surgery, various congenital and acquired defects can be corrected. Heart valves can be repaired or replaced by artificial valves, abnormal openings closed, and blood vessels transposed, bypassed, or replaced. Heart and vascular surgery, once rarely performed, have now become commonplace.

Effects of Valvular Disorders. Two of the most common effects of valvular disorders are interference with venous return and hypertrophy of the heart.

INTERFERENCE WITH VENOUS RETURN. Regurgitation of blood into the atria interferes with venous return of blood from the venae cavae and pulmonary veins into these chambers. This causes stagnation of blood in the capillaries (*passive congestion*).

HYPERTROPHY OF THE HEART. Incompetent valves reduce cardiac output; consequently, the heart must perform more work to maintain adequate blood flow. To accomplish this, the cardiac muscle fibers become longer and thicker, the walls of the heart thicken, and the chambers grow larger. This brings about increased size and bulk (*hypertrophy* or enlargement of the heart). These adjustments are referred to as *cardiac compensation.* They enable an individual to live in spite of severe valvular deficiency, but *the reserve power of the heart is reduced* and activities must be kept at a minimum.

Cardiac Reserve. The ability of the heart to perform work above the basal requirements of the body is known as its *reserve power* or *cardiac reserve.* Under normal conditions, the heart has a reserve that is adequate to meet any demands to which it may be subjected; that is, it cannot be "strained" by overactivity. Even the most strenuous physical activity will not injure a normal healthy heart. It may hypertrophy (enlarge), but, within normal limits, no ill effects will ensue.

There are great individual differences in the reserve power of the heart. In diseased hearts or those with valvular lesions or impaired circulation, the reserve power may be so low that the heart may be unable to maintain circulation adequate to meet the needs of the body. Then signs of heart failure occur: breathlessness, cyanosis, edema (especially in the lower extremities), a tendency to fainting, and the occurrence of chest pains (angina pectoris).

Cardiac Output. The amount of blood leaving a ventricle at each beat is called the *stroke volume.* It averages about 70 ml for the normal heart. The *cardiac output* or *minute volume* is the amount of blood discharged from the heart in 1 min. This is determined by multiplying the heart rate and the volume of blood ejected at each beat. It averages 4 to 5 l/min in a resting adult. During strenuous exercise, it may increase to 20 to 25 l/min. Cardiac output depends upon the *rate* and *force* of the heartbeat.

RATE OF HEARTBEAT. The rate of the heartbeat is usually ascertained by feeling the pulse. The *pulse rate* is extremely variable, owing to many factors. Some of these are as follows:

Sex. The rate is faster in women (72 to 80 beats/min) than in men (64 to 72 beats/min).

Age. The rate is faster at birth, decreasing progressively to old age. The averages are: at birth, 140; at 1 yr, 120; at 10 yr, 90; in adulthood, 70.

Body Position. The rate is faster when standing erect than in a reclining or sitting position.

Physical Training. As a rule, athletes have a slower resting pulse rate than do untrained persons.

Miscellaneous Factors. The pulse rate is increased by a number of *normal* conditions: muscular activity, high altitudes, activity of the digestive tract, and normal emotional excitement; it is also increased by a number of *abnormal* conditions: high body temperature (fever), hemorrhage, surgical shock, cardiac disease, hyperthyroidism, and exaggerated emotional states.

FORCE OF HEARTBEAT. Although the heart follows the all-or-none law, the force of its contractions varies according to the needs of the body. It is thus capable of *graded* contractions, which may be weak or strong. In skeletal muscle, an increase in load increases the force of the contraction; similarly, in cardiac muscle, an increase in the stretch of the muscle, up to a certain point, increases the force of the contraction. This is known as *Starling's law of the heart.*

The volume of blood entering the ventricles during diastole is the primary factor in the stretch of heart muscle. During exercise, venous return is greatly increased, and as a result the heart beats more vigorously. Following hemorrhage or shock, when the volume of blood in circulation is lessened, venous return is reduced, and heart contractions are feeble.

Blood Supply and Nutrition of the Heart. The tissues of the heart receive their blood supply through two *coronary arteries,* which are the first branches of the aorta. Their openings lie in the pockets of the aortic valves. These arteries ramify to all parts of the heart, supplying cardiac muscle. In excessive activity, they are capable of dilating in order that an adequate supply of oxygen and food materials may be maintained. Any interference with blood supply lessens the efficiency of the heart and reduces cardiac reserve.

The energy sources for the heart are glucose and lactate rather than glycogen, as in the case of skeletal muscle. Cardiac muscle is incapable of contracting an oxygen debt. The release of energy in the heart is aerobic, in contrast to the anerobic process that takes place in skeletal muscle. This accounts for the serious effects of oxygen deprivation. A cutting off of the oxygen supply, as occurs in coronary thrombosis, may result in cessation of contractions within 60 sec or less.

REGULATION OF THE HEARTBEAT

The rate and force of the heartbeat are regulated by nerve impulses, chemical substances in the bloodstream, and physical factors, such as temperature and pressure within the heart and the great blood vessels.

Nervous Regulation of the Heartbeat. The rate and force of the heartbeat are regulated by impulses transmitted over the afferent and efferent fibers of the cardiac nerves, operating through the cardiac center located in the medulla oblongata.

CARDIAC NERVES. Nerves that supply the heart tissue are known as *cardiac nerves.* Although, as previously stated, the heartbeat is an automatic action originating within the heart itself, nervous impulses discharged from centers in the brain are capable of altering the heart rate in response to various bodily conditions. This is accomplished reflexly. Cardiac nerves are of two types, afferent and efferent.

Afferent Nerves. These nerves carry sensory impulses from the heart to the central nervous system. Pain impulses thus conducted are of clinical significance, for they are indicative of coronary disturbances or other pathologic conditions. Afferent fibers enter the spinal cord through dorsal roots of spinal nerves by way of sympathetic ganglia. Their cell bodies lie in the dorsal root ganglia.

Afferent fibers also carry impulses from receptor cells lying in the base of the aorta. These cells are sensitive to pressure, which serves as a stimulus for reflex control of blood pressure. Such fibers are called *depressor fibers;* as blood pressure rises, they reflexly slow down the heart rate. Receptors are also present in the large veins and in the right atrium. Distension of these structures by increased venous return initiates impulses that reflexly increase the heart rate. These fibers are called *pressor fibers.* Both pressor and depressor fibers pass through the vagus trunks to the *reflex centers* in the medulla.

Efferent Nerves (Fig. 11-1). The heart receives nerve impulses through two sets of efferent fibers, both of which belong to the autonomic nervous system. These are the *right* and *left vagus nerves* (of the parasympathetic division) and the *accelerator nerves* (of the sympathetic division).

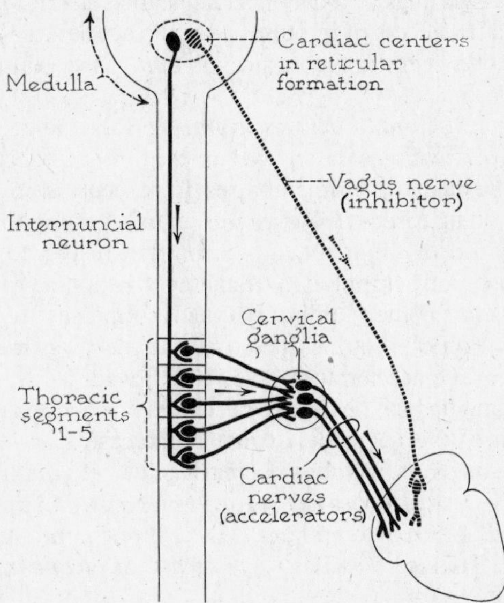

Fig. 11-1. Efferent nerves of the heart. (Reprinted with permission of W. B. Saunders Co., Philadelphia, from B. G. King and M. J. Showers, *Human Anatomy and Physiology,* 6th ed., 1969.)

Efferent fibers of the *vagus* (10th cranial nerve) are preganglionic fibers ending in a small ganglion in the wall of the heart. Here they synapse with postganglionic neurons, the fibers of which are distributed to the SA node, the AV node, and the bundle of His. Impulses carried by these fibers are *inhibitory;* that is, they slow down or stop the beat of the heart. The fibers of the right vagus are distributed principally to the SA node; its impulses affect the rate of the heart more than do impulses of the left vagus, which are distributed principally to the AV node. Nevertheless, both nodes receive fibers from both vagus nerves.

Efferent fibers of the *accelerator nerves* consist of preganglionic and postganglionic fibers. The former are axons of neurons located in the gray matter of the first five thoracic segments of the spinal cord. They synapse with neurons whose cell bodies lie in the ganglia of the sympathetic trunk (the upper thoracic ganglia, stellate ganglia, consisting of the 1st thoracic and inferior cervical ganglia, and the middle and superior cervical ganglia). Axons of these neurons (postganglionic fibers) pass to the heart, some ending near the SA and AV nodes, others in the wall of the ventricle. Impulses over the accelerator nerves *increase* the rate of contraction of both the atria and the ventricles; the *force* of their contractions is also increased.

The vagus nerves bring about their inhibitory effects through the liberation at their endings of a chemical substance, *acetylcholine.* Similarly, the accelerator effects of impulses over the accelerator nerves are due to the release of a chemical substance, *norepinephrine,* at their peripheral ends.

Tonic Action of Cardiac Nerves. Both the vagus and the accelerator nerves exhibit *tone;* that is, they exert a continuous effect on the organ innervated, the heart. Nerve impulses from centers in the brain are continuously being discharged over the vagus, exerting their restraining action to slow down the heart rate. Simultaneously and continuously, the heart is receiving impulses over the accelerator nerves, which tend to speed up its rate. When inhibitory (vagus) impulses are blocked, as by the administration of atropine, the heart rate increases markedly. When the influence of the accelerator nerves is removed, as by excision of the sympathetic ganglia, the heart rate decreases.

Stimulation of the vagus may cause a decrease in rate of beat, or it may decrease the force without decreasing the rate, or it may stop the beat. When the heart has ceased beating as a result of vagal stimulation, it may resume beating temporarily even though the stimulus is still being applied. This is known as *vagal escape* or *escape from vagal inhibition.*

THE CARDIAC CENTER. Nervous control of the heart resides primarily in the *cardiac center,* which is located in the medulla in the floor of the fourth ventricle. From a portion of this center, the *cardioinhibitory center* (Fig. 11-2), impulses are discharged through the vagus nerves to the heart; from another portion, the *cardioaccelerator center,* impulses travel down the spinal cord to the upper thoracic region, then, by way of spinal nerves and sympathetic ganglia, through accelerator nerves to the heart.

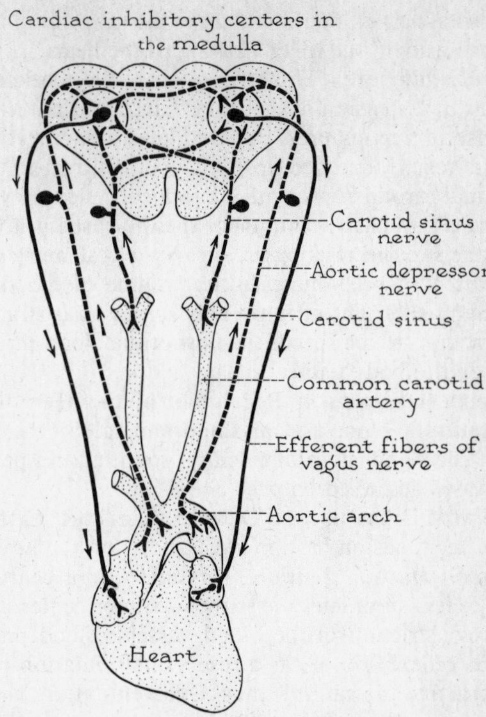

Fig. 11-2. Diagram showing reflex action of aortic depressor and carotid sinus nerves upon heart rate. (Reprinted with permission of W. B. Saunders Co., Philadelphia, from B. G. King and M. J. Showers, *Human Anatomy and Physiology,* 6th ed., 1969.)

CARDIAC REFLEXES. Stimulation of sensory nerves in almost any part of the body may reflexly bring about a slowing of the heart: a blow to the abdomen; manipulation of internal organs, as in an operation; inflammation of the middle ear; application of cold water to the surface of the body; inhalation of an irritant gas; or pressure on the eyeball (*oculocardiac reflex*). Stimuli that give rise to exaggerated emotional states, however, in general tend to speed up the heart rate reflexly (a loud sound, a sudden flash of light).

Pressoreceptors. Changes in blood pressure also reflexly affect the rate of heartbeat, which is inversely proportional to the arterial blood pressure. This principle is known as *Marey's law.* If the blood pressure rises, the heart rate is reflexly slowed; if it falls, the rate is reflexly increased. These effects are brought about through stimulation of *pressoreceptors* (*baroreceptors*) located in the arch of the aorta, the base of the heart, and the carotid sinus. Afferent fibers from these receptors pass through the vagus nerves and the *sinus nerve* (a branch of the glossopharyngeal, the

9th cranial nerve). They conduct to the cardioinhibitory center impulses that bring about the reflex slowing of the heart.

Bainbridge Reflex. This term refers to the acceleration of heart rate as a result of distension of the venae cavae and the right atrium by blood. The rise in venous pressure stimulates afferent vagal endings, and the heart is reflexly speeded up. This is in contrast to the distension of the aorta and carotid sinus, with consequent reflex slowing of the heart rate. This reflex, in part, accounts for the increased heart rate during muscular exercise. The milking action of skeletal muscles and the action of respiratory muscles increase the volume of blood entering the venae cavae and the right atrium. This serves as a stimulus for accelerated heart action, which speeds the flow of the blood through the heart to the lungs and to body tissues.

Chemical Factors in Regulation of the Heartbeat. Chemical substances in the blood play an important role in the control of the heartbeat. (The significance of calcium, sodium, and potassium ions in this regard was described on page 348.)

CARBON DIOXIDE AND OXYGEN TENSION. Carbon dioxide tension and oxygen tension have marked effects on the heart rate.

Carbon Dioxide Tension. This affects the heart in several ways. A *slight* excess stimulates the vasoconstrictor center of the brain, bringing about constriction of the blood vessels. Blood pressure rises and the heart is reflexly slowed as a result of stimulation of pressoreceptors of the aorta and the carotid sinus. Overventilation, with consequent lower carbon dioxide tension in the blood, brings about the opposite effect, that is, an increase in the heart rate. A great excess of carbon dioxide decreases the tone of the cardioinhibitory center and decreases impulse conduction in the atrioventricular bundle, thus slowing the heart. Excessive increase of hydrogen ions, as occurs in acidosis, results in heart block.

Oxygen Tension. Low oxygen tension increases the heart rate but if it is continued and *oxygen want* becomes severe, marked irregularities in heart action occur. The rate becomes slower, and the heart fails. Interference in oxygen supply, as occurs in coronary thrombosis, may damage the heart tissue and lead to heart failure.

EFFECT OF HORMONES ON THE HEART. Cardiac action is affected by hormones. *Epinephrine* and *norepinephrine* from the medulla of the adrenal gland increase both the rate and the amplitude of heart contractions. Injected into the body, they raise blood pressure, which condition reflexly slows the heart. *Thyroxine,* a thyroid hormone that speeds metabolism, increases the rate of the beat.

DRUGS. Many drugs, especially alkaloids, affect heart action. Some produce their effects principally through their action on acetylcholine, which is produced at the vagus endings. *Atropine* increases the rate by neutralizing the action of acetylcholine. *Muscarine,* a poisonous substance found in mushrooms, accentuates vagus action, slowing down the heart or stopping it completely.

Pilocarpine, physostigmine (*eserine*), and *choline* all have an inhibitory effect similar to that of muscarine. Physostigmine acts by inhibiting the action of cholinesterase, the enzyme that destroys acetylcholine. It prolongs any inhibition initiated by vagal stimulation.

Nicotine at first slows the heart by stimulating the vagal ganglion cells of the heart. Later, however, paralysis of these cells results in acceleration of heart rate.

Digitalis, a drug widely used in the treatment of cardiac disease, has a steadying and slowing influence on the heart. It depresses conduction in the atrioventricular bundle, causing a slowing of ventricular activity. Excessive dosage, however, may cause heart block.

Physical Factors in Regulation of the Heartbeat. In cold-blooded animals, the rate of heartbeat is dependent on body temperature; it decreases as this temperature is lowered and increases as it is raised. In warm-blooded animals, the body temperature remains constant under normal conditions, but slight rises in this temperature, as in strenuous exercise or in fever, increases the heart rate.

PERIPHERAL RESISTANCE

With each contraction of the heart, blood is forced into the aorta and its larger branches, the elastic arteries. From these, the muscular arteries distribute the blood to the various organs; smaller branches, the arterioles, distribute blood to the capillaries of specific tissues. The arteries and arterioles contain in their walls smooth muscle fibers so arranged that upon their contraction, the lumen of the blood vessel is constricted. Through such changes in the caliber (inside diameter) of the blood vessels supplying the tissues, the volume of blood supplied to the tissues is controlled.

In the tissues, the capillaries play an important role in peripheral resistance. The total cross-sectional area of the capillaries (the *capillary bed*) in a tissue is much greater than the cross-sectional area of the vessel or vessels supplying that tissue. This increased capillary bed results in a reduced rate of blood flow; consequently, blood flows very slowly through the capillaries. This is the principal factor in the marked drop in pressure in the capillaries shown in tables of pressure gradients. But the extremely small caliber of the capillaries, which resists the flow of the viscous blood, acts to increase the resistance and to counteract the drop in pressure due to the increased capillary bed.

The caliber of the arteries and arterioles is under both nervous and chemical control. Nervous impulses from reflex centers in the brain may constrict or dilate the vessels. Chemical substances may alter the size of blood vessels either by (1) acting locally, (2) circulating and acting directly on nervous centers of control, or (3) stimulating sensory receptors, which initiate reflex control responses. Physical factors, such as temperature, also influence the size of blood vessels.

Local Control. The blood flow to various organs and tissues corresponds in general to the degree of activity of the organ or tissue; that is, with increased activity there is an increase in blood flow; with reduced activity the blood flow is decreased. This is especially true of the heart, skeletal muscle, glands, and the digestive tract. When a skeletal muscle contracts, the arterioles within it dilate, the capillary bed opens up, and the tissue is supplied with an increased amount of blood which is essential for its proper functioning. This increased flow of blood, called *active hyperemia,* is a local phenomenon occurring automatically and independently of nervous impulses or the effects of hormones.

The exact mechanisms involved in this precisely controlled process are not fully known, but the most important factors appear to be local chemical changes resulting from increased activity. In active tissues, there is a decrease in oxygen tension and an increase in the concentrations of carbon dioxide and hydrogen ions. Metabolites are produced, salt balance is altered, and osmotic changes occur. All these changes have been shown experimentally to cause arteriolar dilatation through direct action on the smooth muscle of arterioles. Thus the blood supply is correlated with the physiological activity of an organ or a tissue.

Other chemical substances may be involved in localized responses. Histamine, serotonin, and bradykinin, produced and liberated as a result of injury or inflammatory processes, influence blood flow and the leakage of plasma from the capillaries. All induce arteriolar dilatation and are thought to play a role in regulating blood supply.

If the blood supply to a part is reduced or cut off, as by the tightening of a tourniquet applied to the upper arm, the tissues will be deprived of blood, a condition called *ischemia.* On loosening the tourniquet, there is a marked increase in the flow of blood to the ischemic area, the arm becoming red and warm. This increased flow of blood into the dilated arterioles and capillaries following a temporary cessation of circulation is called *reactive hyperemia.* Dilatation of the blood vessels is the result of the same factors as in active hyperemia. The reaction is of importance in restoring circulation quickly to tissues in which blood supply may have been reduced, as by the occlusion of an artery.

Nervous Control of Blood Vessels. The smaller arteries and arterioles possess a thick layer of smooth muscle fibers, arranged in a circular fashion, which is under nervous control. Contraction and relaxation of the fibers of this layer determine the diameter of the vessels. The muscles are supplied by two sets of nerve fibers: *vasoconstrictors,* which bring about contraction of the muscles, and *vasodilators,* which bring about relaxation of the muscles. Through coordinated reflex mechanisms, supplemented by the action of chemical substances in the blood, the diameters of the vessels are so regulated as to bring about the proper distribution of the blood to the various tissues in response to their physiologic needs.

Vasoconstrictor Nerves. These belong to the sympathetic division of the autonomic nervous system. The nerve fibers arise from cell bodies that lie in the gray matter of the spinal cord. They pass through spinal nerves and white rami to sympathetic ganglia, through which the impulses are relayed to the various parts of the body. Fibers to the head and neck pass from the cervical ganglia to the carotid and other arteries, sending branches to the smaller arterioles in those regions. Fibers to the visceral organs of the abdomen pass through the *greater, lesser,* and *least splanchnic nerves.* Fibers to the blood vessels of the body wall and limbs pass from sympathetic ganglia through the gray rami to the spinal nerves, which distribute them to the smooth muscles of the blood vessels.

A *vasoconstrictor center* (*vasomotor center*) lies in the medulla oblongata. Stimulation of this center brings about a contraction of smooth muscle in the walls of blood vessels. *Vasoconstriction* of the arteries, especially the arterioles, is marked, bringing about a rise in blood pressure. The action of this vasoconstrictor center is *tonic;* that is, it is constantly sending out impulses to the blood vessels, keeping them in a continual state of moderate contraction. If this center is destroyed, or if the spinal cord is severed in the cervical region, the stream of impulses is stopped, and arterial tone is abolished. This brings about *passive vasodilatation* with reduced peripheral resistance, and the blood pressure is greatly reduced. After a time, however, arterial tone begins to be reestablished, and blood pressure may rise, which indicates that there must be secondary reflex centers in the spinal cord capable of inducing vasoconstriction.

Although it is tonic, the vasoconstrictor center *does not act automatically.* Its action depends on the continual reception of impulses from receptor organs. Voluntary control cannot be exercised over it, except under such unusual conditions as biofeedback. The responses evoked are entirely reflex in nature, but the reflexes mediated through the center may result secondarily from conscious states involving the cerebral cortex. Strong emotions (fright, anger, extreme joy or pleasure) usually bring about readily observable circulatory changes (vasoconstriction or vasodilatation). The hypothalamus is also involved in circulatory responses, since it has been found to contain specific areas that when stimulated bring about vasoconstriction or vasodilatation.

Inhibition of the vasoconstrictor center reduces the outflow of impulses, and dilatation occurs as a result of the reduced arterial tone. This is called *passive vasodilatation,* in contrast to *active vasodilatation,* which is brought about by impulses carried by the vasodilator nerves. Cutting or sectioning of nerves carrying the vasoconstrictor fibers produces vasodilatation and a fall of blood pressure to the parts involved.

Vasodilator Nerves. These nerves are much less widely distributed to various part of the body than are the vasoconstrictors. Some of the fibers belong to the *sympathetic division* of the autonomic nervous system; these are distributed to the blood vessels of the muscles and the viscera. Others belong to the *parasympathetic division;* these are distributed through the cranial nerves, especially the chorda tympani branch of the facial nerve, which supplies the blood vessels of the tongue and the salivary glands, and the nerves of the lumbosacral plexus, which supply blood vessels to the urinary bladder, rectum, and external genitalia. Experimental evidence seems to indicate the existence of a third type of vasodilator fibers that pass from the spinal cord via the *posterior roots of spinal nerves* and supply arteries in the skin of the limbs. When these fibers are stimulated, vasodilatation results. There are two views on the nature of these fibers: one, that they are efferent fibers contained in the posterior root; the other, that they are fibers of sensory neurons capable of carrying vasodilator impulses peripherally. If the latter be the case, the impulses must pass in a direction opposite to that of sensory impulses. For this reason they are called *antidromic impulses.* On the basis of this view, these nerve fibers would serve a double function, namely, carrying (1) *afferent* impulses from the sense organs centrally and (2) *efferent* impulses peripherally from the spinal cord to the smooth muscles of blood vessels. An understanding of their function awaits further experimental work.

Impulses carried by vasodilator nerves, then, bring about vasodilatation of blood vessels, with an accompanying increase in flow of blood to a tissue.

It was once postulated that there was a specific vasodilator center in the medulla, since electric stimulation of a lateral area in the vasomotor center produced vasoconstriction, whereas stimulation of a medial area resulted in vasodilatation. It was found, however, that the vasodilatation was the result of inhibition of vasoconstriction impulses, with a reduction in muscle tone of the blood vessels. It was also noted that vasodilator reflexes, such as those involved in the erection of the penis, had their centers in the lower portion of the spinal cord and did not depend upon impulses from the medulla.

Vascular Reflexes. Changes in the caliber of blood vessels are entirely of a reflex nature. Afferent impulses pass to reflex centers in the spinal cord or the brain, where connections are made and impulses are discharged along efferent nerves (vasoconstrictors or vasodilators) to blood vessels. The blood flow to an organ or a tissue then increases or decreases, and the blood pressure rises or falls, depending on which center is involved.

Afferent nerve fibers carrying impulses that result in vasoconstriction are *pressor fibers;* those resulting in vasodilatation are *depressor fibers.* A reflex rise in blood pressure is a *pressor reflex;* a reflex fall, a *depressor reflex.*

PRESSOR REFLEXES. Pressor nerve fibers are widely distributed in all regions of the body; probably all spinal and cranial nerves contain pressor fibers carrying afferent impulses from the skin. Various types of stimuli can bring about vasoconstriction; cold air or cold water striking the skin constricts blood vessels in the skin in parts quite remote from the area stimulated, as well as in the mucous membranes of the respiratory tract (nose, pharynx, trachea). Almost any strong stimulus to the skin may induce vasoconstriction, with an accompanying rise in blood pressure; stimulation of pain receptors also produces this effect.

DEPRESSOR REFLEXES. Reflex dilatation of blood vessels is accomplished primarily by inhibition of the action of the vasoconstrictor center, thus reducing its tonic state. There are relatively few regions in the body where blood vessels have been shown to possess true vasodilator nerves. Two such regions are the genital organs (penis and clitoris) and the salivary glands. There is also evidence that vasodilator fibers may go to the blood vessels of the mucous membranes of the mouth, pharynx, nose, and skin of the face.

Two important sets of afferent depressor fibers are those originating in (1) the aorta and base of the heart and (2) the carotid sinus, a slightly dilated area at the junction of the external and internal carotid arteries. Their endings are called *pressoreceptors* or *baroreceptors* because they respond to a stretch stimulus. The normal stimulus is blood pressure. When the blood pressure rises (as after an increase in heart rate), the increased flow of blood through these structures stretches their walls, stimulating the pressoreceptor endings. Initiated impulses ascend to the vasomotor centers, and efferent impulses are discharged through the vasodilator fibers or vasoconstrictor impulses are reduced. Vasodilatation takes place, and blood pressure is reduced.

The *carotid sinus* can be stimulated by applying pressure to the skin lying over it. Reflex slowing of the heart and vasodilatation with reduced blood pressure result. In some individuals the sinus is hypersensitive, responding to even light pressure applied in the vicinity of the sinus; a tight collar or other undue pressure on the neck may induce a fall in blood pressure, accompanied by dizziness, fainting, or both.

Blood Flow Through Capillaries (Fig. 11-3). It was originally thought that all the capillaries connecting arterioles with venules were simple endothelial tubes, but it has been found that some of the vessels extending beyond the terminal arteriole in the capillary plexus contain smooth muscle fibers in their walls. These vessels that extend from arterioles to venules are called *metarterioles,* and from these vessels the true capillaries are given off. The latter, after following a short course, reenter the metarteriole near its venous end. The metarteriole forms a more or less direct channel, called a *thoroughfare channel,* through which blood can pass into the venules.

At the point of exit of a capillary from the metarteriole, fibers of smooth muscle form a *precapillary sphincter* that, by contracting or di-

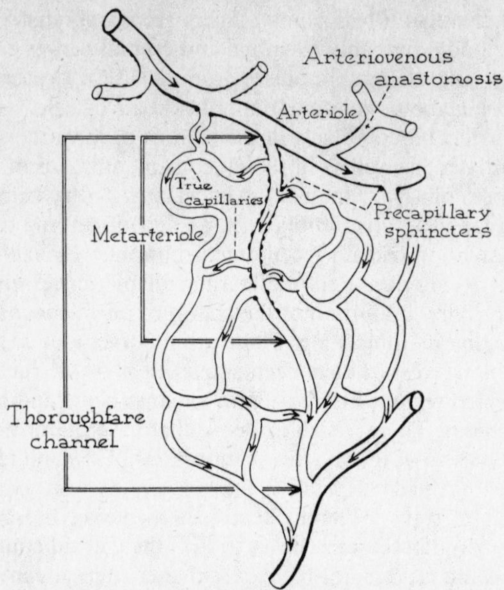

Fig. 11-3. Schematic diagram showing blood flow through capillaries. (After Chambers and Zweifach.) (Reprinted with permission of W. B. Saunders Co., Philadelphia, from B. G. King and M. J. Showers, *Human Anatomy and Physiology,* 6th ed., 1969.)

lating, controls the flow of blood through the capillary. When a tissue is in a resting state, the precapillary sphincters are constricted so that the blood is confined to the thoroughfare channel; when the tissue becomes active, the sphincters dilate and the capillaries open up, increasing the flow of blood to the tissue involved.

The smooth muscle cells of the metarterioles and precapillary sphincters are under nervous control, since stimulation of sympathetic nerves induces their constriction. Local capillary responses, however, are due principally to humoral factors, that is, chemical substances resulting from metabolic activities in the tissues or chemical substances in the blood. Histamine, a substance released from injured tissues, is a powerful capillary dilator, acting on the metarterioles and capillary sphincters. Metabolites (carbon dioxide and other substances) that lower the pH, as well as local heat, cause vasodilatation. Hormones, such as epinephrine, and serotonin released from platelets induce constriction, especially in injured tissues.

That capillaries are capable of independent activity is shown by the various reactions of the skin to different types of stimulation. If a blunt instrument is drawn gently across the skin, a well-defined white line appears after an interval of several seconds. This is known as the "*white reaction*" and is due to contraction of the capillaries in response to the stimulus. No nervous factors are involved. If the blunt instrument is drawn more forcibly across the skin, a distinct red line (instead of a white line) is seen. This result of localized capillary dilatation is known as the "*red reaction.*" If a stronger stimulus is applied, about the red line for some distance a red area appears. This area, designated *erythema* or *flare,* is due to arteriolar dilatation and is dependent upon local nervous factors (the axon reflex). In normal individuals, if an intense stimulus (such as the lash of a whip) is applied, an elevated area, called a *wheal* or *welt,* develops in the injured area. In sensitive individuals, this response may occur from a slight stimulus. A wheal is due to the escape of fluid from the capillaries into the tissue spaces; it is a localized edema resulting from increased permeability of the capillary walls.

These three reactions to injury (the red reaction, flare, and wheal) comprise the "triple response," which can be induced by various types of skin injury such as scratching, burning, freezing, or applying injurious chemical agents. It is thought to be brought about by the release of histamine, or a substance similar to it from injured cells. The responses of tissue to inflammation (redness, heat, swelling) can be accounted for in this way.

Influence of Higher Brain Centers on Blood Vessels. The vasomotor centers, especially the vasoconstrictor center, are highly responsive to conditions that involve the higher brain centers. Various emotions are registered by changes in the size of blood vessels. Blushing, from embarrassment or shame, and turning pale, from fright, are common reactions. The former is brought about by dilatation, the latter by constriction of the vessels of the face and neck. Anger may bring about either constriction or dilatation of blood vessels; which it will be depends on the personality or life experiences of the individual.

Chemical Control of Blood Vessels. Chemical substances, especially oxygen, carbon dioxide, and acids, play an important role in the maintenance of blood pressure through their effects on the size of blood vessels. Their effects are mediated through the *carotid* and *aortic bodies,* chemoreceptors that respond to chemical changes in the blood. Each *carotid body* is located on a branch of the carotid artery near the carotid sinus; the *aortic bodies* are located in the arch of the aorta and at the junction of the right subclavian and right carotid arteries. These structures, at their strategic locations, sample the blood leaving the heart for its oxygen and carbon dioxide content and its pH. They are stimulated by a decrease in oxygen (*hypoxia*), an increase in carbon dioxide (*hypercapnia*), or a decrease in pH (*acidosis*). Impulses arising in these receptors are carried to the brain, where, in the vasomotor and cardiac centers, they reflexly evoke responses that alter blood pressure. Some of the effects of specific chemical substances are as follows:

OXYGEN. A marked decrease in oxygen tension of the blood causes vasoconstriction, with resulting reduced blood supply, especially in pulmonary vessels and vessels in the extremities.

CARBON DIOXIDE. Carbon dioxide is thought to be the normal stimulus for the tonic action of the vasoconstrictor center. It may act directly on the medullary center, or it may act through the carotid body and aortic body chemoreceptors. Either way, an increase in carbon dioxide tension in the blood brings about generalized peripheral vasoconstriction. Locally, however, opposite effects may result, since an increase in carbon dioxide tension causes peripheral vasodilatation.

HYDROGEN ION CONCENTRATION. In the tissues, an increase in the production of acid substances, especially lactic acid, may bring about a fall in the pH of the blood, with resulting vasoconstriction. The effect is brought about through direct action on the vasomotor center and stimulation of receptors in the carotid and aortic bodies, especially the latter. Locally, the effect of excess hydrogen ions on peripheral vessels is dilatation.

HORMONES. *Epinephrine* and *norepinephrine* from the adrenal medulla, when injected into the bloodstream, bring about a marked constriction of visceral and cutaneous arterioles. This results in a rise in blood pressure. In skeletal muscle, however, dilatation of the arterioles occurs. *Vasopressin,* a hormone from the posterior pituitary, is a vasoconstrictor.

MISCELLANEOUS SUBSTANCES. Two substances of importance are histamine and angiotensin. *Histamine* is produced in excess in certain inflammatory conditions and especially in allergies or hypersensitivity reactions. It is one of the most powerful dilators known. It dilates capillaries and increases their permeability, constricts bronchioles, and increases mucus secretion. Because of its potent hypotensive effects, it is an important causative factor in shock involving circulatory collapse. *Angiotensin* is a powerful vasoconstrictor. When blood pressure in the kidney falls, a substance, *renin* (not rennin) is liberated from the juxtaglomerular cells. This substance, sometimes considered to be a hormone, reacts with *angiotensinogen* in the blood to form *angiotensin,* which constricts peripheral arterioles, with a resulting increase in blood pressure. It also stimulates the production of aldosterone by the adrenal cortex.

Other substances having vasoactive effects include kinins, serotonin, and prostaglandins. *Kinins* comprise a group of vasodilator peptides (bradykinins) formed during the active secretion of the salivary glands, sweat glands, and the exocrine portion of the pancreas. They resemble histamine in that they cause contraction of visceral smooth muscle, but they relax vascular smooth muscle and increase vascular permeability. They are potent vasodilators and participate in normal inflammatory responses and some allergic reactions. *Serotonin* released from blood platelets induces local vasoconstriction. The effect of *prostaglandins* on the musculature of blood vessels is uncertain. Prostaglandins A and E lower blood pressure; prostaglandin F increases it.

EXTRANEOUS SUBSTANCES. *Alcohol* causes a pronounced dilatation of the arterioles of the skin and has a depressant effect on the vasoconstrictor center. This results in a lowering of peripheral resistance and a decrease in arterial blood pressure. The use of *tobacco* in smoking increases the heart rate and constricts the peripheral blood vessels. As a result, blood pressure is increased.

ELASTICITY OF THE ARTERIES

The elasticity of the arteries is the primary factor responsible for the maintenance of blood pressure between the contractions of the ventricles (diastolic pressure) and for the steady continuous flow of blood through the capillaries. Normally, the arteries are always overfilled with blood, and their walls are slightly stretched. At each contraction of the ventricles, a quantity of blood is discharged into the aorta and its branches, further distending their walls. At the cessation of the contraction phase (systole), the stretched walls, by virtue of the elastic tissue contained within them, recoil. This exerts pressure on the blood, forcing it onward in the arteries during the interval between the ventricular contractions. As a result, the blood flows through the capillaries steadily and continuously, rather than in spurts, as would be the case if the arteries possessed rigid instead of elastic walls. The elasticity of the arterial walls prevents the blood pressure from falling to excessively low levels between the beats of the heart.

BLOOD VOLUME AND VISCOSITY

In addition to its response to nervous and chemical influences, blood pressure is affected by the volume of the blood and its viscosity.

Blood Volume. Under normal conditions, blood volume is fairly constant. If, however, there is a significant reduction in the volume of the blood in circulation, the pressure will fall. This may occur as a result of severe *hemorrhage* or in *shock*. In the latter, although the blood is still within the body, the plasma is lost from the circulation to the tissues, and there is stagnation of blood in the greatly dilated capillaries. Changes in *osmotic pressure* may also alter from time to time the volume of blood in circulation.

Blood Viscosity. The pressure required to force any fluid through a narrow tube varies with the viscosity of the fluid, a greater pressure being required to propel a viscous or "thick" fluid. The relative high viscosity of the blood (five to six times that of water) is due to the presence of blood proteins and corpuscles in the plasma. Conditions that alter the viscosity affect the ease of flow and consequently affect blood pressure. Alterations in the blood proteins or their loss, as in kidney disease, or a reduction in the number of red blood cells, as in anemia or following a hemorrhage, result in reduced viscosity accompanied by a fall in blood pressure. Conversely, an increase in the number of blood cells, as in polycythemia or leukemia, or a decrease in the fluid portion of the blood (anhydremia) increases the viscosity of the blood and brings about an increase in blood pressure.

VENOUS RETURN

Blood pressure, which is high in the aorta and the larger arteries, drops significantly in the capillaries and is gradually reduced until it approaches zero in the large veins. This is due principally to the fact that when a person is standing, the blood in the lower portion of the body must rise against the force of gravity in the large veins. Under ordinary conditions, the force of the heartbeat is not strong enough to maintain adequate venous return to the heart without the aid of certain actions, the more important of which are as follows:

1. *Suction force of the atrium,* when it is expanding following contraction.
2. *Respiratory movements,* which bring about changes of pressure within the thorax. Increase in intrathoracic pressure during inspiration acts on the venae cavae, exerting an aspirating effect on the veins of the abdomen. In addition, increased intra-abdominal pressure caused by downward movement of the diaphragm forces blood in the abdominal veins upward.
3. *Contraction of skeletal muscles,* especially those of the limbs, during muscular activity. This acts to compress the veins, exerting a massaging action on them. The presence of valves that prevent backflow enables this pressure to propel the blood onward toward the heart. Contraction of the abdominal muscles tends to force the blood in the abdominal veins upward.
4. *Movements in the visceral organs,* such as peristaltic action of the intestines and vasoconstriction in the splanchnic area.

Fainting *(syncope)* may result from inadequate venous return. Shock or a strong emotional reaction may induce dilatation of splanchnic veins, with reduced blood return to the heart and subsequent loss of consciousness. Standing still for a prolonged period of time (as when a soldier stands at attention) may result in fainting, owing to lack of muscular activity. Tenseness of antagonistic muscles may also impede circulation. The alternate contraction and relaxation of limb muscles produce a "milking" effect of the blood vessels lying between them; this favors the flow of blood to the tissues and venous return to the heart.

PRACTICAL CONSIDERATIONS

Velocity of Blood Flow (Fig. 11-4). Blood flows very rapidly in the aorta and the large arteries (about 40 cm/sec). Its velocity decreases in the smaller arteries and arterioles, and in the capillaries its flow is very slow (about 0.5 mm/sec). This is due to (1) the extremely narrow caliber of the lumen of the capillaries and the added resistance they offer to the flow of blood, a viscous fluid, and (2) the great increase in the cross-sectional area of the *capillary bed,* which is estimated to be 800 times that of the aorta.

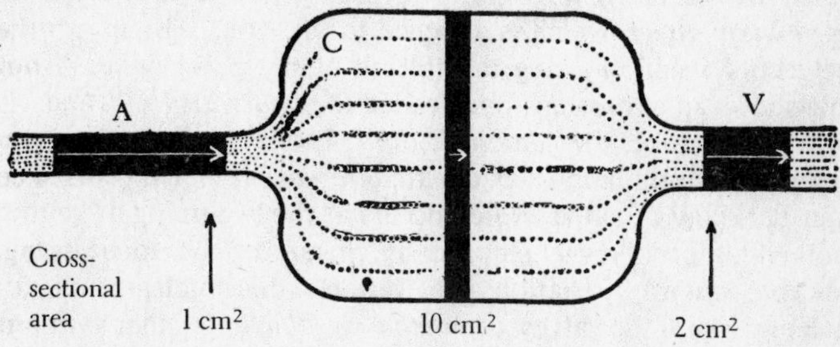

Fig. 11-4. Blood flow velocity. Lengths of white arrows represent rates of flow (10 cm/sec in *A,* 1 cm/sec in *C,* and 5 cm/sec in *V*). (Reprinted with permission of the University of Chicago Press from A. J. Carlson and V. E. Johnson, *The Machinery of the Body,* 4th ed., 1953.)

When the blood has passed through the capillaries and entered the smaller veins, the velocity picks up somewhat, and it increases progressively as the blood passes through the larger veins toward the heart. But the flow in the venae cavae is never so rapid as that in the aorta.

The Pulse. When a finger is placed over an artery lying near the surface of the body, a beat or pulsation is felt. This is due to the alternate expansion and contraction of the arterial walls owing to the beat of the heart. When the blood is forcibly ejected from the heart into the aorta, its impact on the elastic walls of the aorta sets up a pressure wave that travels along the arteries. This is known as the *arterial pulse,* or simply the *pulse.* It is not due to the passage of blood, since the pulse wave travels much faster than the blood flows. The velocity of the pulse wave averages about 7m/sec; that of the blood ranges from 10 to 15 cm/sec. It takes about 3 sec for the blood discharged from the heart to reach the wrist (where the pulse is commonly felt); but the pulse wave originating in the aorta traverses the same distance in about 0.1 sec.

The speed of the pulse wave varies with age. In children it averages 5 m/sec. As age increases and the arteries lose more and more of their elasticity, the speed increases, until in old age it may reach 8 to 10 m/sec; in atherosclerosis the velocity of the pulse wave may be even higher.

WHERE THE PULSE IS FELT. All arteries exhibit a pulse, but the pulse is most readily felt at points where an artery lies near the surface, especially when it is located over a bone. Following are the places where the pulse is most readily distinguished: *radial artery* (at the wrist), *temporal artery* (temporal region of the skull), *facial artery* (where it crosses the mandible), *carotid artery* (side of the neck), *brachial artery* (inner side of the biceps muscle), *femoral artery* (where it crosses the pelvic bone), *popliteal artery* (posterior surface of the knee), and *dorsalis pedis artery* (at the instep).

RECORDING OF THE PULSE. A graphic record of the pulse wave can be obtained with a *sphygmograph,* a recording device connected with a *sphygmomanometer,* which is attached to the wrist. The *sphygmogram* is the record itself, showing the tracing of the pulse wave. A normal tracing shows an abrupt upstroke, called the *anacrotic limb,* and a more gradual downstroke, the *catacrotic limb.* The latter normally reveals a secondary elevation, known as the *dicrotic wave* or *notch.* This is due to a slight backflow of blood, which occurs at the beginning of ventricular diastole. This backflow is stopped by closure of the aortic semilunar valves; consequently, variations in elevation of this notch are indicative of irregular action of the valves, or *valvular dysfunction.* Other variations in the pulse wave are characteristic of certain pathologic conditions.

Usually the arterial pulse disappears in the arterioles, but occasionally it may spread through the capillaries and appear in the veins. In the large veins near the heart, especially the jugular vein, a distinct venous pulse can be detected. The latter is due primarily to pressure exerted by the atrial systole and by the protrusion of the atrioventricular valves into the atria during ventricular systole. It is of importance in interpreting irregularities in heartbeat. A venous pulse is also observed in the retina of the eye. It is due to the transmission of the impulse from the artery through the vitreous body to the veins.

Blood Pressure. This is the pressure that is exerted by the blood against the walls of the vessels within which it is contained. Blood pressure varies in different parts of the system, being highest in the aorta, lower in the arteries, and progressively lower in the capillaries and veins. The higher pressure in the arteries can be noted when an artery is cut, for the blood flows out forcibly and in spurts. On the other hand, the flow from a cut vein is continuous and steady.

KINDS OF BLOOD PRESSURE. The term *blood pressure,* when unqualified, refers to arterial pressure (that is, pressure in the large arteries), which is usually determined by taking the pressure of the blood in the left brachial artery. Pressure of blood in the veins is called *venous pressure;* that in the capillaries, *capillary pressure.*

Blood pressure is highest in the arteries at the time of contraction of the ventricles *(ventricular systole)*; this is known as *systolic pressure.* Pressure during ventricular diastole is known as *diastolic pressure.* The latter is due principally to the force exerted by the elastic rebound of the arterial walls.

RANGES OF BLOOD PRESSURE. Blood pressure is usually expressed as a fraction; for example, 120/80, which is read "one-twenty over eighty." The first of these figures is the systolic pressure, the second the diastolic. They record the mark on the scale of the sphygmomanometer that is reached by a column of mercury that responds to the force of the pressure. For the young adult in good health, the average blood pressure in different parts of the circulatory system is:

Brachial Artery	
systolic	110–120 mm Hg
diastolic	65–80 mm Hg
Capillaries	20–30 mm Hg
Veins	0–20 mm Hg (in veins near the heart)

Pulse pressure is the difference between systolic and diastolic pressure.

Mean blood pressure is usually considered to be halfway between systolic and diastolic values which, for a person with a blood pressure of 120/80, would be 100 mm Hg. However, the actual value is usually lower, approaching the diastolic value. Another method of determining it is taking the diastolic value and adding one-third of the pulse pressure. By this method, mean pressure would be 93 mm Hg (80 + 13).

METHODS OF DETERMINING BLOOD PRESSURE. Two methods are used to determine blood pressure: the direct method, used principally on experimental animals, and the indirect method, which is used clinically on humans.

Direct Method. A glass or metal tube called a *cannula* is inserted into an artery, and the cannulated vessel is connected to a manometer (Fig. 11-5), a U-shaped tube filled with water or mercury. The manometer is

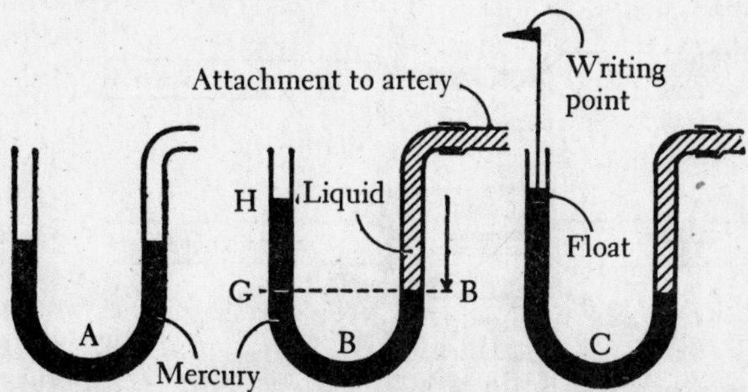

Fig. 11-5. Mercury manometer. (*A*) Condition with both arms of U-tube open to the air. (*B*) Condition with one arm of manometer attached to artery by rubber tubing. (*C*) Manometer equipped with float and writing point to transcribe blood-pressure record. (Reprinted with permission of the University of Chicago Press from A. J. Carlson and V. E. Johnson, *The Machinery of the Body,* 4th ed., 1953.)

usually equipped with a recording apparatus *(kymograph)* so that a permanent record will be obtained.

Indirect Method. The apparatus used in the indirect method is called a *sphygmomanometer.* The principle underlying the method involves the balancing of pressure within the artery with an external pressure exerted by air contained within a cuff applied around the arm. Actually, what is measured is the air pressure in the cuff. The manometer may be of either the mercury or the aneroid type. (*Aneroid* means "without fluid"—in place of the mercury, there is a delicate metal spring that is sensitive to changes in pressure).

The steps in determining blood pressure in man through this indirect method (Fig. 11-6) are as follows:

1. The cuff or armlet is wrapped securely about the arm *above* the elbow.

2. Air is pumped into the cuff by means of a rubber bulb until the pressure in the cuff is sufficient to stop the flow of blood in the *radial artery.* The radial pulse disappears at this point. The amount of pressure within the cuff is exhibited on the scale of the sphygmomanometer.

3. The observer then places a stethoscope over the brachial artery just *below* the elbow and gradually releases the air from within the cuff. When the air pressure falls to a point that permits the blood to flow and

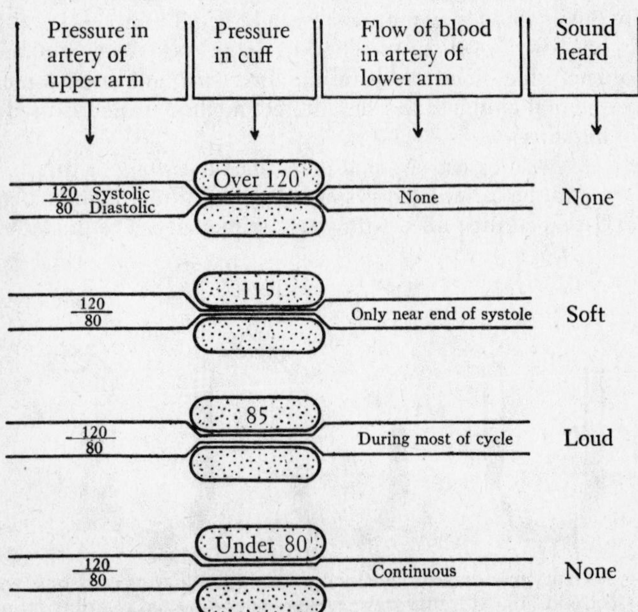

Fig. 11-6. Determination of blood pressure. Pressures in millimeters of mercury. (Reprinted with permission of the University of Chicago Press from A. J. Carlson and V. E. Johnson, *The Machinery of the Body,* 4th ed., 1953.)

fill the artery below the cuff, faint tapping sounds will be heard in the stethoscope. These reflect the heartbeat. At the appearance of the first sound, the pressure is read on the scale; this is the *systolic pressure.*

4. As more air in the cuff is released, the sounds become progressively louder, then disappear. Just before they disappear, they change in quality from very loud to very soft. At this point the manometer reading is again taken; this is the *diastolic pressure.*

The foregoing method is called the *auscultatory method.* It is the method most generally employed. In the *palpatory method* (*palpation*), the flow of blood in the radial artery is determined by digital pressure on that artery at the wrist. It is desirable to check auscultatory readings with palpatory readings.

FACTORS INFLUENCING BLOOD PRESSURE. Several factors, aside from the structure and functioning of the heart and the circulatory system, influence the blood pressure: age, muscular activity, sleep, emotional states, sex, weight, and pathologic conditions.

Age. The average systolic pressures at different ages are as follows:

	mm Hg
At birth	40
At end of 12th month	80
At age 12	100
At age 15	110
At age 20	120
At age 40	125
At age 65	134
After age 65	tendency to rise more rapidly

(Blood pressure above 140 mm Hg indicates hypertension.)

Muscular Activity. Blood pressure rises with increased muscular activity. The increase may be 60 to 80 mm Hg above normal. After cessation of activity, normal pressure is resumed.

Sleep. Systolic pressure falls during quiet sleep, owing to the prone position and reduced body activities. It rises slowly just before awakening. In disturbed sleep (e.g., during nightmares) systolic pressure may become very high.

Emotional States. Excitement, fright, anger, worry, and other extreme emotional states usually bring about a rise in systolic pressure. Grief, feelings of hopelessness and deprivation, and depression tend to lower the blood pressure.

Sex. Blood pressure averages 8 to 10 mm lower in women than in men. For women, this is applicable up to the time of menopause, when the pressure beings to rise. At age 60, blood pressure in women usually equals, and may even exceed slightly, the average for men.

Weight. An increase in body weight after age 60 is usually accompanied by a rise in blood pressure.

Pathologic Conditions. In arteriosclerosis and certain diseases of the kidneys, liver, and heart, blood pressure may become elevated (*hypertension*). In certain other diseases of the heart and under conditions of shock, blood pressure may be below normal (*hypotension*).

Recording the Heartbeat. The *electrocardiograph* is used to record changes in electrical potential originating from the action of a beating heart. It is, in effect, a very sensitive galvanometer. The record obtained from it is called an *electrocardiogram* (ECG or EKG). Deflections of the tracings on this record are produced in response to the passage of the heart current through the galvanometer. A normal electrocardiogram tracing is shown in Fig. 11-7. Irregularities of heart action revealed in the electrocardiogram are of value in the diagnosis of certain heart disorders, especially disturbances in impulse conduction and changes in the integrity of heart muscle.

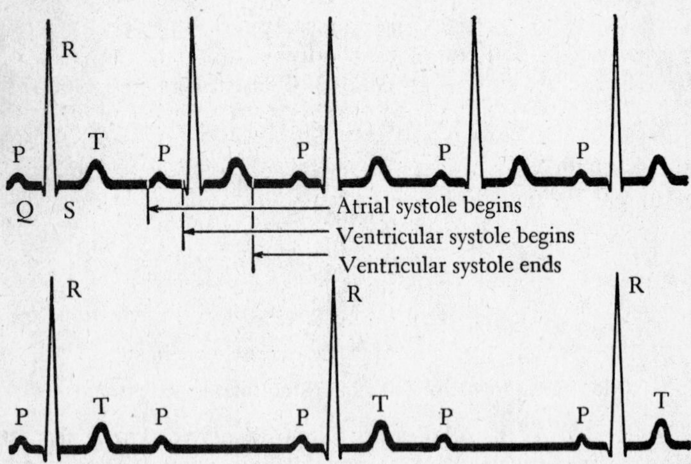

Fig. 11-7. A normal electrocardiogram. The P wave corresponds to atrial contraction; the QRS and T waves correspond to ventricular contraction. (Reprinted with permission of the University of Chicago Press from A. J. Carlson and V. E. Johnson, *The Machinery of the Body,* 4th ed., 1953.)

DISEASES AND DISORDERS OF THE HEART AND CIRCULATORY SYSTEM

Angina Pectoris. Severe thoracic pain that tends to radiate from the region of the heart to the shoulder and down the left arm, accompanied by a feeling of suffocation and syncope. It is due to an inadequate supply of blood to the myocardium of the heart, generally the result of coronary spasm or thrombosis. Attacks are frequently precipitated by muscular activity or excitement.

Aneurysm. A dilatation of an arterial wall, forming a sac that fills with blood. It is due to a weakening of the artery wall, usually the result of injury or inflammatory changes. Aneurysms may occur in any artery, but they are most common in the aorta because of the high blood pressure there. Aortic aneurysms are commonly brought on by syphilitic infections.

Arteriosclerosis. "Hardening" of the arteries. A condition in which there is thickening and loss of elasticity of the layers of the artery wall, especially the tunica intima, associated with degenerative changes in the muscular coat. Arteriosclerosis is common in advanced age.

Atherosclerosis. A type of arteriosclerosis in which fatty deposits or *atheromas* develop within and beneath the intima. Degenerative changes occurring in the elastic and muscular tissue, accompanied by hyperplasia of the tissue, result in reduction or obliteration of the lumen of the vessel.

Bradycardia. Abnormally slow rate of heartbeat (less than 60 beats/min).

Coronary Artery Disease. Obstruction of or interference with the flow of blood through an artery, owing to either narrowing of the lumen of the vessels from arteriosclerosis or presence of a thrombus or an embolus. The deprivation of oxygen and accumulation of metabolic substances stimulate the pain endings of afferent nerves, giving rise to agonizing chest pains (*angina pectoris*). Attacks are usually precipitated by muscular exertion or emotional excitement. The pain may be referred to the left shoulder and arm. Relief may be obtained by inhalation of amyl nitrate which dilates the coronary arteries.

Coronary artery disease and angina pectoris are also being treated by coronary bypass graft surgery. In this procedure, a vessel from elsewhere in the body, usually a portion of the saphenous vein from the leg, is connected surgically with the aorta and the portion of the occluded coronary artery located distal to the obstructing lesion. Such is especially effective in relieving angina.

Embolism. The sudden blocking of an artery by an embolus.

Embolus. A mass of material carried within the bloodstream to a point where it lodges in a vessel and obstructs the flow of blood. An embolus may consist of a blood clot (thrombus), air or gas bubbles, fat globules, tissue debris, cancer cells, or clumps of bacteria or animal parasites.

Fibrillation (Delirium Cordis). A condition in which the heart beats with such rapidity (400 to 600 times/min) that the beats are not distinguishable one from the other. It may involve the atria (*atrial fibrillation*) or the ventricles (*ventricular fibrillation*). Atrial fibrillation usually occurs in association with other cardiac disorders (e.g., *mitral stenosis*). Ventricular fibrillation may be induced by lack of oxygen (as in coronary occlusion), mechanical injury to the chest wall, a strong electric shock, chloroform or cyclopropane anesthesia, or the use of certain drugs.

Fibrillation can be stopped by the use of an electronic defibrillator or by cardiac massage. *External cardiac massage* is accomplished by forcibly depressing the sternum 4 to 5 cm at a rate of 60 times/min. *Internal cardiac massage* involves manually squeezing the ventricles, if the chest is open.

Flutter (Atrial Flutter). A condition in which the atria beat at an excessively fast rate (250 or more beats/min.). There is functional impairment in impulse conduction, which causes the ventricles to beat much slower, usually at a ratio of 1 to 2 or 3 atrial beats.

Heart Block. The condition in which the passage of impulses over the impulse-conducting system of the heart is inhibited, slowed, or interrupted. It includes (1) *sinoatrial (SA) block,* in which impulses fail to leave the SA node; (2) *intra-atrial (IA) block,* in which there is interference in the passage of impulses through the atrial myocardium; (3) *atrioventricular (AV) block,* in which impulses are blocked in the AV node or bundle, and (4) *intraventricular (I-V)* or *bundle branch block,* in which impulses are blocked in the branches of the bundle of His. It may result from congenital defects, disease (rheumatic fever, diphtheria, syphilis), myocardial infarction, the action of various drugs (e.g., digitalis), an excess of potassium or calcium salts, or excessive vagal stimulation.

Heart Failure. The condition in which the heart is unable to pump blood at a rate sufficient to meet the needs of the body. Since congestion in either the pulmonary or systemic circuits or both usually occurs, the condition is commonly referred to as *congestive heart failure.* Basically, heart failure is due to (1) the inability of the heart to perform its normal work load, (2) an increase in the work load of the heart, or (3) a combination of these conditions.

Some factors that may be responsible for the inability of the heart to function normally are coronary insufficiency, a diseased myocardium, inadequate venous return, defective origin and transmission of the heart impulse, or reduced oxygen-carrying capacity of the blood.

Some conditions that may increase the work load of the heart are hypertension, which increases arterial resistance; diseased or defective heart valves; congenital defects, such as a patent foramen ovale or ductus arteriosus; and increased metabolic demands, as from hyperthyroidism, extreme obesity, or excessive physical activity.

In congestive heart failure, the heart is unable to empty completely one or more of its chambers. Progressive dilatation of the chamber involved and hypertrophy of the heart muscle usually occur. Signs of a failing heart are the development of arrhythmias, tachycardia, abnormal heart sounds, fatigue, muscular weakness, pulmonary edema, and edema in the lower extremities (swelling of the ankles).

Hypertension (High Blood Pressure). Hypertensive vascular disease is the persistent elevation of blood pressure above that which is normal for a given age. For adults, blood pressure above 140/90 is indicative of hypertension, although in some cases higher readings may be encountered without ill effects. An elevated diastolic pressure is of greater significance. Hypertension is primarily the result of a spasm of the muscles in the arterioles. Arteriosclerosis may be a factor, but it is difficult to determine whether it is a cause or an effect. Some of the effects of hypertension are cardiac hypertrophy, with eventual cardiac failure; further hardening of the arteries (arteriosclerosis); possible rupture of blood vessels, especially in the brain (cerebral hemorrhage, "apoplexy"); and kidney dysfunction due to degenerative changes in renal vessels.

The causes of essential (chronic) hypertension are unknown. Predisposing factors include heredity; excess ingestion of sodium chloride; renal disorders, especially obstruction to renal blood flow, which induces the production of renin; excessive production of aldosterone by the adrenal cortex; and the effects of trace elements, especially cadmium. Neurogenic factors may also be involved. About 10 percent of the adult population is affected, females more frequently than males.

Hypotension (Low Blood Pressure). A condition in which the blood pressure is persistently below that which is normal for a given age. In an adult, systolic pressure below 90 mm Hg is regarded as indicative of hypotension. But low blood pressure, in the absence of other unfavorable findings, is not regarded as necessarily indicating a pathologic condition; indeed, it may favor longevity.

Phlebitis. Inflammation of a vein, usually accompanied by formation of pus. It frequently leads to the formation of a thrombus within a vein (thrombophlebitis) that may break loose and result in the distribution of infective emboli to other parts of the body. Phlebitis occurs most commonly in the veins of the lower extremities, often following long confinement in bed, abdominal operations, or childbirth.

Phlegmasia Alba Dolens ("Milk Leg"). Acute edema that begins in the ankle and ascends or begins in the groin and descends, resulting from obstruction of venous return. It usually occurs as a consequence of septic infection following childbirth or surgery.

Shock (Circulatory Shock). A condition of profound circulatory collapse characterized by inadequate return of blood to the heart and the pooling of blood in the peripheral vessels due to vasodilatation. Shock may result from reduced volume of blood in circulation, increased capacity of peripheral vessels resulting from vasodilatation and venous pooling, myocardial failure, or a combination of these factors. It commonly occurs after a massive hemorrhage, major surgery, trauma (as from burns or extensive soft tissue injury), acute myocardial failure, dehydration, or overwhelming infection. It may also result from acute adrenal insufficiency, drug intoxication or poisoning, and anaphylactic or hypersensitivity reactions. A patient in shock is apprehensive; the skin is cold, pale, and ashen gray; the blood pressure is below 90/60; the pulse rate is fast; respirations are rapid and shallow; and urine production is reduced. Blood sugar level falls, and metabolic acidosis may set in. Unless prompt and vigorous treatment is employed, shock may reach an irreversible stage, leading to death. Treatment involves sedation, restoration of blood volume by injection of normal saline or a balanced electrolyte solution, and administration of vasodepressors and cardiac stimulants. Inhalation oxygen therapy is sometimes required. Neurogenic shock may follow cerebral trauma or hemorrhage.

Tachycardia. Excessively rapid heartbeat. In paroxysmal tachycardia, the heart begins suddenly to beat at an abnormally high rate, as fast as 150 or more beats/min. The causes are numerous and varied.

Tetralogy of Fallot. A congenital anomaly of the heart characterized by (1) the presence of a permanent interventricular foramen; (2) stenosis of the pulmonary artery, restricting blood flow to the lungs; (3) dextroposition of the opening of the aorta so that it receives blood from both ventricles; and (4) hypertrophy of the right ventricle. These conditions result in little blood going to the lungs; consequently, cyanosis ensues. This is one of the possible causes of the "blue baby syndrome."

Varicocele. Dilatation of the veins of the spermatic plexus in the spermatic cord, usually on the left side. It is a common cause of male infertility.

Varicose Veins. Swollen, knotted, and tortuous veins, most commonly seen in the lower extremities. They result from a weakening of the walls of the veins or an interference in venous return. Blood tends to stagnate in the vessels, and the valves become incompetent. Varicosities are more common in women than in men. Standing for long periods of time, injury, and obesity are predisposing factors. Treatment is the wearing of elastic stockings or bandages, injection of sclerosing solutions, or surgery involving ligation and stripping.

INDEX